The Handbook of Environmental Chemistry

Volume 90

Founding Editor: Otto Hutzinger

Series Editors: Damià Barceló · Andrey G. Kostianoy

In over three decades, *The Handbook of Environmental Chemistry* has established itself as the premier reference source, providing sound and solid knowledge about environmental topics from a chemical perspective. Written by leading experts with practical experience in the field, the series continues to be essential reading for environmental scientists as well as for environmental managers and decision-makers in industry, government, agencies and public-interest groups.

Two distinguished Series Editors, internationally renowned volume editors as well as a prestigious Editorial Board safeguard publication of volumes according to high scientific standards.

Presenting a wide spectrum of viewpoints and approaches in topical volumes, the scope of the series covers topics such as

- local and global changes of natural environment and climate
- anthropogenic impact on the environment
- water, air and soil pollution
- remediation and waste characterization
- environmental contaminants
- biogeochemistry and geoecology
- chemical reactions and processes
- chemical and biological transformations as well as physical transport of chemicals in the environment
- environmental modeling

A particular focus of the series lies on methodological advances in environmental analytical chemistry.

The Handbook of Envir onmental Chemistry is available both in print and online via http://link.springer.com/bookseries/698. Articles are published online as soon as they have been reviewed and approved for publication.

Meeting the needs of the scientific community, publication of volumes in subseries has been discontinued to achieve a broader scope for the series as a whole.

The Seine River Basin

Volume Editors: Nicolas Flipo · Pierre Labadie · Laurence Lestel

With contributions by

S. Alligant · F. Alliot · J. Anglade · P. Ansart · S. Ayrault · S. Azimi ·
F. Baratelli · I. Barjhoux · S. Barles · N. Beaudoin · J. Belliard ·
M. F. Benedetti · T. Berthe · S. Beslagic · A. Bigot-Clivot · G. Billen ·
M. Blanchard · H. Blanchoud · J. Boé · M. Bonnard · R. Bonnet ·
D. Boust · A. Bressy · C. Briand · J.-M. Brignon · H. Budzinski ·
C. Carré · M. Chevreuil · N. Chong · F. Collard · M.-A. Cordier ·
O. Dedourge-Geffard · S. Derenne · J.-F. Deroubaix · A. Desportes ·
R. Dris · D. Eschbach · F. Esculier · A. Euzen · N. Flipo · N. Gallois ·
J. Garnier · J. Gaspéri · D. Gateuille · A. Geffard · A. Gelabert ·
F. Gob · A. Goffin · A. Goutte · A. Groleau · E. Guigon · S. Guillon ·
Y. Guo · F. Habets · J.-P. Haghe · P. Labadie · B. Labarthe · J. Le Noë ·
C. Le Pichon · J. Léonard · L. Lestel · C. Lorgeoux · Y. Louis ·
F. S. Lucas · A. Marescaux · B. Mary · Z. Matar · M. Meybeck ·
C. Mignolet · R. Moilleron · J.-M. Mouchel · L. Moulin · B. Muresan ·
P. T. Nguyen · L. Oziol · M. Palos-Ladeiro · E. Parlanti · P. Passy ·
F. Petit · T. Puech · W. Queyrel · M. Raimonet · A. Rivière ·
V. Rocher · C. Schott · J. Schuite · L. Seguin · P. Servais · M. Silvestre ·
C. Soares-Pereira · E. Tales · G. Tallec · B. Tassin · M.-J. Teil ·
S. Théry · D. Thevenot · V. Thieu · J. Tournebize · R. Tramoy ·
R. Treilles · G. Varrault · P. Viennot · L. Vilmin · S. Wurtzer

 Springer

Editors
Nicolas Flipo
Geosciences Department
MINES ParisTech, PSL University
Fontainebleau Cedex, France

Pierre Labadie
CNRS, UMR 5805 EPOC, Equipe LPTC
Université de Bordeaux
Talence Cedex, France

Laurence Lestel
Sorbonne Université CNRS
UMR 7619 METIS
Sorbonne Université
Paris Cedex 05, France

ISSN 1867-979X   ISSN 1616-864X   (electronic)
The Handbook of Environmental Chemistry
ISBN 978-3-030-54262-7   ISBN 978-3-030-54260-3   (eBook)
https://doi.org/10.1007/978-3-030-54260-3

This book is an open access publication.

This Springer imprint is published by the registered company Springer Nature Switzerland AG.
The registered company address is: Gewerbestrasse 11, 6330 Cham, Switzerland

Series Preface

With remarkable vision, Prof. Otto Hutzinger initiated *The Handbook of Environmental Chemistry* in 1980 and became the founding Editor-in-Chief. At that time, environmental chemistry was an emerging field, aiming at a complete description of the Earth's environment, encompassing the physical, chemical, biological, and geological transformations of chemical substances occurring on a local as well as a global scale. Environmental chemistry was intended to provide an account of the impact of man's activities on the natural environment by describing observed changes.

While a considerable amount of knowledge has been accumulated over the last four decades, as reflected in the more than 150 volumes of *The Handbook of Environmental Chemistry*, there are still many scientific and policy challenges ahead due to the complexity and interdisciplinary nature of the field. The series will therefore continue to provide compilations of current knowledge. Contributions are written by leading experts with practical experience in their fields. *The Handbook of Environmental Chemistry* grows with the increases in our scientific understanding, and provides a valuable source not only for scientists but also for environmental managers and decision-makers. Today, the series covers a broad range of environmental topics from a chemical perspective, including methodological advances in environmental analytical chemistry.

In recent years, there has been a growing tendency to include subject matter of societal relevance in the broad view of environmental chemistry. Topics include life cycle analysis, environmental management, sustainable development, and socio-economic, legal and even political problems, among others. While these topics are of great importance for the development and acceptance of *The Handbook of Environmental Chemistry*, the publisher and Editors-in-Chief have decided to keep the handbook essentially a source of information on "hard sciences" with a particular emphasis on chemistry, but also covering biology, geology, hydrology and engineering as applied to environmental sciences.

The volumes of the series are written at an advanced level, addressing the needs of both researchers and graduate students, as well as of people outside the field of "pure" chemistry, including those in industry, business, government, research establishments, and public interest groups. It would be very satisfying to see these volumes used as a basis for graduate courses in environmental chemistry. With its high standards of scientific quality and clarity, *The Handbook of Environmental Chemistry* provides a solid basis from which scientists can share their knowledge on the different aspects of environmental problems, presenting a wide spectrum of viewpoints and approaches.

The Handbook of Environmental Chemistry is available both in print and online via www.springerlink.com/content/110354/. Articles are published online as soon as they have been approved for publication. Authors, Volume Editors and Editors-in-Chief are rewarded by the broad acceptance of *The Handbook of Environmental Chemistry* by the scientific community, from whom suggestions for new topics to the Editors-in-Chief are always very welcome.

Damià Barceló
Andrey G. Kostianoy
Series Editors

Preface

Among the six hydrographic basins of metropolitan France, the Seine-Normandy basin is the most human-impacted. This territory of 76,000 km^2 and 17 million people receives the highest anthropogenic pressure, due to its industrial and agricultural activities linked to the development of the urban area of Paris, which has been and still is the economic and social heart of France. These pressures have gradually impacted the hydrological, chemical, and ecological functioning of the basin, leading to a maximum chemical degradation between the 1960s and the 1990s.

The very poor chemical and ecological status of water bodies in the 1980s led a small group of researchers to propose a PIREN-Seine, i.e. an interdisciplinary environmental research program created by the French National Centre for Scientific Research (CNRS). This program was launched in 1989 in the context of insufficient wastewater treatment in the Paris conurbation. Its first achievement consisted in developing models to better understand the river hydrological and biogeochemical functioning. These tools have made it possible to bring together research teams on a common object of study, the whole Seine watershed; the program has also been a forum for dialogue between the basin's institutional partners and researchers, enabling the latter to make management proposals to establish investment priorities based on the results of these models.

Over the past 30 years, the PIREN-Seine program has grown up, has attracted social scientists, and has generated a vast number of disciplinary and interdisciplinary publications, more than 100 PhD theses, hundreds of publications in scientific journals, and as many communications in international workshops and conferences. Nevertheless, the collective visibility of this group of scientists and institutional partners is still relatively low internationally, since most publication credits are given to individuals, to their laboratories, or to their research institutions. Moreover, the names of these laboratories and institutions have also evolved over time, making the recognition of this collective effort even more difficult. This book is the opportunity to present the most salient or recent results of the program, presented here as trajectories that relate environmental changes, societal changes, and the state of the Seine River basin waterbodies, which is nowadays largely controlled by the balance between pressures, water and river uses, and social responses.

The book covers a broad range of topics such as (1) the estimation of fluxes transported from headwaters to the coastal zone, at a very fine spatial scale, using models, (2) long-term analyses (50–200 years) of the socio-ecosystem Seine, using archives, retro and prospective modeling, (3) the identification and quantification of sources and transfer of a wide variety of elements and pollutants (nutrients, carbon, trace metals, POPs, pharmaceuticals, pesticides, microplastics, etc.), (4) the study of microbial contaminations, and (5) the analysis of the impact of water quality and contaminations on biota.

Studies considering the Seine River basin as a socio-ecological system are increasingly present within the PIREN-Seine program, which has been included in the Zone Atelier Seine (ZA Seine). The latter coordinates the research activities on the Paris city, the Seine River basin, and the Seine River estuary. It is also part of the European Long-Term Socio-Economic and Ecosystem Research (LTSER) programs. The large size of the basin and the longue-durée approach (up to 200 years) make this territory a rare and fully documented example of the multiple and evolving interrelations between a river, its large basin and their society which characterize the Anthropocene era.

Fontainebleau, France   Nicolas Flipo
Talence, France   Pierre Labadie
Paris, France   Laurence Lestel
Paris, France   Michel Meybeck

Contents

Trajectories of the Seine River Basin.......................... 1
Nicolas Flipo, Laurence Lestel, Pierre Labadie, Michel Meybeck,
and Josette Garnier

The Evolution of the Seine Basin Water Bodies Through Historical
Maps... 29
Laurence Lestel, David Eschbach, Michel Meybeck, and Frédéric Gob

Pluri-annual Water Budget on the Seine Basin: Past, Current and
Future Trends.. 59
Nicolas Flipo, Nicolas Gallois, Baptiste Labarthe, Fulvia Baratelli,
Pascal Viennot, Jonathan Schuite, Agnès Rivière, Rémy Bonnet,
and Julien Boé

The Seine Watershed Water-Agro-Food System: Long-Term
Trajectories of C, N and P Metabolism......................... 91
Gilles Billen, Josette Garnier, Julia Le Noë, Pascal Viennot,
Nicolas Gallois, Thomas Puech, Celine Schott, Juliette Anglade,
Bruno Mary, Nicolas Beaudoin, Joël Léonard, Catherine Mignolet,
Sylvain Théry, Vincent Thieu, Marie Silvestre, and Paul Passy

Past and Future Trajectories of Human Excreta Management Systems:
Paris in the Nineteenth to Twenty-First Centuries................ 117
Fabien Esculier and Sabine Barles

How Should Agricultural Practices Be Integrated to Understand
and Simulate Long-Term Pesticide Contamination in the Seine
River Basin?... 141
H. Blanchoud, C. Schott, G. Tallec, W. Queyrel, N. Gallois, F. Habets,
P. Viennot, P. Ansart, A. Desportes, J. Tournebize, and T. Puech

Mass Balance of PAHs at the Scale of the Seine River Basin   163
D. Gateuille, J. Gasperi, C. Briand, E. Guigon, F. Alliot, M. Blanchard,
M.-J. Teil, M. Chevreuil, V. Rocher, S. Azimi, D. Thevenot, R. Moilleron,
J.-M. Brignon, M. Meybeck, and J.-M. Mouchel

**Ecological Functioning of the Seine River: From Long-Term
Modelling Approaches to High-Frequency Data Analysis**   189
J. Garnier, A. Marescaux, S. Guillon, L. Vilmin, V. Rocher, G. Billen,
V. Thieu, M. Silvestre, P. Passy, M. Raimonet, A. Groleau, S. Théry,
G. Tallec, and N. Flipo

**Aquatic Organic Matter in the Seine Basin: Sources, Spatio-Temporal
Variability, Impact of Urban Discharges and Influence on
Micro-pollutant Speciation**   217
G. Varrault, E. Parlanti, Z. Matar, J. Garnier, P. T. Nguyen, S. Derenne,
V. Rocher, B. Muresan, Y. Louis, C. Soares-Pereira, A. Goffin,
M. F. Benedetti, A. Bressy, A. Gelabert, Y. Guo, and M.-A. Cordier

**Experience Gained from Ecotoxicological Studies in the Seine River
and Its Drainage Basin Over the Last Decade: Applicative Examples
and Research Perspectives**   243
M. Bonnard, I. Barjhoux, O. Dedourge-Geffard, A. Goutte, L. Oziol,
M. Palos-Ladeiro, and A. Geffard

**Sedimentary Archives Reveal the Concealed History of Micropollutant
Contamination in the Seine River Basin**   269
Sophie Ayrault, Michel Meybeck, Jean-Marie Mouchel, Johnny Gaspéri,
Laurence Lestel, Catherine Lorgeoux, and Dominique Boust

**Changes in Fish Communities of the Seine Basin over a Long-Term
Perspective**   301
Jérôme Belliard, Sarah Beslagic, and Evelyne Tales

**Bathing Activities and Microbiological River Water Quality
in the Paris Area: A Long-Term Perspective**   323
Jean-Marie Mouchel, Françoise S. Lucas, Laurent Moulin,
Sébastien Wurtzer, Agathe Euzen, Jean-Paul Haghe, Vincent Rocher,
Sam Azimi, and Pierre Servais

**Contaminants of Emerging Concern in the Seine River Basin: Overview
of Recent Research**   355
Pierre Labadie, Soline Alligant, Thierry Berthe, Hélène Budzinski,
Aurélie Bigot-Clivot, France Collard, Rachid Dris, Johnny Gasperi,
Elodie Guigon, Fabienne Petit, Vincent Rocher, Bruno Tassin,
Romain Tramoy, and Robin Treilles

**River Basin Visions: Tools and Approaches from Yesterday
to Tomorrow** 381
Catherine Carré, Michel Meybeck, Josette Garnier, Natalie Chong,
José-Frédéric Deroubaix, Nicolas Flipo, Aurélie Goutte, Céline Le Pichon,
Laura Seguin, and Julien Tournebize

Correction to: The Seine River Basin C1
Nicolas Flipo, Pierre Labadie, and Laurence Lestel

Index 415

Trajectories of the Seine River Basin

Nicolas Flipo, Laurence Lestel, Pierre Labadie, Michel Meybeck, and Josette Garnier

Contents

1 Introduction: River Systems in the Anthropocene 2
2 Multiple and Heavy Pressures on the Seine River System 4
   2.1   The Hydrological Features of the Seine River Basin 4
   2.2   Evolution of the Basin Population 7
   2.3   The Land Cover 9
   2.4   Industries and Navigation 10
3 PIREN-Seine Research on the River Basin Trajectories 10
   3.1   Circulation of Material Within the Basin 11
   3.2   Multiple Long-Term Trajectories of River Control Factors 13
   3.3   Trajectories of River State and Societal Response to River Issues 16
4 Structuration of the Socioecological Research in the Context of the PIREN-Seine 17
   4.1   Evolution of Main Research Themes 18
   4.2   Present Structuration of the PIREN-Seine Research Programme 18
5 The Spatio-Temporal Scales of the Research Themes Selected in This Publication 22
References 23

Abstract The Seine River basin in France (76,238 km^2, 17 million (M) people) has been continuously studied since 1989 by the PIREN-Seine, a multidisciplinary programme of about 100 scientists from 20 research units (hydrologists, environmental chemists, ecologists, biogeochemists, geographers, environmental historians). Initially PIREN-Seine was established to fill the knowledge gap on the river functioning, particularly downstream of the Paris conurbation (12 M people), where the pressure and impacts were at their highest in the 1980s (e.g. chronic summer hypoxia). One aim was to provide tools, such as models, to manage water resources

N. Flipo (✉)
Geosciences Department, MINES ParisTech, PSL University, Fontainebleau, France
e-mail: Nicolas.Flipo@mines-paristech.fr

L. Lestel, M. Meybeck, and J. Garnier
SU CNRS EPHE UMR 7619 Metis, Paris, France

P. Labadie
University of Bordeaux, CNRS, UMR 5805 EPOC, Talence, France

Nicolas Flipo, Pierre Labadie, and Laurence Lestel (eds.), *The Seine River Basin*,
Hdb Env Chem (2021) 90: 1–28, https://doi.org/10.1007/698_2019_437,
© The Author(s) 2020, Published online: 3 June 2020

and improve the state of the river. PIREN-Seine gradually developed into a general understanding and whole-basin modelling, from headwater streams to the estuary, of the complex interactions between the hydrosystem (surface water and aquifers), the ecosystem (phytoplankton, bacteria, fish communities), the agronomic system (crops and soils), the river users (drinking water, navigation), and the urban and industrial development (e.g. waste water treatment plants). Spatio-temporal scales of these interactions and the related state of the environment vary from the very fine (hour-meter) to the coarser scale (annual – several dozen km). It was possible to determine the trajectories (drivers-pressures – state-responses) for many issues, over the longue durée time windows (50–200 years), in relation to the specific economic and demographic evolution of the Seine basin, the environmental awareness, and the national and then European regulations. Time trajectories of the major environmental issues, from the original organic and microbial pollutants in the past to the present emerging contaminants, are addressed. Future trajectories are simulated by our interconnected modelling approaches, based on scenarios (e.g. of the agro-food system, climate change, demography, etc.) constructed by scientists and engineers of major basin institutions that have been supporting the programme in the long term. We found many cumulated and/or permanent hereditary effects on the physical, chemical, and ecological characteristics of the basin that may constrain its evolution. PIREN-Seine was launched and has been evaluated since its inception, by the National Centre for Scientific Research (CNRS), today within its national Zones Ateliers (ZA) instrument, part of the international Long-Term Socio-Economic and Ecosystem Research (LTSER) network.

Keywords LTSER, PIREN-Seine, Seine basin, Seine River, Socio-ecosystem, Zone Atelier Seine

1 Introduction: River Systems in the Anthropocene

Water resources are indispensable for social development. Human needs shaped rivers and river basins, leading to a drastic modification of the global water cycle. Changes due to human activities are so large nowadays that they override natural processes, leading to the end of a geological era and the beginning of a new one called "the Anthropocene" first conceptualized by Crutzen [1, 2]. Using this concept first proposed by the Earth science community, and increasingly considered by the human sciences [3, 4], makes it possible to highlight multiple and profound changes in the watersheds: (1) the physical environment is largely modified compared with its initial state; (2) terrestrial, aquatic, and estuarine environments are increasingly managed by societies, according to their interests and their representation of nature: flows, river morphology, summer temperature, water quality, and biodiversity are now largely controlled by dedicated institutions according to specific criteria; and (3) some of the factors influencing such management are located within the catchment area, but the external components have been increasing steadily for 50 years (national and European regulations, national and European markets for agricultural products, decline or recovery of industry and mining, as well as the evolution of international trade, etc.).

River system refers primarily to the hydrosphere, i.e. the water circulation over a well-delineated area, the basin watershed. It includes the atmospheric inputs and water flow components of the drainage network and is separated into (1) the surficial hydrographic network, from headwater streams to the estuary, including stagnant systems such as ponds, wetlands, lakes, reservoirs, and canals, (2) shallow and deep aquifers and their related unsaturated zone. It also includes the terrestrial biosphere and the pedosphere, which regulate water circulation and provide the river-borne and groundwater materials, and the aquatic biosphere, from micro-organisms to fish populations. Finally, the system also includes all the controlling factors that regulate these fluxes of water and materials and their composition: internal factors, either natural (e.g. hydrological regime, river morphology) or anthropogenic (e.g. water abstraction, pollution, hydrological control, and artificialization of river course), and external factors (e.g. climate change, trans-basin trade, species introduction, etc.). It can therefore be considered as a socio-ecosystem in the sense of Haberl [5].

Among the six hydrographic basins of metropolitan France, the Seine-Normandy basin is the most human-impacted (see Sect. 2 of this chapter). It receives the highest anthropic pressure, due to its industry and agriculture linked to the development of the urban area of Paris, which has been and still is the economic and social heart of France. The very poor chemical and ecological status of water in the 1980s led a small group of researchers to propose a PIREN-Seine, i.e. an interdisciplinary environmental research programme launched by the French National Centre for Scientific Research (CNRS), as had already been put in place for the Rhône River, the Garonne River, and the Alsace plain in 1979 [6]. It was created in 1989 in a context of insufficient wastewater treatment in the Paris conurbation and new investment projects in sanitation facilities [7]. Its first achievement consisted in developing a model, Riverstrahler [8], to dynamically represent the biogeochemical fluxes of carbon, nitrogen, phosphorus and then silica, for each body of water in the basin, from headwaters to the Seine outlet, according to constraints such as geomorphology, hydrography, agricultural diffuse sources, and urban discharges. Another modelling tool, ProSe, was also developed with a transient hydrology on the lower Seine more dedicated to the Paris conurbation domestic load [9–12]. These tools have made it possible to bring together research teams on a common object of study, the entire Seine watershed; the programme has also been a forum for dialogue between the basin's institutional partners and researchers, enabling the latter to make management proposals to establish investment priorities based on the results of these models [13]. Despite the fact that PIREN programmes have been replaced by other interdisciplinary programmes, the PIREN-Seine programme has continued to exist, financially supported by the institutions in charge of the Seine basin management. Over the past 30 years, it has generated a vast number of publications, more than 100 PhD theses, hundreds of publications in scientific journals and as many communications in international workshops and conferences, as well as special issues [14–16] and several booklets [17]. Nevertheless, the collective visibility of this group of scientists and institutional partners is relatively low internationally, since most publication credits are given to individuals, to their laboratories, or to their research institutions. Moreover, the names of these laboratories and institutions have also evolved over time, making the recognition of this collective effort even more difficult. This book provides the opportunity to present a selection of some of the most salient results of the programme, obtained within the framework of our current socioecological approach of relevant environmental issues in the Seine River basin. Most of the results will be presented as trajectories that relate environmental changes and societal changes. A number of the trajectories identified within the Seine basin will be presented in Sect. 3 of this chapter.

These studies of socioecological systems are by nature interdisciplinary; the proposed conceptual frameworks such as material flow analysis (MFA) [18, 19], the driver-pressure-state-impact-response (DPSIR) concept recommended by the European Environment Agency [20–22], the human-environment systems (HES) framework [23], or the social-ecological systems framework (SESF) [24] are means of structuring this research or proposing a common language for the different disciplinary components used to analyse the complexity of these human-environmental relationships, with each of the methods having its objectives and limitations [25]. These approaches are increasingly present within the PIREN-Seine, which has been included in the Zone Atelier Seine (ZA Seine) and the Long-Term Socio-Economic and Ecosystem Research (LTSER) programmes, as presented in Sect. 4 of this chapter. The spatial and temporal scales of the trajectories described in the following 14 chapters are presented at the end of this introduction.

2 Multiple and Heavy Pressures on the Seine River System

In a basin like that of the Seine River which has long been populated, deforested, and industrialized and where the large urban conurbation of Paris and highly productive agriculture areas coexist, human control of the system, its water, and the material fluxes often exceeds natural controls. This makes the Seine River system a case study in which complex interactions between societal and biophysical processes can be examined. The next subsections summarize the present situation in terms of land use, water use, as well as human pressures and their dynamics since the beginning of the twentieth century.

2.1 The Hydrological Features of the Seine River Basin

The Seine River basin extends over 76,000 km^2, of which 65,000 km^2 are upstream of its estuary, with the outlet of the basin located at Poses (Fig. 1). It lies 97% within the sedimentary Paris basin, the largest groundwater reservoir in Europe (Triassic to tertiary, Fig. 2a). The lithology of the basin encompasses carbonates (69.6%) and sandy formations (13.6%), interbedded by poorly permeable clayey and marl units (9.1%) and covered by alluvial deposits (5.4%) (Fig. 2b) [27]. The sedimentary basin is covered by carbonated loess in its western and central parts [28].

The river network can be described using the Strahler stream order concept [29], which is used throughout the morphological, hydrological, and biogeochemical modelling of the Seine basin. The total length of the river network is 27,500 km, mostly composed of first and second Strahler orders. The main tributaries of the Seine River are the Yonne River, the Marne River, and the Oise River. The Seine River reaches the seventh Strahler order after its confluence with the Yonne River.

The hydrological regime of the Seine River basin is pluvial/oceanic. The mean rainfall over the basin is 800 mm year^{-1} and exhibits some spatial variability, with a maximum of around 1,200 mm year^{-1} along the coastal shoreline and in the Morvan mountainous range and only 650 mm year^{-1} in its central part [30–32]. For the past 50 years, the Seine discharge at the last gauging station before the estuary at Poses has had an average value of 485 m^3 s^{-1} but reached 2,280 m^3 s^{-1} in winter, with summer minimums at 80 m^3 s^{-1} (low flows sustained by reservoirs). In addition to

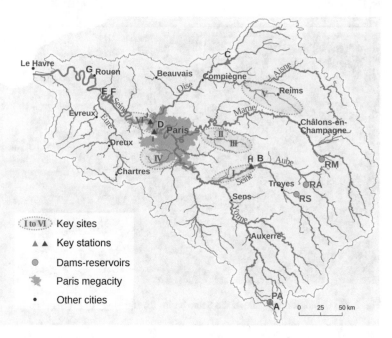

Fig. 1 The hydrological network (Strahler orders 3–7) of the Seine and the main experimental sites that have been studied. I Bassée floodplain, II Orgeval stream experimental area, III Grand Morin River, IV Orge River, V Vesle River, VI Lower Seine sector. Core sites [26]: A Pannecières reservoir, B Troyes Seine canal, C Chauny Oise River, D Chatou Seine sluice, E Muids Seine floodplain, F Bouafles Seine floodplain, G Rouen Seine estuary dock. Sampling stations: H Marnay, I Bougival, J Triel. Reservoirs: PA Pannecières, RA Aube reservoir, RS Seine reservoir, RM Marne reservoir

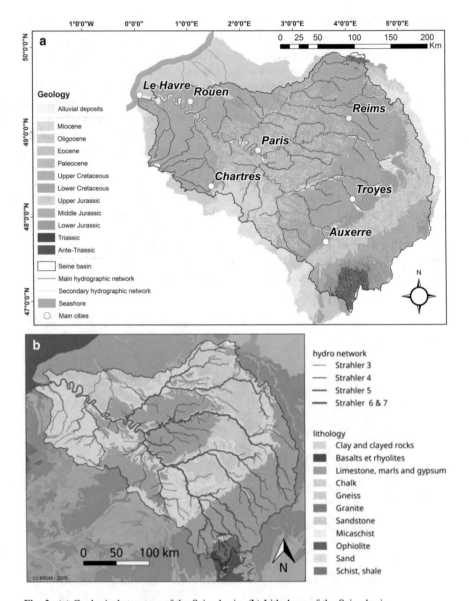

Fig. 2 (**a**) Geological structure of the Seine basin. (**b**) Lithology of the Seine basin

these seasonal river flow variations, a 17-year cycle associated with the North
Atlantic Oscillations (NAO) structures the long-term temporal variability of stream
discharges [33], as well as groundwater levels [34].

The rainfall exhibits virtually no seasonality, meaning that the river flow regime
derives from seasonal variations in real evapotranspiration, thereby resulting into
winter high flows and summer low flows; these are naturally sustained by numerous

sedimentary aquifers, and, for the middle and lower reaches of the Seine and Marne rivers, by large reservoirs that were built between 1931 and 1990 (Fig. 1) [35]. These reservoirs, totalling 841 Mm^3, regulate floods and can provide more than 50% of the river flow in summer, thus contributing to dilute wastewater inputs of the Paris conurbation (i.e. 25 $m^3 s^{-1}$ discharge downstream of Paris). Water management is therefore a key factor of the current Seine River discharge profile [32]. It should be noted that the relative natural dilution power of the Seine River in relation to Parisian wastewaters is very limited, making the Seine River sensitive to point source pollution compared with most large European rivers.

The PIREN-Seine programme focuses mainly on the Seine River basin upstream of its estuary, the latter also being studied by the long-term programme, Seine-Aval, which started in 1995 [36]. The main experimental sites studied within the Seine River basin and the coring sites presented in Ayrault et al. [26] are shown in Fig. 1 and will be discussed in further detail in the following chapters.

2.2 Evolution of the Basin Population

Paris, the largest megacity in Europe, has developed on the Seine River in the central part of the basin, near the confluence of the Marne River and upstream of the confluence with the Oise River. This location once represented a favourable factor for the long-distance transport of food, timber, and construction materials. The urbanized part of the basin has been growing over the past two centuries. Paris megacity increased from 75 km^2 in the 1850s (Paris area) to 2,850 km^2 today [37]. In 2015, it was home to 12.4 million (M) of the 16.7 M inhabitants (inhab.) within the basin. The population density of the Seine basin in 1901, 1954, and 2016 is provided in Fig. 3, showing the increasing population in the Ile-de-France region, around Paris, and downstream along the lower Seine. The population density near urban tributaries in the Paris area ranges between 1,000 and 5,000 inhab km^{-2}, while it is much lower (<20 inhab km^{-2}) in the upstream areas, with an average of 250 inhab km^{-2} for the whole basin.

The urban pollution generated by the city of Paris, and later by the Parisian conurbation, which impacts the lower Seine and the estuary has been a major concern of the authorities for over a century [37–39]. Water quality has been severely degraded in terms of oxygen levels, ammonia, and nitrite concentrations, as well as the occurrence of faecal bacteria over a section of river extending from 100 to 250 km downstream from Paris [40]. In the 1870s, the collected wastewater started to be spread over sewage farms near Paris. Wastewater treatment plants (WWTPs) were then gradually built but insufficiently to treat the volume of wastewater generated by the Paris conurbation [41]. One of the sewage farms at Achères, located at 60 km downstream from Paris, was converted into the Seine-Aval WWTP between 1930 and the 1980s, treating up to 8 M inhab. equivalents in the 1970s [35]. The lag between sewage collection and its adequate treatment was only bridged in the 1990s [42].

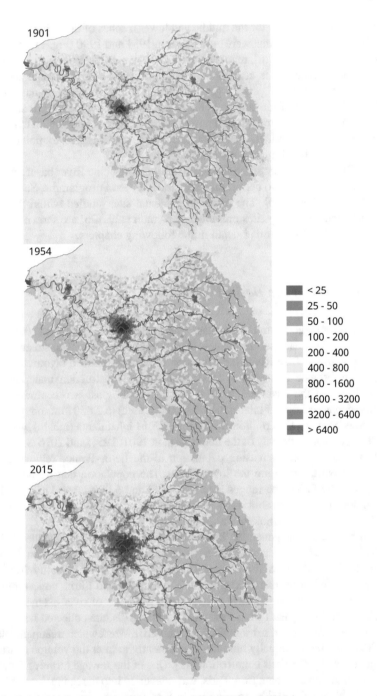

Fig. 3 Population density of the Seine basin in 1901, 1954, and 2015

2.3   The Land Cover

Nowadays, the Seine River basin is characterized by a relatively high proportion of urbanized area (7.6%), compared with grasslands (9.5%), forested areas (25.6%), and croplands (56.5%). Wetlands and water bodies represent low surface areas (Fig. 4).

Paris is surrounded by areas of high-intensity agriculture, oriented towards specialized intensive production of cereal and industrial crops. The development of agriculture initially paralleled the growth of Paris and its food demand in recent centuries [43]. Then a significant turning point in land use took place in the 1960s, when nitrogen fertilizers were applied intensively to croplands and converted grasslands [44, 45]. A major socioecological transition of the Seine system is thus identified at that time.

The PIREN-Seine has documented these changes in great detail (e.g. crop rotation, fertilizers applications, livestock density, etc.) from agricultural censuses conducted since 1850 at the level of the French departments and also by enquiries at the scale of small homogeneous agricultural areas (PRA, petites régions agricoles; originally defined in the 1950s). This information supported the reconstruction of nutrient circulations (nitrogen and phosphorus) in the basin at a fine spatio-temporal resolution [46]. Changes in land cover since the beginning of the twentieth century were also used to reconstruct the evolution of the Seine basin water budget [32].

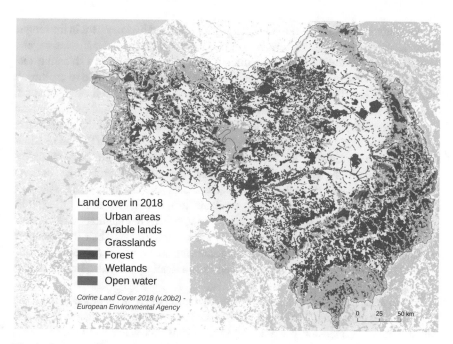

Fig. 4 Land cover distribution (2018) of the Seine River basin

2.4   Industries and Navigation

Heavy industrialization began in Paris and its suburbs on the eve of the nineteenth century. Until the end of the "Thirty Glorious Years" (1946–1975), industries were mostly located in Paris megacity, along the Lower Seine industrial corridors and along one of its main tributaries, the Oise River (see Fig. 1). Industrial wastewaters were barely treated on site and were discharged directly to the closest rivers, with the official assumption that they would be diluted and that rivers had enough self-cleaning potential. Deindustrialization and industrial wastewater treatment promoted by the Seine-Normandy Basin Agency (AESN) led to an improvement in the quality of the Seine River's water. Until the late 1980s, the level of toxic substances in the river, the fluxes released by both the city and its industries, and their effects on receiving waters were largely ignored by French scientists and authorities [26, 42, 47].

A further factor affecting the river has been the demand for deeper, larger, and more extended navigated reaches in the basin, specifically after the 1837 law for the improvement of navigation on the Seine River, allowing increased sand and gravel extraction in the floodplain, transported by waterway and used for Paris urban growth [48].

3   PIREN-Seine Research on the River Basin Trajectories

An investigation of the trajectories of this highly modified hydrosystem in the longue durée is the main focus of this book. The study of river systems over several centuries using historical data or sediment core analysis reveals a dynamic that contemporary observation – over a few years or decades – is not able to capture. It allows for the construction of trajectories that link societal evolution and changes in the overall state of the river. These trajectories represent simplified expressions of our understanding at a given stage of (1) the spatial description of the system state and its functioning, (2) its temporal evolution in the longue durée, (3) the relationships between river/water actors, and (4) the evolving position of river basins within the Earth system, particularly within the new Anthropocene concept [42, 49]. Trajectories bridge the gap between natural scientists and social scientists and make some results more readily available to our partners and to the public [50]. When possible, the usage of numerical models leads to quantitative results that help to define trajectories. A few examples are provided here. Their time window exceeds the 50 years of environmental monitoring in France and extends to two centuries and even more, owing to particularly abundant archives and historical data for the Seine River basin.

Within the PIREN-Seine, the ways of studying the trajectories of the Seine basin have evolved over time. The first representations were spatial: a catchment area in which a human dimension was introduced to reconstruct water and material flows within the whole catchment (e.g. metals) or to model the biogeochemical cycles of selected major elements (e.g. nitrogen). The opening of the biogeochemical cycles due to multiple factors has led to more systemic approaches combining the following trajectory components: (1) state indicators; (2) controlling factors and pressures indicators, generally economic; and (3) a set of social indicators, such as scientific knowledge and social awareness, inclusion of issues on the political agenda, environmental surveys, as well as regulatory and technical responses.

This work is complex, since both water and river systems are at the same time a resource, an economic good, and a cultural and symbolic asset, subject to different regulations according to these various functions. Each function generates specific actors or sets of actors and is perceived differently by each of these actors according to the numerous and variable reading grids over time.

3.1 Circulation of Material Within the Basin

The main advantage of river systems is the possibility to carry out material balances by monitoring material fluxes at the outlet of the basin. This approach provides integrated information for the entire territory of the basin and its population, at various time steps. Deciphering all the complexity and heterogeneity of the multiple fluxes requires specific studies for the past and present situations. For almost 30 years, river systems have been recognized for this complexity and studied in an interdisciplinary way owing to it, such as in the PIREN-Seine programme since 1989 or in other programmes [51].

For geologists and geochemists, natural material fluxes within river basin fluxes are derived from the erosion and weathering products of surficial rocks and from the uptake of atmospheric carbon and nitrogen occurring within the basin. For environmental chemists, the river-borne material contamination results from waste discharges and runoff and from the erosion of soils contaminated by diffuse sources. For environmental economists, material flow analysis over a given territory reveals the metabolism of the anthroposphere, the storage of left-over products, the growing pattern of infrastructures, and the recycling of products and goods [16]. The comparison of additional river fluxes – compared with estimated natural material fluxes – reveals that the circulation of many economic products used in the Seine River basin is between one and two orders of magnitude greater than the natural material fluxes, as for the heavy metals [52]. This new view of territorial functioning is schematized in Fig. 5, in which the system is described by a set of natural and/or anthropogenic reservoirs between which there is a continuous circulation.

The river receives a share – between 0.1 and 40% depending on products and periods – of this economic circulation (e.g. 0.1% for metals [55], 7% for phosphorus [56], and 40% for nitrate [57]). The specific circulations and their impacts on river exports vary over time and may relate to different locations in a river basin (see Fig. 6 for metals).

The reconstruction of these fluxes and of their evolution over time makes it necessary to correlate the knowledge on the circulation of substances in the anthroposphere (manufacturing, importation, consumption, and evolution of uses)

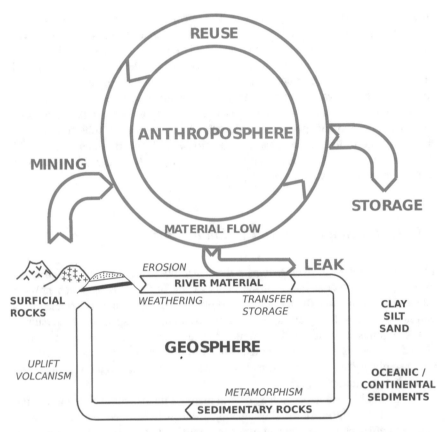

Fig. 5 Schematic circulation and storage of material within an impacted river basin: material flow within the river basin territory and its leaks into the aquatic system (based on [53, 54])

with field data provided by official water surveys or collected by the teams involved in the PIREN-Seine programme. Mass balances at the scale of the Seine River basin were undertaken for nitrogen [58], non-ferrous metals [59], and polycyclic aromatic hydrocarbons (PAHs) [60].

The lack of available sources, either spatial or temporal, makes a retrospective modelling approach to fill the gaps essential. Widely developed within the PIREN-Seine, e.g. for water budget [32], oxygen budgets [41], or nitrogen circulation at the watershed level [46], this approach remains original within the scientific community.

For centuries, the Seine River basin system was relatively closed, excepted for some imported products carried through the river waterway. Nowadays, the socio-economic system has been widely opened through massive imports and exports. The basin exports a great quantity of agricultural, food, and manufactured products, and it consumes a large quantity of imported animal feed, fossil fuels, mining, and manufactured products. It also emits and receives atmospheric pollutants to and from other basins through long-range transport.

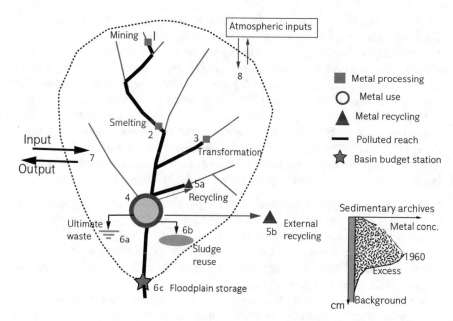

Fig. 6 Circulation of products and materials within impacted basins: (1) mining (very limited, occurring in the 1700s); (2) metal smelting, mostly in Paris and along the Seine and Oise; (3) metal transformation, mostly in Greater Paris; (4) use of metal and metal-containing products; (5) recycling of metal products such as batteries and pipes, first carried out in Paris suburbs prior to 1950 (5a), then outside the basin, and finally outside France (5b); and (6) storage in contaminated soils and landfills (6a); recycling of Paris WWTP sludge, once highly contaminated (<1990) (6b); and storage in floodplain sediments providing sedimentary archives of contamination (6c) (from [49])

This link with other river basins has been evidenced for the nitrogen and phosphorus circulation; it is also obvious for heavy metals, as there are no mines of non-ferrous metal in the basin. The strong impact of the Seine basin on the eutrophication of the Bay of Seine to the English Channel, and of the North Sea, has also been well documented [41], representing an example of the interest in considering the land-to-sea continuum, from headwaters to the coastal zone and eventually to oceans. These studies also showed the growing role of society in the control of rivers during the Anthropocene era [49]. These controlling factors are detailed in the following section.

3.2 Multiple Long-Term Trajectories of River Control Factors

For nearly 150 years, the social response to manage the uses of the river (navigation, flood control, and recreational purposes) or of its waters (drinking or agricultural water resources, dilution of the residual pollution from wastewater treatment plants) has taken place in successive stages [61]. It followed changes in paradigms (all-sewer sanitation), triggers (catastrophic floods), increasing water needs, new uses (cooling of nuclear power plants, agricultural irrigation), new awareness of water quality, wetland functionality, and aquatic biodiversity. A marker of these social responses are the laws concerning aquatic environments. The main laws for France are those of 1807, 1858, and 2014–2017 (flood prevention); 1829, 1865, and 1984 (river fishing laws); 1837, 1846, 1878, and 1880 (Seine River navigation improvement) and 1956 (Code for Waterways and Inland Navigation); and 1898, 1964, 1992, and 2006 (water quality and general river management).

It has been possible to undertake such a study on the Seine River basin due to the fact that it is an exceptionally rich territory in terms of environmental archives:

1. We have been able to reconstruct the nitrogen circulation and the related water quality, going as far as back in the Middle Ages, based on quantitative economic archives from ecclesiastic records [62].
2. A cartographic database on the river course and its corridor showing its evolution since the end of the eighteenth century has been set up, thanks to the archives of Ponts et Chaussées engineers [48].
3. We benefit from historical data (e.g. the nutrient chemistry and microbiological analyses; river hydrological records), which started in the second half of the nineteenth century – the chemical and bacteriological analyses of Seine water are one of the earliest ever made and have been regularly surveyed since the late nineteenth century [63].
4. We have also used sediment archives to reconstruct the river contamination by numerous and diverse legacy pollutants (heavy metals, PCBs, etc.), over 50–80 years, while the reliable surveys of these substances only started after the 2000s [26].
5. We have analysed historical agricultural censuses from 1854 to the present [45] and their associated greenhouse gas emissions [64].
6. We have also reconstructed the evolution of the fish population, a key integrative ecological indicator within the basin, from historical and archaeological archives collected over more than two centuries [65, 66].

The Seine River basin, therefore, provides an exceptional array of quantitative information, since it has provided water, energy, and food to Paris, France's capital city, for centuries. Reliable documentation dating back to the First Empire (early nineteenth century), when systematic mapping, population censuses, and economic statistics were started, is available.

To further address the interactions between the Seine basin society and its river over the longue durée and to understand these changes, we need access to the evolution of several controlling factors of the river basin system such as hydro-climate evolution (so far addressed from the second half of the nineteenth century [67]), changes in agricultural practices, industrial activities, population distribution and sanitation history, river navigation structures and other river uses, the introduction of aquatic species, etc. A schematic view of these trends is presented in Fig. 7. They show a wide spectrum of evolution, rarely linear and regular, varying from a uniform increase in population over the past 200 years to plateau features, bell shapes

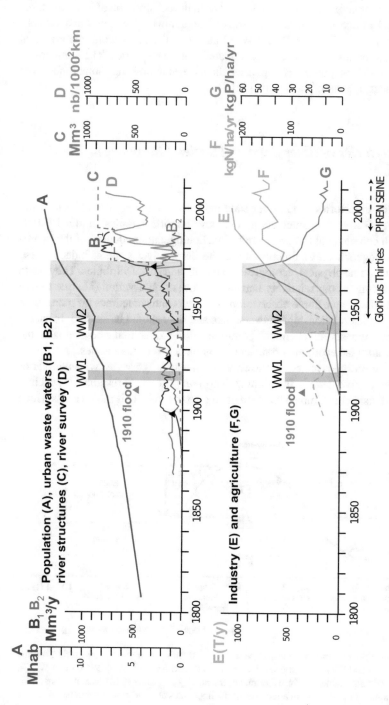

Fig. 7 Elements of the Seine River trajectories construction: evolution of controlling factors on the Seine River basin (1800–2010). A, total population; B1 and B2, collected and treated volumes of wastewaters from Paris [38] (triangles: tipping points at curve intersects); C, total capacity of reservoirs; D, density of river quality survey stations (nb/1000 km²); E, total metal use in France (Cu + Pb + Zn) (from [52, 55]), extrapolated before 1950 from industrial indicators; F, Nitrogen fertilizers applications; and G, Phosphorus fertilizer. Chronological markers: 1910, centennial flood; WWI and WWII conflicts; 1944–1974, period of rapid post-war development; 1989, start of the PIREN-Seine programme

(urban inputs), ruptures, and stepwise variations in the construction of infrastructures (WWTPs and reservoirs) but also in terms of regulations (laws, pollutants bans), notch patterns particularly for World War I and II or the 1929 economic crisis, as well as tipping points (e.g. the collected/treated wastewater ratio). The comparison of the trajectories of pressures/responses shows a temporal lag, sometimes over several decades, typical of each issue [42].

3.3 Trajectories of River State and Societal Response to River Issues

Analysing the trajectories of the river state indicators and their control factors and investigating river-society interactions are now important topics of the PIREN-Seine, which complement the description, functioning, and modelling of the hydrological and biogeochemical functioning of the Seine River system. Indeed, these trajectories combine physical, ecological, chemical, and social attributes. They make it possible to convert our relatively narrow window of observation (i.e. one to five decades) into the longue durée observation (100 years and sometimes far more) with which the complex relationships between man and river should be studied. In such a time window, the severe chemical pollution of the river that characterized the twentieth century is becoming a transient phenomenon for many issues.

Knowledge on the river basin shows a non-linear progression, with periods of disinterest in the river and its functioning. This complex social response is schematized in Fig. 8 using the "impair-then-repair" model [61, 68]. This model starts with a

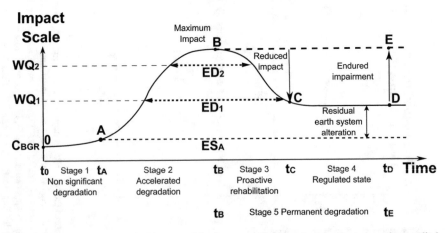

Fig. 8 The impair-then-repair scheme and the five stages defining river quality trajectories, applied to North America and Western European river basins (adapted from [61, 68]). WQC$_1$ and WQC$_2$, water quality criteria established for water management. C$_{BGR}$, pristine state concentrations. ES$_A$, ED$_1$, ED$_2$, duration of Earth system alteration, of the impaired state, and severe degradations of the river, respectively, as defined by river basin societies

period of *non-significant impact* on the Earth system or on water resources (OA, stage 1). The next stage is a period of *accelerated degradation* of the aquatic environment and of water resources (AB, stage 2), often faster than population growth in the river basin, first reaching the level of water quality (WQC_1) at which water resources are impaired, often followed by a very poor level of water quality (WQC_2). When the technical and regulatory measures taken by a society become efficient, a *proactive rehabilitation phase* is observed after a stage of maximum impact (BC, stage 3). When a satisfactory state is finally achieved, reaching the desired level of quality WQC_1, the *regulation stage* ensures a stable quality even though the population and economy of the basin continue to grow (CD, stage 4). If environmental management is insufficient, the impact can reach a *permanent degradation* (BE, stage 5) stabilized at an altered level ($>WQC_2$), which can be considered as irreversible.

The duration of the moderate environmental degradation (ED_1) from the societal perspective is defined here as exceeding WQC_1, and the duration of severe degradation (ED_2) as exceeding WQC_2 (Fig. 7). Water quality scales, WQC_1 and WQC_2, defined for each of the issues recognized by the society may evolve over time, thereby changing the environmental assessments made by societies. From an Earth system perspective, the analysis may be quite different: Any significant deviation from the pristine state, as defined by the background concentrations (C_{BGR}), is expressing an alteration in the Earth system (ES_A) and may generate a change on receiving waters – for instance, along the coastal zone.

4 Structuration of the Socioecological Research in the Context of the PIREN-Seine

At the beginning of the programme in 1989, originally launched by hydrologists, environmental chemists, biochemists, and engineers, only the natural processes, as modified and sometimes regulated by present and past human activities, were considered. After the seven stages of the programme, the interactions between the society and its environment now account for a major aspect of our research. In some projects we even aim to understand the generation and regulation of these activities, a turning point that has been taken gradually in the second and third stages of the programme. This comprehensive study of the Seine basin territory, a hydrological and a social system, over a well-defined spatial entity, enables us to generate river functioning models, which now implicitly include a number of socio-economic variables (e.g. agricultural practices, consumers habits, and water use) and technical changes or innovations (e.g. for wastewater treatment). Finally, in the last sixth and seventh stages of the programme, the future state of the river has been explored on the basis of narrative scenarios indicating the evolution of controlling factors, sometimes coupled with the hydrological response to various climate change scenarios. These scenarios have been implemented in the modelling approaches.

4.1 Evolution of Main Research Themes

Although the PIREN-Seine programme was not initially planned to last as long as it has or to encompass so many different topics, it gradually became apparent that it was very well suited to study many environmental issues far beyond the initial water resources issues.

The gradual changes in the main themes, as presented in annual reports to our financial sponsors, are schematized in Fig. 9. The colour coding makes it possible to represent the permanent themes, such as hydrology and water circulation, biogeo-chemistry, ecology, basin history, sediment contamination, as well as agricultural and urban impacts. The main models (Figs. 9 and 10), Riverstrahler [8, 70], ProSe [9, 12, 71, 72], STICS-MODCOU [73, 74], and ANAQUALAND [75], integrate many of these subjects. All these models are mechanistic in nature and evolved towards the next generation for Riverstrahler and MODCOU, now called pyNuts-Riverstrahler [76] and CaWaQS [32], respectively. They have been continuously improved, by integrating new processes, new interactions with the basin society, and refining the space or time resolution. These models can be considered as general tools that encapsulate most of the knowledge gained by the programme at a given stage (see [50]).

In fact, this representation conceals highly specialized scientific work: There is a great variety and specificity to the research topics published in scientific journals, such as surface water and groundwater motion, behaviour and transfer of microbial and chemical contaminants across the river continuum, denitrification processes in riparian zones, greenhouse gas emissions, analytical improvements for trace contaminant analysis, ecotoxicology or microbiological activity in field-based and laboratory experiments, databases construction, water use and treatment history, evolution of environmental concerns, changes in agricultural practices, etc.

4.2 Present Structuration of the PIREN-Seine Research Programme

The PIREN-Seine programme includes about 100 researchers from 22 research teams, belonging to research institutions (universities, CNRS, INRA, IRSTEA, MINES ParisTech, École des Ponts ParisTech, EPHE, to mention the major ones) that ensure their remuneration and evaluate their research (typically every 1–5 years).

The PIREN-Seine programme by itself is evaluated every 4 years by its financial partners. The permanent partners since the beginning of the programme in 1989 are basin actors for water quality management such as the Seine-Normandy Basin Agency (AESN) coupled with French authority DRIEE, or the EPTB Seine Grands Lacs, and services for drinking water supply (SEDIF – Syndicat des Eaux d'Ile-de-France-, Eau de Paris, SUEZ Eau France) or wastewater treatment (SIAAP, Sewage

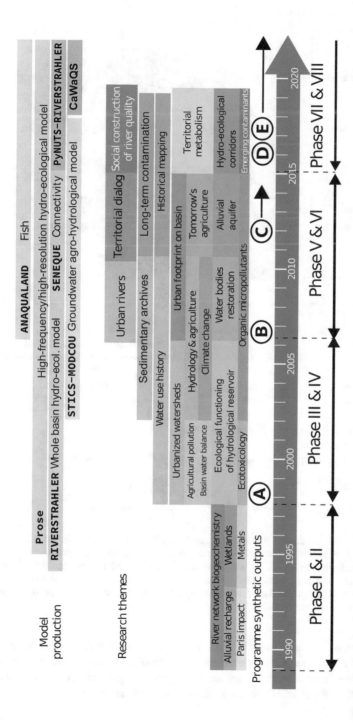

Fig. 9 Gradual development of research themes and modelling tools of the PIREN-Seine programme (1989–2019). A, [14]; B, [15]; C, [17]; D, [16]; E, [69]. For model references, see Sect. 4.1

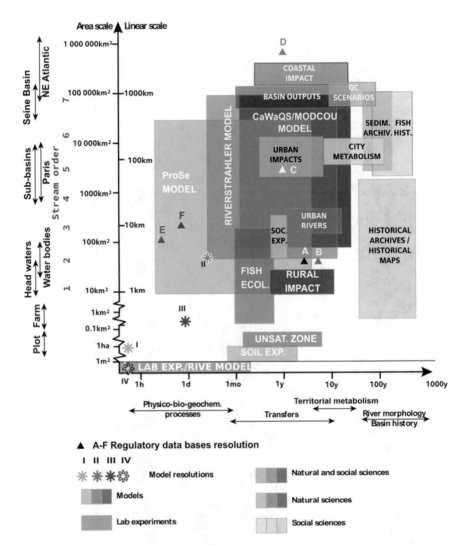

Fig. 10 Main research themes mobilized to describe, understand, and model the Seine River basin and its social interactions, presented in a spatio-temporal diagram. The resolutions of models (I, ProSe; II, Riverstrahler; III, CaWaQS/MODCOU; IV, RIVE) and of regulatory databases (A, population census; B, agricultural census; C, economic statistics (administrative districts); D, economic statistics (national); E, meteorological data; F, hydrological data) are distinguished from the scales of the research items

Public Company of the Greater Paris), VNF, Voies Navigables de France, and the CNRS via the Zones Ateliers network [77]. Funding is used to support specific projects and scholarships to address the main scientific questions collectively arising for the PIREN-Seine scientific and actor partners and stakeholders.

The PIREN-Seine has gradually managed to build up permanent communication gateways and meeting places where all these parties can meet, communicate, and exchange their experience. Over 30 years, numerous scientists, trained under the programme, have reached the public or private sectors dealing with environmental management; some of them are now involved, together with their former PhD advisors, in the stakeholders' meetings of the programme, sharing their views and knowledge, as well as their concepts and tools.

Another set of products comprising leaflets and booklets that have been produced since 2009 is directly targeted at French end users: technicians of the water and/or river institutions, environmental engineers, associations, and educators. These are written by PIREN-Seine scientists and were first published jointly with the AESN and now with ARCEAU Idf [78]. They are freely available [17, 69]. Each one deals with a specific topic. The 40–60-page booklets – a total of 18 in 2019 – can be considered as a whole book on the Seine River basin and its evolution, on river functioning, and on river-society interactions.

The PIREN-Seine is now embedded in the interdisciplinary instrument Zone Atelier (ZA) Seine of the CNRS [79], which comprises two other projects on Parisian urban waters and on the estuary. These ZAs, including the ZA Seine, are part of the national and international LTSER network [80]. Through the network of ZAs, the Seine River programme is connected to other collaborative river research programmes (Rhône, Moselle, Loire, and Garonne).

In addition to these written products, the PIREN-Seine scientists also deliver an annual oral reporting of their main results to their partners. These meetings, established at the outset and opened to associations and journalists, are attended by around 200 people over a 2-day period. They have greatly contributed to establishing mutual trust and shared visions among scientists and partners. These meetings are also attended by the doctoral and post-doctoral students of the programme who also present their main results. The great variety of presentations covering the entire spectrum of our research promotes the interdisciplinary vision built up by the PIREN-Seine community. The programme has also carried out social experiments: They bring together PIREN-Seine social scientists and modellers with groups of citizens, association members, and water institutions technicians, who meet several times to present their visions of river functioning and uses. These experiments, which last at least 1 year, allow the social scientists to explore how the territorial dialogue can be built. Some of the model tools that have been adapted to the territorial conditions can be used to explore local-scale scenarios (e.g. for conflicting river uses, restoration of fish migration, reshaping the agricultural water-scape) (see examples in [50]). At the scale of the whole basin, other scenarios are being developed between scientists and partners, exploring possible future perspectives, particularly in terms of sanitation, agriculture, water demand, and flood security demand (future water storage) [32, 64, 81, 82].

5 The Spatio-Temporal Scales of the Research Themes Selected in This Publication

The selected research themes presented here encompass a wide range of space and time scales schematized in Fig. 10. In addition to laboratory analyses (e.g. characterization of organic matter [83], analysis of contaminants [60, 84], ecotoxicology [85], microbiology [40]), the field activities range from agricultural plot studies to interannual river exports assessed at the basin outlet. Various intermediate time and spatial scales are considered, depending on the scientific or societal question being addressed. They include numerous experimental sites (Fig. 1) that are selected to cover a wide range of impacts, which are generally studied over periods of 3–30 years. At the process scale, the functioning is, for instance, studied in first-order catchments of the rural Orgeval site (II on Fig. 1), an experimental site established in the early 1960s, initially for hydrology and drainage impacts and then for nutrient budgets and pesticides [86]. Recently, continuous records have been set up for hydrological, and especially stream aquifer interactions [87, 88], and biogeochemical monitoring as part of innovative national equipment for the critical zone [89]. Other experimental sites include mid-order streams (3 and 4), such as the Grand Morin River (III on Fig. 1), for which social experiments among river users have been undertaken [50, 90], and the Orge River, a suburban river in the Paris region with a high population density, where many analyses of legacy and emerging contaminants have been carried out (IV in Fig. 1) [85]. The Paris station on the Middle Seine (order 7), just downstream of the Seine-Marne confluence, is located near the laboratory facilities and is used in particular to study the seasonal variability of river chemistry, with a handful of other key stations on upper stream orders, including the Seine reservoirs (5–7, Fig. 1). The Lower Seine course, downstream of Paris (VI on Fig. 1), corresponds to the maximum degradation sector of the Seine River. The impact of the Paris conurbation on this section of the river is present in almost all the following chapters (see, e.g. [37, 40]). The basin output is considered at the Poses station (Fig. 1), just upstream of a weir preventing tide propagation from the estuary. This station at Poses integrates all information on the whole basin upstream. An original feature of the PIREN-Seine programme was to take into account the whole continuum from small streams to the Seine mouth since the outset of the programme [32, 41, 46, 66].

The PIREN-Seine has gradually developed numerous models [50]. These are schematized in Fig. 10 into items with different space and time resolutions: (1) the biogeochemical RIVE model embedded in the Riverstrahler model, which describes the river continuum (10 days, 1 km river reach) [41], and in the ProSe model devoted to a two-dimensional description of the Parisian river sector (15 min, 0.1 km) [12, 91], and (2) the CaWaQS model (formerly MODCOU model), devoted to groundwaters (1 day; 1–5 km^2) [32, 92]. These models are actually interconnected and are also connected to another set of estuarine and coastal models. The trajectory of the Seine River has been modelled for retrospective and climate change scenarios for both a hydrological [32] and a biogeochemical perspective [41, 76].

The longue durée evolution of the Seine basin is studied over 50–500 years. Sedimentary archives [26] are used to reconstruct the records of dozens of metals and persistent contaminants that were not adequately surveyed before the twenty-first century. Their interpretation is aided by the historical research on archives for past pressures, water quality analyses, and social responses dating back more than 150 years. The historical maps [48] enable a description of geomorphological modifications in the river course since the eighteenth century, which can be very slow (as with most natural processes) or stepwise (channelization, damming).

All these areas of research are presented here as belonging to the natural sciences (e.g. hydrology, environmental chemistry, biogeochemistry, etc.), the social sciences (history, geography, economics), or their combinations (agronomy, social experiments based on river models, scenarios of basin evolution, etc.). These distinctions are now less and less marked, in particular with regard to the scenarios run with the models and the analysis of socioecological trajectories.

The field, laboratory, and archive research activities are jointly analysed and processed with many regulatory databases and geographic information system (GIS) representations, thereby allowing a description of the entire basin and its evolution. Many of these databases were not set up for environmental studies, but for economic purposes, and, apart from the river discharge measurements, none of them is generated with hydrological boundaries, but rather at administrative levels [52, 53]. All are established at various space and time resolutions, from the daily-subdaily scale and 10 km^2 for the hydro-meteorological data (Fig. 10, database A) to the yearly statistics of the national economy (database D). These databases are a precious asset of the Seine River basin, with some of them starting at the beginning of the nineteenth century.

Despite the diversity of the spatial and temporal scales shown here, the joint work carried out over 30 years has one common goal: to show the evolution of the Seine basin and its possible future, given the construction of the past trajectories and possible future scenarios that are presented in the following chapters.

Acknowledgements This work is a contribution to the PIREN-Seine research programme (www. piren-seine.fr), which belongs to the Zone Atelier Seine part of the international Long-Term Socio-Economic and Ecosystem Research (LTSER) network.

References

1. Crutzen PJ (2002) Geology of mankind. Nature 415:23. https://doi.org/10.1038/415023a
2. Crutzen PJ, Steffen W (2003) How long have we been in the Anthropocene era? An editorial comment. Clim Chang 61:251–257. https://doi.org/10.1023/B:CLIM.0000004708.74871.62
3. Hamilton C, Gemenne F, Bonneuil C (eds) (2015) The Anthropocene and the global environmental crisis: rethinking modernity in a new epoch. Routledge, London
4. Latour B (2015) Face à Gaïa: Huit conférences sur le nouveau régime climatique. La Découverte, Paris

5. Haberl H, Fischer-Kowalski M, Krausmann F et al (eds) (2016) Social ecology. Society-nature relations across time and space. Springer, Heidelberg. https://doi.org/10.1007/978-3-319-33326-7
6. Jollivet M (2001) Un exemple d'interdisciplinarité au CNRS: le PIREN (1979-1989). Rev Hist CNRS 4. https://journals.openedition.org/histoire-cnrs/3092
7. Bouleau G, Fernandez S (2012) Trois grands fleuves, trois représentations scientifiques. In: Gautier D, Benjaminsen T (eds) L'approche Political Ecology: pouvoir, savoir et environnement. Editions Quae, Paris, pp 201–217
8. Billen G, Garnier J, Hanset P (1994) Modelling phytoplankton development in whole drainage networks: the RIVERSTRAHLER model applied to the Seine river system. Hydrobiologia 289:119–137
9. Even S, Poulin M, Garnier J et al (1998) River ecosystem modelling: application of the ProSe model to the Seine River (France). Hydrobiologia 373:27–37
10. Even S, Poulin M, Mouchel JM et al (2004) Modelling oxygen deficits in the Seine River downstream of combined sewer overflows. Ecol Model 173:177–196
11. Even S, Mouchel JM, Servais P et al (2007) Modeling the impacts of combined sewer overflows on the river Seine water quality. Sci Total Environ 375:140–151. https://doi.org/10.1016/j.scitotenv.2006.12.007
12. Vilmin L, Flipo N, Escoffier N et al (2016) Carbon fate in a large temperate human-impacted river system: focus on benthic dynamics. Glob Biogeochem Cycles 30:1086–1104. https://doi.org/10.1002/2015GB005271
13. Billen G (2001) Le PIREN-Seine: un programme de recherche né du dialogue entre scientifiques et gestionnaires. Rev Hist CNRS 4. http://journals.openedition.org/histoire-cnrs/3182
14. Meybeck M, de Marsily G, Fustec E (eds) (1998) La Seine en son bassin. Elsevier, Paris
15. Billen G, Garnier J, Mouchel JM et al (2007) The Seine system: introduction to a multidisciplinary approach of the functioning of a regional river system. Sci Total Environ 375(1):1–12
16. Mouchel JM (2018) Spatial and temporal patterns of anthropogenic influence in a large river basin? A multidisciplinary approach. Environ Sci Pollut R 25(24):23373–23594
17. PIREN Seine booklets. https://www.piren-seine.fr/fr/fascicules/. Accessed 15 July 2019
18. Ayres RU, Ayres LW (eds) (2002) A handbook of industrial ecology. Edward Elgar, Cheltenham
19. Baccini P, Brunner PH (2012) Metabolism of the anthroposphere: analysis, evaluation, design. MIT Press, Cambridge
20. Lammers PEM, Gilbert AJ (1999) Towards environmental pressure indicators for the EU: indicator definition. EUROSTAT, Brussels
21. Svarstad H, Petersen LK, Rothman D et al (2008) Discursive biases of the environmental research framework DPSIR. Land Use Policy 25(1):116–125
22. Fernandez S, Bouleau G, Treyer S (2014) Bringing politics back into water planning scenarios in Europe. J Hydrol 518:17–27
23. Scholz RW, Binder CR, Lang DJ (2011) The HES framework. In: Scholz RW (ed) Environmental literacy in science and society. From knowledge to decisions. Cambridge University Press, Cambridge, pp 453–462
24. Ostrom E (2007) A diagnostic approach for going beyond panaceas. Proc Natl Acad Sci U S A 104:15181–15187
25. Binder CR, Hinkel J, Bots PWG et al (2013) Comparison of frameworks for analyzing social-ecological systems. Ecol Soc 18(4):26. https://doi.org/10.5751/ES-05551-180426
26. Ayrault S, Meybeck M, Mouchel JM et al (2020) Sedimentary archives reveal the concealed history of micropollutant contamination in the Seine River basin. In: Flipo N, Labadie P, Lestel L (eds) The Seine River basin, Hdb Env Chem. Springer, Cham. https://doi.org/10.1007/698_2019_386

27. Guillocheau F, Robin C, Allemand P et al (2000) Meso-Cenozoic geodynamic evolution of the Paris Basin: 3D stratigraphic constraints. Geodin Acta 13:189–245. https://doi.org/10.1080/09853111.2000.11105372
28. Guerrini MC, Mouchel JM, Meybeck M et al (1998) Le bassin de la Seine: la confrontation du rural et de l'urbain. In: Meybeck M, de Marsily G, Fustec E (eds) La Seine en son bassin. Elsevier, Paris, pp 29–75
29. Strahler AH (1957) Quantitative analysis of watershed geomorphology. EOS Trans Am Geophys Union 38(6):913–920
30. Quintana-Seguí P, Moigne PL, Durand Y et al (2008) Analysis of near-surface atmospheric variables: validation of the SAFRAN analysis over France. J Appl Meteorol Clim 47:92–107
31. Vidal JP, Martin E, Franchistéguy L et al (2010) Multilevel and multiscale drought reanalysis over France with the Safran-Isba-Modcou hydrometeorological suite. Hydrol Earth Syst Sci 14:459–478
32. Flipo N, Gallois N, Labarthe B et al (2020) Pluri-annual water budget on the Seine basin: past, current and future trends. In: Flipo N, Labadie P, Lestel L (eds) The Seine River basin, Hdb Env Chem. Springer, Cham. https://doi.org/10.1007/698_2019_392
33. Massei N, Laignel B, Deloffre J et al (2010) Long-term hydrological changes of the Seine River flow (France) and their relation to the North Atlantic oscillation over the period 1950-2008. Int J Climatol 30:2146–2154
34. Flipo N, Monteil C, Poulin M et al (2012) Hybrid fitting of a hydrosystem model: long term insight into the Beauce aquifer functioning (France). Water Resour Res 48:W05509
35. Garnier J, Meybeck M, Ayrault S et al (2020) Continental Atlantic rivers: the Seine basin. In: Tockner K, Zarfl C, Robinson C (eds) Rivers of Europe, 2nd edn. Elsevier, London. ISBN: 9780081026120 (in press)
36. GIP Seine-Aval. https://www.seine-aval.fr/. Accessed 15 July 2019
37. Esculier F, Barles S (2020) Past and future trajectories of human excreta management systems - the case of Paris 19th-21st centuries. In: Flipo N, Labadie P, Lestel L (eds) The Seine River basin, Hdb Env Chem. Springer, Cham. https://doi.org/10.1007/698_2019_407
38. Barles S, Guillerme A (2014) Paris: a history of water, sewers, and urban development. In: Tvedt T, Oestigaard T (eds) Water and urbanization. A history of water, series III, vol 1. Tauris, London, pp 384–409
39. Lestel L, Carré C (eds) (2017) Les rivières urbaines et leur pollution. Quae, Paris
40. Mouchel JM, Lucas F, Moulin L et al (2020) Bathing activities and microbiological water quality in the Paris area: a long-term perspective. In: Flipo N, Labadie P, Lestel L (eds) The Seine River basin, Hdb Env Chem. Springer, Cham. https://doi.org/10.1007/698_2019_397
41. Garnier J, Marescaux A, Guillon S et al (2020) Ecological functioning of the Seine River: from long term modelling approaches to high frequency data analysis. In: Flipo N, Labadie P, Lestel L (eds) The Seine River basin, Hdb Env Chem. Springer, Cham. https://doi.org/10.1007/698_2019_379
42. Meybeck M, Lestel L, Carré C et al (2018) Trajectories of river chemical quality issues over the Longue Durée: the Seine River (1900S–2010). Environ Sci Pollut R 25(24):23468–23484
43. Billen G, Barles S, Chatzimpiros P et al (2012) Grain, meat and vegetables to feed Paris: where did and do they come from? Localising Paris food supply areas from the eighteenth to the twenty-first century. Reg Environ Chang 12(2):325–335
44. Mignolet C, Schott C, Benoît M (2007) Spatial dynamics of farming practices in the Seine basin: methods for agronomic approaches on a regional scale. Sci Total Environ 375:13–32
45. Le Noë J, Billen G, Esculier F et al (2018) Long-term socioecological trajectories of agro-food systems revealed by N and P flows in French regions from 1852 to 2014. Agric Ecosyst Environ 265:132–143. https://doi.org/10.1016/j.agee.2018.06.006
46. Billen G, Garnier J, Le Noë J et al (2020) The Seine watershed water-agro-food system: long-term trajectories of C, N, P metabolism. In: Flipo N, Labadie P, Lestel L (eds) The Seine River basin, Hdb Env Chem. Springer, Cham. https://doi.org/10.1007/698_2019_393

47. Bouleau G, Marchal PL, Meybeck M et al (2017) La construction politique d'un espace de commune mesure pour la qualité des eaux superficielles. L'exemple de la France (1964) et de l'Union Européenne (2000). Développement durable et territoires 8(1). https://doi.org/10.4000/developpementdurable.11580

48. Lestel L, Eschbach D, Meybeck M et al (2020) The evolution of the Seine basin water bodies through historical maps. In: Flipo N, Labadie P, Lestel L (eds) The Seine River basin, Hdb Env Chem. Springer, Cham. https://doi.org/10.1007/698_2019_396

49. Meybeck M, Lestel L (2017) A Western European River in the Anthropocene. The Seine, 1870-2010. In: Kelly JM, Scarpino P, Berry H et al (eds) Rivers of the Anthropocene. University of California Press, Oakland, pp 84–100

50. Carré C, Meybeck M, Garnier J et al (2020) River basin vision: tools and approaches, from yesterday to tomorrow. In: Flipo N, Labadie P, Lestel L (eds) The Seine River basin, Hdb Env Chem. Springer, Cham. https://doi.org/10.1007/698_2019_438

51. Schwarz HE, Emel J, Dickens WJ et al (1990) Water quality and flows. In: Turner II BL, Clark WC, Kates R et al (eds) The earth as transformed by human action. Global and regional changes in the biosphere over the past 300 years. Cambridge University Press, Cambridge, pp 253–270

52. Lestel L (2012) Non-ferrous metals (Pb, Cu, Zn) needs and city development: the Paris example (1815–2009). Reg Environ Chang 12(2):311–323

53. Lestel L, Meybeck M, Thévenot D (2007) Metal contamination budget at the river basin scale: an original Flux-Flow Analysis (F2A) for the Seine River. Hydrol Earth Syst Sci 11(6):1771–1781

54. Meybeck M (2013) Heavy metal contamination in rivers across the globe: an indicator of complex interactions between societies and catchments. In: Arheimer B, Collins A, Krysanova V et al (eds) Understanding freshwater quality problems in a changing world. IAHS Publ., vol 361. IAHS Press, Wallingford

55. Meybeck M, Lestel L, Bonté P et al (2007) Historical perspective of heavy metals contamination (Cd, Cr, Cu, Hg, Pb, Zn) in the Seine River basin (France) following a DPSIR approach (1950–2005). Sci Total Environ 375(1):204–231. https://doi.org/10.1016/j.scitotenv.2006.12.017

56. Garnier J, Lassaletta L, Billen G et al (2015) Phosphorus budget in the water-agro-food system at nested scales in two contrasted regions of the world (ASEAN-8 and EU-27). Global Biogeochem Cycles 29(9):1348–1368. https://doi.org/10.1002/2015GB005147

57. Billen G, Thieu V, Garnier J et al (2009) Modelling the N cascade in regional watersheds: the case study of the Seine, Somme and Scheldt rivers. Agric Ecosyst Environ 133(3–4):234–246

58. Billen G, Garnier J, Némery J et al (2007) A long-term view of nutrient transfers through the Seine river continuum. Sci Total Environ 375(1):80–97

59. Thévenot DR, Moilleron R, Lestel L et al (2007) Critical budget of metal sources and pathways in the Seine River basin (1994–2003) for Cd, Cr, Cu, Hg, Ni, Pb and Zn. Sci Total Environ 375 (1):180–203

60. Gateuille D, Gaspery J, Briand C et al (2020) Mass balance of PAHs at the scale of the Seine River basin. In: Flipo N, Labadie P, Lestel L (eds) The Seine River basin, Hdb Env Chem. Springer, Cham. https://doi.org/10.1007/698_2019_382

61. Vörösmarty CJ, Meybeck M, Pastore CL (2015) Impair-then-repair: a brief history & global-scale hypothesis regarding human-water interactions in the Anthropocene. Daedalus 144 (3):94–109

62. Benoit P, Berthier K, Billen G et al (2004) Agriculture et aménagement du paysage hydrologique dans le bassin de la Seine aux XIVe et XVe siècles. In: Burnouf J, Leveau P (eds) Fleuves et marais, une histoire au croisement de la nature et de la culture. CTHS, Paris, p 235

63. Meybeck M, Lestel L, Briand C (2017) La Seine sous surveillance: Les analyses des impacts de l'agglomération parisienne par l'Observatoire de Montsouris de 1876 à 1937. In: Lestel L, Carré C (eds) Les rivières urbaines et leur pollution. Quae, Paris, pp 32–43

64. Garnier J, Le Noë J, Marescaux A et al (2019) Long-term changes in greenhouse gas emissions from French agriculture and livestock (1852-2014): from traditional agriculture to conventional intensive systems. Sci Tot Environ 660:1486–1501. https://doi.org/10.1016/j.scitotenv.2019.01.048

65. Belliard J, Beslagic S, Delaigue O et al (2018) Reconstructing long-term trajectories of fish assemblages using historical data: the Seine River basin (France) during the last two centuries. Environ Sci Pollut R 25(24):23430–23450

66. Belliard J, Beslagic S, Tales E (2020) Changes in fish communities of the Seine basin over a long-term perspective. In: Flipo N, Labadie P, Lestel L (eds) The Seine River basin, Hdb Env Chem. Springer, Cham. https://doi.org/10.1007/698_2019_380

67. Bonnet R (2018) Variations du cycle hydrologique continental en France des années 1850 à aujourd'hui. Dissertation, Université de Toulouse 3 Paul Sabatier

68. Meybeck M (2002) Riverine quality at the Anthropocene: propositions for global space and time analysis, illustrated by the Seine River. Aquat Sci 64(4):376–393

69. PIREN Seine leaflets. https://www.piren-seine.fr/content/fiches-4-pages. Accessed 15 3July 2019

70. Garnier J, Billen G, Coste M (1995) Seasonal succession of diatoms and Chlorophyceae in the drainage network of the river Seine: observations and modelling. Limnol Oceanogr 40:750–765

71. Flipo N, Even S, Poulin M et al (2004) Biogeochemical modelling at the river scale: plankton and periphyton dynamics – Grand Morin case study, France. Ecol Model 176:333–347

72. Vilmin L, Flipo N, De Fouquet C et al (2015) Pluri-annual sediment budget in a navigated river system: the Seine River (France). Sci Total Environ 502:48–59. https://doi.org/10.1016/j.scitotenv.2014.08.110

73. Ledoux E, Gomez E, Monget JM et al (2007) Agriculture and groundwater nitrate contamination in the Seine basin. The STICS-MODCOU modelling chain. Sci Total Environ 375:33–47

74. Beaudoin N, Gallois N, Viennot P et al (2018) Evaluation of a spatialized agronomic model in predicting yield and N leaching at the scale of the Seine-Normandie basin. Environ Sci Pollut R 25:23529–23558. https://doi.org/10.1007/s11356-016-7478-3

75. Le Pichon C, Gorges G, Boët P et al (2006) A spatially explicit resource-based approach for managing stream fishes in riverscapes. Environ Manag 37:322–335. https://doi.org/10.1007/s00267-005-0027-3

76. Raimonet M, Thieu V, Silvestre M et al (2018) Landward perspective of coastal eutrophication potential under future climate change: the Seine River case (France). Front Mar Sci 5:136. https://doi.org/10.3389/fmars.2018.00136

77. PIREN Seine partners. https://www.piren-seine.fr/fr/partenaires. Accessed 15 Sept 2019

78. ARCEAU Idf. http://arceau-idf.fr/. Accessed 15 Oct 2019

79. Zones Ateliers. https://www.inee.cnrs.fr/fr/zones-ateliers. Accessed 15 Oct 2019

80. Long-term ecosystem research in Europe. https://www.lter-europe.net/. Accessed 15 Oct 2019

81. Passy P, Le Gendre R, Garnier J et al (2016) Eutrophication modelling chain for improved management strategies to prevent algal blooms in the Bay of Seine. Mar Ecol Prog Ser 543:107–125. https://doi.org/10.3354/meps11533

82. Billen G, Le Noë J, Garnier J (2018) Two contrasted future scenarios for the French agro-food system. Sci Total Environ 637–638:695–705. https://doi.org/10.1016/j.scitotenv.2018.05.043

83. Varrault G, Parlanti E, Matar Z et al (2020) Aquatic organic matter in the Seine basin: sources, spatio-temporal variability, impact of urban discharges and influence on micro-pollutant speciation. In: Flipo N, Labadie P, Lestel L (eds) The Seine River basin, Hdb Env Chem. Springer, Cham. https://doi.org/10.1007/698_2019_383

84. Labadie P, Alligant S, Berthe T (2020) Contaminants of emerging concern in the Seine River basin: overview of recent research. In: Flipo N, Labadie P, Lestel L (eds) The Seine River basin, Hdb Env Chem. Springer, Cham. https://doi.org/10.1007/698_2019_381

85. Bonnard M, Barijhoux L, Dedrouge-Geffard O et al (2020) Experience gained from ecotoxi-cological studies in the Seine River and its drainage basin over the last decade: applicative examples and research perspectives. In: Flipo N, Labadie P, Lestel L (eds) The Seine River basin, Hdb Env Chem. Springer, Cham. https://doi.org/10.1007/698_2019_384

86. Blanchoud H, Schott C, Tallec G et al (2020) How agricultural practices should be integrated to understand and simulate long-term pesticide contamination in the Seine River basin? In: Flipo N, Labadie P, Lestel L (eds) The Seine River basin, Hdb Env Chem. Springer, Cham. https://doi.org/10.1007/698_2019_385

87. Mouhri A, Flipo N, Rejiba F et al (2013) Designing a multi-scale sampling system of stream-aquifer interfaces in a sedimentary basin. J Hydrol 504:194–206. https://doi.org/10.1016/j.jhydrol.2013.09.036

88. Cucchi K, Rivière A, Baudin A et al (2018) LOMOS-mini: a coupled system quantifying transient water and heat exchanges in streambeds. J Hydrol 561:1037–1047. https://doi.org/10.1016/j.jhydrol.2017.10.074

89. Gaillardet J, Braud I, Hankard F et al (2019) OZCAR: the French network of critical zone observatories. Vadose Zone J. https://doi.org/10.2136/vzj2018.04.0067

90. Carré C, Haghe JP, De Coninck A et al (2014) How to integrate scientific models to switch from flood river management to multifunctional river management. Int J River Basin Manag 12 (3):231–249. https://doi.org/10.1080/15715124.2014.885439

91. Vilmin L, Aissa-Grouz N, Garnier J et al (2015) Impact of hydro-sedimentary processes on the dynamics of soluble reactive phosphorus in the Seine River. Biogeochemistry 122:229–251. https://doi.org/10.1007/s10533-014-0038-3

92. Pryet A, Labarthe B, Saleh F et al (2015) Reporting of stream-aquifer flow distribution at the regional scale with a distributed process-based model. Water Resour Manag 29:139–159. https://doi.org/10.1007/s11269-014-0832-7

The Evolution of the Seine Basin Water Bodies Through Historical Maps

Laurence Lestel, David Eschbach, Michel Meybeck, and Frédéric Gob

Contents

1 Introduction 30
2 The Physical Anthropogenic Transformation of the Seine River (1800–2010) 31
  2.1 Transformations of Seine Water Bodies in the Early 1800s and Their Drivers 31
  2.2 Current State of the Seine River Water Bodies 34
3 Historical Maps: A Tool for Quantifying the Physical Changes of Water Bodies 36
  3.1 The Inventory and Critical Analysis of River Maps 38
  3.2 General Use of Maps to Document River Environmental Changes 40
4 Historical Trajectories of Selected Water Bodies in the Seine River Basin 45
  4.1 Stream Network Modification on the Versailles-Saclay Plateau (1670–1860) 46
  4.2 Man-Made Heterogeneity of the Floodplain: Channelisation and Sandpits
     in the Bassée Floodplain 47
  4.3 Simplification of Seine River Channel and Loss of Islands 50
5 Conclusion and Perspectives 52
References 54

Abstract The Seine River basin (65,000 km^2) is extremely rich in cartographic documents generated over the past two centuries: general maps describing the territory, fiscal land registries, navigation charts (e.g. bathymetric profiles and maps), etc. After 1830 river engineers (Ponts et Chaussées) started to develop a huge network of waterways, which were charted with precision and accuracy. These documents, retrieved from various archives, have been checked, selected, geo-referenced and digitalised within an open-access database (ArchiSeine). It has allowed researchers to fully quantify the state of rivers, often in their lateral, longitudinal and vertical dimensions, their long-term and slow natural dynamics

The copyright year of the original version of this chapter was corrected from 2019 to 2020. A correction to this chapter can be found at https://doi.org/10.1007/698_2020_667

L. Lestel (✉), D. Eschbach, and M. Meybeck
Sorbonne Université, CNRS, EPHE, UMR Metis, Paris, France
e-mail: laurence.lestel@sorbonne-universite.fr

F. Gob
Université Paris 1 Panthéon Sorbonne, Laboratoire de Géographie Physique, Meudon, France

Nicolas Flipo, Pierre Labadie, and Laurence Lestel (eds.), *The Seine River Basin*,
Hdb Env Chem (2021) 90: 29–58, https://doi.org/10.1007/698_2019_396,
© The Author(s) 2020, corrected publication 2020, Published online: 3 June 2020

(e.g. meander movement) and their abrupt modifications by man-made river works due to various and evolving river use (water supply, wood rafting, navigation, hydropower, sand extraction, flood protection), all closely connected to Paris growing demands, and the adjustments of the fluvial system to these changes. From headwaters to the estuary, the physical attributes of the Seine River system have been substantially modified. Examples of such environmental trajectories are provided for the Versailles plateau headwaters, the Bassée alluvial plain and the Lower Seine sector.

Keywords Environmental trajectories, Geo-history, Historical maps, Secular evolution, Waterbodies

1   Introduction

Rivers and their physical evolution are prime examples of the longue durée (multi-secular) spatial evolution of the environment, particularly as concerns the hydro-morphology of the fluvial system – the river channel(s) and the floodplain at the maximum flood stage [1], ecology [2] and the river landscape [3].

More recently river management and restoration operation have also considered the historical evolution of the fluvial system [4], for instance, to restore the ecological continuity or to take regulatory measures to face extreme events such as droughts and floods. To achieve these objectives, historical data, textual and cartographic, have been increasingly developed since the 1960s. This has been facilitated over the last three decades by the new geographic information system (GIS) techniques and the digitalisation of maps, which allows the analysis of selected cartographic elements represented on a given map, and the comparison of maps, which makes it possible to evaluate spatial dynamics [5–8].

Historical cartographic documents have already been analysed for environmental purposes on many European fluvial systems such as the Po [9], the Rhône [10], the Danube [11] and the Rhine [12]. This approach has also been used by river historians [13–16] and to compare the river-city relations for four main European cities (Paris, Brussels, Berlin and Milan) [17].

In natural conditions, numerous factors may change the hydrology and hydro-morphology of rivers, which in turn have an impact on the river hydraulics and on sediment transport, including the floodplain [18, 19]. The response of the fluvial system to historical man-made changes is less known. In many contemporaneous environmental studies, focused on annual to decadal time scales, the slow secular river evolution is not taken into account: the physical attributes of the river are considered invariant, particularly as concerns the river morphology, and the past natural- or human-induced transformations of the system are not considered. Actually the Seine River system has been gradually modified since the Middle Ages, particularly between the mid-1800s and the 1970s for navigation purpose, from the headwaters to the estuary [20], by sand and gravel extraction in alluvial plains [21] and evolving floodplains and wetlands [22].

Our objectives here were to illustrate the secular evolution of the Seine River and its related water bodies in order to quantify the trajectories of their physical evolution in their triple dimension – longitudinal, transversal and vertical – based on historical maps. Herein the trajectories are considered as the combination of (1) a set of quantified indicators of changes, natural, man-made or both and (2) the analysis of the historical causes and the context of these changes.

The first section of the chapter is a general overview of the multiple river uses in the Seine basin and the related man-made physical transformations of water bodies, from the headwaters to the estuary. Then we present the historical cartographic sources and the methodology that have been developed to transform them into GIS databases. In the last section we present a selection of trajectories in headwater streams, a river floodplain and the Seine River channel, based on cartographic archives covering the last 200 years.

2 The Physical Anthropogenic Transformation of the Seine River (1800–2010)

The river hydrological network is schematised here using the Strahler representation of stream orders, from the first permanent headwater streams (stream order 1) to the mouth of the Seine River (stream order 7) (Fig. 1). The stream order increases when two streams of similar n orders meet to form a $n + 1$ order. In the Seine basin, the 5,610 stream order 1 rivers are on average 0.07 m deep and 1.76 km long and drain 5.7 km^2 [23]. These parameters reach their maximum values for the 438-km-long Seine River from its confluence with the Yonne River to the sea (stream order 7). We have considered two schematic pictures of the basin, one around 1800 and the other represents the contemporary period (2010s).

2.1 Transformations of Seine Water Bodies in the Early 1800s and Their Drivers

Before the French Revolution, the river basin was already somehow affected by human activities (Fig. 1, upper part).

In agricultural land, the headwater network was not significantly modified (Fig. 1a). However, in the headwaters of the forested Morvan mountainous massif (Yonne tributary headwaters, altitude 600–900 m), many medium-sized ponds (<3 ha) regulated water reservoirs for timber rafting. High flows were generated in November to allow the transportation of "free" logs, 1.14 m long, from the Morvan to the city of Clamecy, where they were stored for several months and eventually assembled and released downstream as 36- then 72-m-long rafts by the spring high flows. This usage had a considerable effect on rivers, due to bank erosion and regular high discharges [24]. Timber rafting, which started in the sixteenth century, was generated by Paris's demand for construction and fuel wood, peaked

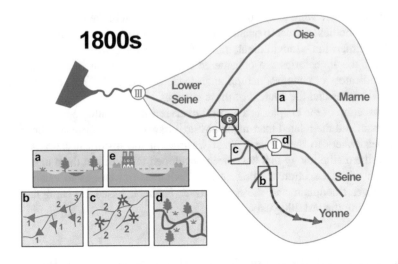

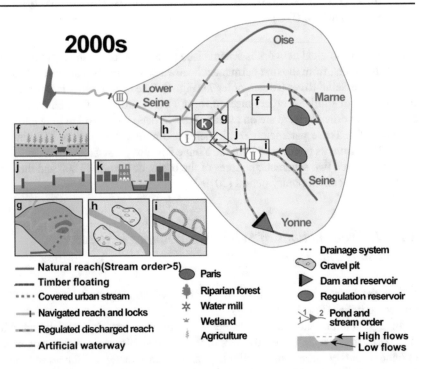

Fig. 1 Schematic presentation of the Seine River basin and its main tributaries (orders 6–7) with zooms on specific man-made physical transformations. *Situation in the early 1800s*: (**a**) unregulated plateau streams; (**b**) fish ponds (plateaus) and water storage ponds for timber rafting; (**c**) fixed water mills; (**d**) free meandering in alluvial plain (e.g. Bassée); (**e**) Seine within Paris city. *Situation in the 2000s*: (**f**) tile drains, irrigation and ditches in cropland; (**g**) deepened and/or covered urban and peri-urban rivers; (**h**) leftover water-filled sandpits in floodplains; (**i**) artificial navigation channel cut

around 1800 with a total volume of 1–1.5 M steres per year and declined rapidly after 1900. In the rest of the basin, other ponds, mostly for fish production, had been established on headwaters since the Middle Ages [25]. An analysis of Cassini's maps showed that, in the late 1700s, 2,550 ponds were present in the whole Seine catchment (65,000 km^2), mostly on first-order streams (69%) [26]. In some regions their density approached 0.5/km^2 (Fig. 1b). Fish production was for local use (e.g. abbeys) or exported to nearby cities and more distant cities as Paris, via waterways. The impact of urban water demand on headwaters is fully documented for Versailles, which started to develop in 1663: the construction of the chateau and city involved a major diversion of stream waters (see Sect. 4, Fig. 5).

Fixed water mills were found on all stream orders, principally on orders 2–3 (Fig. 1c). They resulted in the partial diversion of streams through a new course with regulated slopes, ending with a small water fall (0.5–1 m), while the "old course" was mostly active at the high-water stage. Around Paris the mill density on small rivers was on average 0.5/km^2 (6,000 over 12,000 km^2) [27] or one mill per 5 km of river course. Within Paris and on navigated rivers, floating mills were established.

In the alluvial plain of the stream orders 3–4, an artificial network of ditches was used for centuries to flood grasslands when needed [25]. In greater floodplains (Fig. 1d), a few kilometres wide such as the Bassée (see below in case studies, Fig. 6), wetlands and riparian forest dominated. In these free-meandering lowlands, side arms of the main river, slack waters and abandoned oxbows were common; they were mostly activated and connected during major floods (>5-year return period).

Stream orders 5–7 were generally navigated, which created a major regulatory difference with non-navigated reaches: the former were managed by the state, the others by local owners, a distinction that continues today [28]. These navigated rivers were not yet regulated, either concerning the water discharge or the river level (Fig. 1e), with the noted exception of the Yonne, used for timber rafting.

Many of the modifications of the river course were already generated by Paris's demand for goods, food, fuel wood and construction wood, largely transported by waterways. In 1800 Paris was already a metropolis located in the centre of the Seine basin, with 11% of the basin's total population (0.5 M for a total of 4.5 M inhabitants) [29], which relied heavily on the basin hinterland for food and energy [30]. However, the Seine River was not regulated and only slightly artificialised; banks were largely natural and sandy slopes were still used for landings (Fig. 1e).

During the Napoleon era, the state began to develop the navigation facilities on the Seine within Paris, with harbours, artificial banks and docks [31]. Numerous institutions were related to the river transport, particularly the state Ponts et Chaussées corps of engineers in charge of managing the state-owned navigated rivers, corresponding to orders 5–7.

Fig. 1 (continued) across floodplains; (**j**) navigated sectors with stable levels between locks; (**k**) regulated river level and artificial banks within the urban sectors. *Case studies*: I, Versailles plateau; II, Bassée alluvial plain; III, Poses lock and weir

2.2 Current State of the Seine River Water Bodies

From 1830 to the present day, most of the water bodies of the Seine River have been physically modified, from the headwaters to the estuary. These changes have been gradual and planned for water resources management, navigation development, agricultural drainage, etc. (Fig. 1, lower part).

In the marl and limestone agricultural parts of the Seine basin (e.g. Beauce and Brie), the tile drainage of cropland reaches its maximum proportion (20–40%) [23]. It has been developed since the nineteenth century by specific engineers and peaked between the 1960s and 1990s (Fig. 1f). Intensification of agriculture has also led to the transformation of some first-order streams into ditches. Meanwhile summer sprinkler irrigation was developed. As a result, the wetland area was reduced, and many order 1 streams are no longer in natural conditions, except in the forested parts of the basin. Most ponds along stream orders 1–3 have now been converted into cropland. The Morvan forest ponds linked to rafting have not been maintained.

Water mills are no longer in operation, but orders 2–4 often keep their heritage of side channels and weirs. Historical maps allow to connect these past river works to the river's cultural history, which gives them a patrimonial value, although they are now considered by the Water Framework Directive as fish obstacles and devices blocking the sediment transit of the river, which should be removed.

Small rivers around Paris (orders 2 and 3) have greatly suffered from urban sprawl: their lower parts have been covered and transformed into sewers and then into storm runoff sewers (e.g. Bièvre, southwest of Paris, and Croult, north of Paris) (Fig. 1g). Urbanisation also results in the water sealing of some parts of catchment surfaces and therefore to riverbed erosion related to peak discharges during heavy rains [32].

In the floodplain of the Middle and Lower Seine, as far as 150 km from Paris, demands for sand and ballast for construction of the expansion of Paris, new railways and a nuclear power plant in the Bassée resulted in the excavation, from 1950 to 2000, of dozens of sandpits, 10–100 ha wide, filled with alluvial groundwaters, 2–5 m deep, some connected with the main channel (Fig. 1h; see also Fig. 6). Given that the sandpit sedimentation rate is low to medium (less than 1 cm year^{-1}), the filling of these water bodies will take centuries.

In 1830–1840, the French state initiated a major effort to develop river transport management for the needs of Paris. The navigated reaches of the Seine, Marne and Oise rivers are those in which the river course has been completely modified in its three dimensions. The trajectory of the river works has been uniform for nearly 200 years: the navigation channel is now narrower; the course is straightened, particularly in floodplains, and equipped with dikes or levees (Fig. 1i); the channel is deeper and regularly dredged; the connection with floodplain water bodies is decreased; and the river banks are steep. This has resulted in a major change on these river reaches, with new types of river works, docks and weirs (Fig. 1j) termed *recalibration* by river engineers, a positive expression during this period of river taming and dedication of the river to human uses. These changes were made in several steps, corresponding to political decisions and technical innovations [33]. River reaches were mapped in three dimensions with great precision and accuracy prior to any construction works, leaving behind precious archives that are used today to access the initial river course morphology.

Until the 1940s, the river level within Paris was not regulated, and extreme low levels were observed in summer (Fig. 1k). In 1910, an exceptional event occurred: the flooding of Paris and its suburbs by a 100-year return period flood. Another phase of river management ensued. The new objective was to maintain the river level at a fixed value ±35 cm, with minimum river discharge variations [34]. Flood protection and low-water discharge regulation for Paris led to the creation of four main reservoirs, which were constructed in the 1930s–1980s on the Yonne, Upper Seine, Aube and Marne rivers (schematically represented in Fig. 1), some 250 river km upstream of Paris. These facilities, financed by the city of Paris, increased the summer low flows from 25 m^3/s (extreme value at Paris) up to 100 m^3/s, thus reducing the impact of the Paris region on the water quality of the Lower Seine.

When comparing the 1800 and 2000 figures, the weight of Paris on the Seine River course is enormous, considering the local and proximal changes such as the artificialisation of the river courses within Paris, the high regulation of discharge and the water level, the loss of many suburban river reaches or the modification of the water balance over the urbanised sectors. But Paris also controls the river in a distal mode, either upstream by the regulation reservoirs for flood protection and low-water enhancement or 100–250 km around Paris on regulated and navigated river reaches and by the sandpit excavation. Many of these river transformations can be documented with historical maps. Cartographic document analysis can also be complemented by other approaches to assess the slow longue durée evolution of the river, such as the use of former furnace slags to measure the very slow river bedload movement (Box 1).

Box 1 Historical Slag Residues: Another Approach to Slow River Dynamics

In lowland meandering rivers such as the Seine, riverbed transport is very slow and the deposited sand-gravel material might only move during medium rare to rare events. In these rivers with relatively low energy and therefore with low morpho-sedimentary adjustments, the impact of obstacles such as weirs, river mill diversions, small reservoirs and navigation sluices is relatively unknown.

Coarse sediment transport in the Seine basin occurs at a slow pace, which can be quantified by the analysis of slag residues from historical high-furnace iron smelting, even if mill weirs have temporarily or permanently interrupted or disturbed sedimentary transit for several centuries.

(continued)

Box 1 (continued)

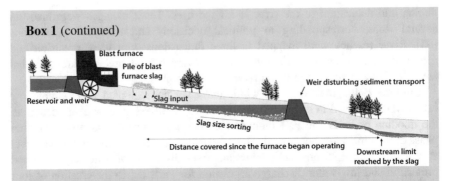

These small industries, using river power, were developed from sedimentary ore deposits in the Upper Marne region (Fig. 1), beginning in the fifteenth century until the early 1900s. From the knowledge of the injection site and its period of operation, by taking the largest slag present in the bed from upstream to downstream, we can assess the competence of the river and the rate of progression of the load background (see above). Their slag residues, of cm size, still found today in the river bed, have a density close to that of the natural gravel which characterises these upper basin reaches (2.2–2.5 g/cm^3 vs 2.6 g/cm^3). It can be postulated that they are transported at the same rate as natural riverbed particles [35].

A specific field study carried out on the Rognon, a medium-sized tributary to the Upper Marne, downstream of a well-dated and well-located former blast furnace, gives a minimum gravel transport rate of only 2.16 km/100 years. Slag residues have been detected in a large number of streams in the Seine basin, which make it possible to determine which grain size particles are regularly transported through the watercourse and their velocity displacement, as well as the role of transverse obstacles on their displacement.

3 Historical Maps: A Tool for Quantifying the Physical Changes of Water Bodies

Historical cartographic documents, maps, charts, plans, river profiles and drawings are stored in archive depositories, generally within administrative services, in thousands of closed boxes, each containing a dozen of maps in average, that must be opened as the preliminary step of the research, followed by the extraction of their cartographic documents, their scanning and the recording of their related information (objectives, techniques, producers, etc.). Documents are then selected according to their scale, coverage, represented elements and then geo-referenced and all river-related elements (punctual, linear, surface) digitalised in a GIS. Finally, their analysis gives quantitative information on the state of the river and its associated water bodies and their dynamics, over the last two centuries (Fig. 2). Several fields of river

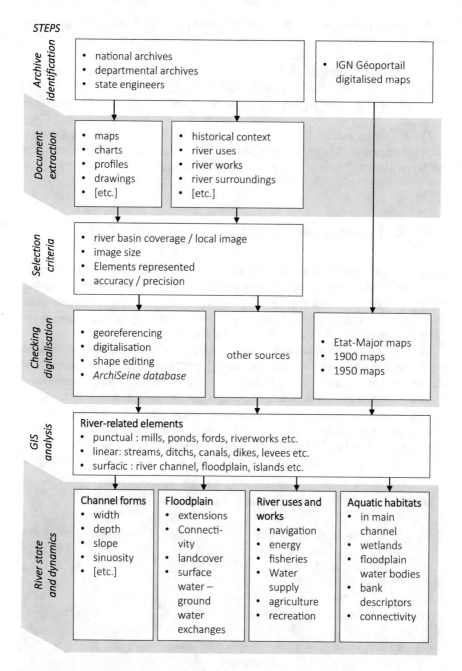

Fig. 2 Schematic steps of the analysis of historical cartographic documents on water bodies to quantify the their morphology, uses and water works as well as aquatic habitats and their dynamics

research may benefit from this quantified spatial and temporal analysis: hydro-morphology, hydro-ecology, socio-economic uses of the fluvial territory, river management and river restoration.

These treated documents related to rivers of the Seine basin were put in an open-access database (ArchiSeine) (Box 2), which complements the Geoportail open-access digitalised historical maps established by the IGN (Institut Géographique National).

The following section focuses on hydro-morphology, particularly for the navigated river channel. A brief overview of the history of the cartographic documents of the Seine River and their critical analysis is provided first.

3.1   The Inventory and Critical Analysis of River Maps

Historical cartographic documents provide area-wide information that is not available in classical archives. They give representations of river-related objects such as islands, river-land interfaces, hydrological networks, man-made infrastructures, river uses, river regulatory status, etc., as well as the territory (relief, pathways and villages) aside from the river network and its corridor.

In the 1700s maps had become an essential means of describing and understanding the territory and its evolution and an indispensable strategic and geopolitical instrument [36]. In France, Colbert's establishment in 1666 of the Royal Academy of Sciences allowed the development of cartographic methods, particularly triangulation by Jean Picard and then by a dynasty of cartographers founded by Jean-Dominique Cassini who achieved the first complete coverage of France at a scale of 1/86,400 in less than 100 years, from 1747 to 1818, but the representation of rivers was not always accurate and insufficiently precise for further quantitative analysis.

Box 2 ArchiSeine: An Open-Access Cartographic Database on River History

In 2013 the PIREN-Seine program launched a joint project with the French National archives to identify, restore and digitalise historical maps, charts and drawings related to the Seine River and its major tributaries. From 1,000 documents already identified, covering a period from the end of the eighteenth century to the 1930s, and chartered by the state's Pont et Chaussées corps of engineers, one-third have already been geo-referenced and presented on the ArchiSeine website [37, 38]. Each document is supplemented by metadata providing its historical context (authors, institution, objectives, etc.).

These charts and drawings are mainly (1) local maps related to planned and/or completed river works and (2) large-scale maps or drawings related to complaints and other judiciary acts concerning river works. These documents

(continued)

The second push for cartography was given by the creation of the Ponts et Chaussées corps of engineers, who were responsible for drawing up these maps. Progressively map status was transformed into a territorial management and decision-making instrument. For Verdier, this was "a profound mutation of cartography... for Ponts et Chaussées engineers, mapping was already part of solving a problem" [39]. These engineers also used standardised methods, initiated by Buchotte in the first specific cartographic manual published in France in 1722, for code colour by land use type or relief representation [40].

All land use and water use development projects were validated by Ponts et Chaussées technical advice, and the related maps were sent by engineers in each *département* and archived in the centralised Maps and Drawings Depository in the Ministry of Public Works, established under the First Empire [41]. The development of river navigation in 1830–1850 [33, 42, 43] generated a great number of

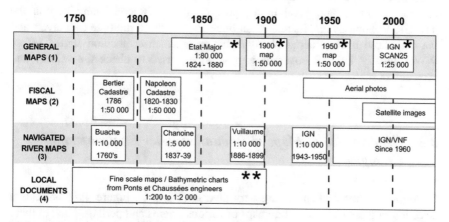

Fig. 3 Main sources of historical cartographic documents on the Seine River used in quantified trajectory analyses. *1*, General medium-scale mapping of the French territory; *2*, general fine-scale mapping for raising taxes (land registry); *3*, maps and charts produced by state engineers for navigated rivers; *4*, other local documents and bathymetric profiles. ∗ Open-access geo-referenced and digitalised documents available on IGN-Geoportail. ∗∗ ArchiSeine documents (see Box 2)

cartographic and textual documents which fulfill the different criteria for potential quantitative analysis. Many documents drawn by river engineers present the original state of the river, prior to their planned works, as well as the future situation after the *recalibration* (rectification, dredging, dikes and levee construction): this data set can therefore be used to approach the natural conditions of the river channel of orders 5–7, which were not yet greatly affected by human activities (see Sect. 2). The 1910 flood which damaged Paris resulted in an exceptional depository of these documents in the national archives.

The Seine River basin is covered by a great variety of maps (Fig. 3), in terms of spatial coverage, from local to basin-wide (65,000 km^2) spatial scales, extending over the last 200 years. In addition to the set of maps and drawings made by the Ponts et Chaussées engineers, one also finds the fine-scale land registry established for taxation purposes, such as the land registry supervised by Bertier [44], the Napoleonic Land Registry [45], or the Etat-Major Map (1/50,000) drawn by the War Ministry (1840–1860s).

Navigation charts give numerous details on the river channels. The first ones were drawn from 1731 to 1766 by Philippe Buache (1700–1773), considered as the first French geographer [46] (see Fig. 4c). They illustrate the many navigation difficulties (shoals, rocks) encountered on the Lower Seine reach, before the implementation of the first river management structures [42, 47]. The 19th July 1837 Law for navigation improvement on the Seine led to the drawing of the Seine River by Ponts et Chaussées engineers and the construction of the first set of six locks on the Seine River. Another set of charts was made between 1886 and 1899 by Raoul Vuillaume, the director of a new magazine dedicated to recreational navigation, *Le Yacht* (see Fig. 4e). These charts will be continued by institutions in charge of the navigation as the Seine Navigation Service (SNS, today Voies Navigables de France, VNF) and published by the IGN after the Second World War.

Around 2000 a major break in map production was observed stemming from the new GIS technology and its related on-line geo-referenced multiple data layers and the digitalised historic aerial photographic sets, which acquisition started before WWII. Finally, LIDAR coverage is now widely used to check the validity and accuracy of maps.

3.2 General Use of Maps to Document River Environmental Changes

Cartographic fluvial elements are very diverse (Table 1). Figure 4a–e presents a selection of these elements from maps of various origins and periods. Each of them provides specific information on the state of the river system at a given period and its potential changes in hydro-morphology, ecology, hydrology, aquatic resources, river uses and bio(geo)-chemical functioning. It is noteworthy that this state is the result of

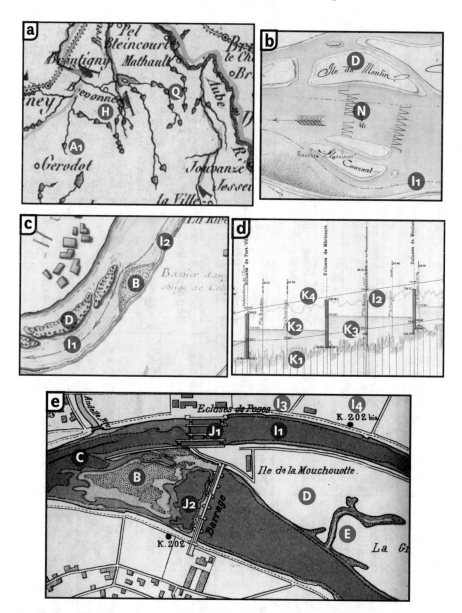

Fig. 4 Examples of fluvial elements represented on historical maps and listed in Table 1 (A–Q). (**a**) Timber rafting in 1828: ports (H), ponds (Q), first-order stream (A1) [Source: AN CP/F/14/10078, file 1, 67a]. (**b**) Fisheries on the Seine River in 1843 (N), islands (D), navigation channel (I1) [AN F/14/6577, file 5, 4]. (**c**) Shoals (B), towpath (I2) in 1766 on a Buache map [AN CP/F/14/10078, file 1, 2 h]. (**d**) 1883 River profile by Lagrené and Caméré: river bed (K1), regulated level (K2), low-water stage (K3), high-water stage (K4) [AN CP/F/14/10078, file 2, 51a]. (**e**) Lock (J1) and weir (J2), isobaths (C), disconnected channel (E), dikes (I3) and navigation mark (I4) in 1899 on a Vuillaume navigation map [Musée de la Batellerie et des voies navigables]

Table 1 Natural and man-made river elements represented on historical maps and charts and their deduced river properties and functionalities

Elements represented / Associated elements	Objectives of representation	Actual and potential consequences on waterbodies			
		Physical	Ecological	Socio-economic	Chemical
Natural entities					
A. River network (A1) (Fig. 4a) and flooded area (A2) ([38], *Fig. 9*)	Territorial/flood risk protection	Groundwater recharge/sediment deposition area	Land-river interface, wetlands, spawning areas	Water sources	C storage, denitrification, poll Archives
B. Fords/shoals (Fig. 4c)	River crossing	Coarse material accumulation	Spawning area	River crossing, navigation obstacle	
C. Bathymetric maps and profiles *Isobaths* (Fig. 4e)	Navigation	Dredged area, river channel	Fish habitat mapping (refuges, reproduction, feeding)	Reduction of navigation risks	
D. Island and/or gravel bar (Fig. 4b, c, e)	Navigation	Channel type and pattern [2]	Habitat diversity		
E. Disconnected lateral channels ([38], *Fig. 21*) *Lentic waters*	Land reclamation	Sediment trapping	Fish habitats (spawning areas, refuges)		Pollutant archives
Anthropogenic entities and engineering works					
F. Water mills *River cut-off, local bypass, canal, weir* ([38], *Fig. 12*)	Energy production	Modified flow, sediment discontinuity, modified surface/gw exchanges	Obstacle to fish movement habitat modification	Water management conflicts	Reoxygenation
G. Waterway port *Docks* ([38], *Fig. 11*)	Navigation	Bank artificialisation, dredging	Destruction of fish habitat area	Improved river access (+)	Deposition of particulate contaminants
H. Timber rafting (Fig. 4a)	Fuel/construction wood	Hydrological modifications and bank erosion [24]	River bank destruction, reduction in fish habitat complexity	Land loss due to erosion	Tannin release
I. Navigation channel (I1), towpath (I2), dikes and levees (I3), navigation marks (I4) (Figs. 4c and 9)	Navigation	Riverbank stabilisation/artificialisation, disconnection of lateral channels	Modification of fish habitat area/riparian vegetation Loss of habitat diversity	"Levee effect" [48, 49] Bank protection, gain of property	

J. Locks (J1) and weirs (J2) (Figs. 4e and 9)	Navigation	Upst./downst. modifications of flow and river/gw connexions	Fish habitat modification and simplification of modified riparian vegetation	Island loss, loss of property	Reoxygenation
K1–K4. Topographic profiles (Fig. 4d)	Navigation	Velocity regulation	Increased depth/decreased light penetration	Groundwater table modification/impact on adjacent wells	
L. Dams and reservoirs (Fig. 1)	River regulation	Modified river regime, upstream sediment trapping, modified surface water/gw dynamics, water evaporation	Facies replacement [4], source of non-native species. Obstacle to fish migration	Flood mitigation, loss of property, new recreation area (+)	
M. Bridges ([38], *Fig. 3*)	River crossing	Hydraulics modification, downstream erosion	Possible obstacle for fish		
N. Fisheries (Fig. 4b)	Food resource		Presence of fish (+)		
O. Gravel pits (Figs. 6 and 8)	Construction, public works	Changes in surface water/gw exchanges; water evaporation	Terrestrial/aquatic habitat shift	Landscape changes. New recreation area (+)	Groundwater exposure
P. Ditches *Tile drainage* (Fig. 5)	Agriculture	Increased water and fine sediment transfer	Loss of wetlands habitats	Increased croplands	
Q. Ponds (Fig. 4a)	Food resource/ Energy	Sediment trapping	Lentic facies. Source of non-native species	Food production	

gw groundwater

the cumulated natural and man-made transformations that have shaped the territory: any map reflects heritages of past dynamics.

Many natural (Table 1, A–E) and man-made (F–Q) elements were represented for navigation purposes, river network (A1) (see Fig. 4a) and navigation channel (I) (Fig. 4b, c, e), river depth (C) and topographic profiles (K) (Fig. 4d), obstacles (B, D, Fig. 4c), harbours (G) and locks (J) (Fig. 4e), or for river regulation, dams (L) or flooded area (A2). Other maps show places where energy is produced (water mills, F) or ponds (Q) and starting areas for timber rafting (H) (Fig. 4a), elements related to agriculture (ditches on very-fine-scale maps (P), Fig. 5) and food production (fisheries, N, Fig. 4b).

The physical river state and dynamics can be studied on numerous elements: identification of channel type and pattern from the presence and shape of the islands [4], modified flows (watermill derivations on a lateral canal (0.1–1 km) while the main river course is left active only at high flows) and changes in river regime by dams and weirs; dredging area (navigation channel and ports); loss of lateral connectivity due to the artificialisation of the banks and raising them to create towpaths; modified groundwater/surface water exchanges due to the change in the river's water level (dams and weirs) and to the presence of gravel pits (O); bank erosion in timber rafting reaches [24]; and sediment trapping within reservoirs (L), disconnection of lateral channels (E), delineation of flooded areas (A2) and ponds (Q).

The ecological state of the river can be addressed particularly through its fish habitats. Presence of islands (D), disconnected lateral channels and oxbows (E) provides potential refuges, spawning grounds and feeding grounds for fish [2]. The deepening of the river and the bank re-profiling seen on bathymetric maps (C) (e.g. Vuillaume, Fig. 4e) result in the loss of fish habitat areas and diversity. Dams (L), water mills (F) and weirs (J) are obstacles to fish migration. In the 1890s, weirs on the Lower Seine (Fig. 4d, e) were equipped with fish ladders, but they were not sufficiently effective to allow salmon, once common in the Seine basin [50, 51], to reach their spawning grounds 400 km upstream in the basin headwaters. The connectivity and the ecology of fluvial wetlands are also affected by river works, mainly dikes and levees (I) and more recently by sandpits.

The socio-economic uses of the river, the potential water conflicts and the flood risks can be analysed from river maps: the river network (A1) shows where to access surface water resources; water mills (F) mark where river power was used and potential water conflicts; navigation and river access are facilitated by river works; but new reservoirs (L) or erosion areas (H) result in loss of property. In contrast, river channeling (I) and the related disconnection/filling of backwater wetlands lead to a possible gain of property for the riparian landowners. Dikes and levees (I) were considered until recently as effective protection against flooding, leading to economic development close to the river. This positive "levee effect" ([48], Cited by [49]) is now being revised [49, 52, 53] because, although dikes prevent the natural inundation of the floodplain, they also limit the water storage at high flows and therefore increase the flood risk for downstream settlements.

Finally, some chemical effects on rivers can be related to fluvial elements represented on maps. Water mills (F), dams (L) and weirs (J) have downstream reoxygenation effects but upstream deoxygenation in the reservoirs, which are also more affected by eutrophication. Timber rafting (H) may release tannins. Gravel pits expose groundwaters to contamination. Docks, disconnected lateral channels and flooded areas (A2) are areas of pollutant storage which can be analysed to reconstruct former pollution levels and trends [54].

In addition, the comparison of successive maps provides insights into the dynamics of the fluvial system (Table 2). Examples of these dynamics and the methodology for quantifying these changes are provided in Sect. 4.

This diachronic analysis can be performed through planimetric (surface evolution) or vertical (depth evolution) comparisons (Sect. 4.2). The river network and the river channel can be completely modified, from the smallest stream orders, whose modification can lead to changes in the shape and size of the catchment area (Sect. 4.1), to the channelised river where many transformations can be demonstrated by comparing maps: the disappearance or modification of islands, often to facilitate navigation (Sect. 4.3), the so-called channel rectification, the channelisation by engineering works and the resulting channel adjustments, i.e. its natural evolution following an engineering structure downstream or upstream.

The comparison of maps also provides information on the evolution of land use in the catchment areas and the possible impact of this evolution on the physical attributes of the rivers, for instance, the evolution of agriculture towards more intensive practices, especially in first-order streams, which can lead to conflicts for water use, the appearance of gravel pits in the alluvial plain which modify its hydraulics (Sect. 4.2) or the waterproofing of urban soils, which results in the "urban stream syndrome", i.e. an increase of rain storm runoff, and riverbed erosion [32, 55].

Ecological responses to changes in fluvial dynamics can be assessed through changes in river fish habitat distribution. They can range from immediate (effect of damming on fish migration) to decadal (effects of minor bed deepening on the floodplain groundwater dynamics and its ecology).

Visualising and quantifying these changes can also help to understand the conflicts reported in the archives, such as water resource problems, property conflicts or increased flood risks.

4 Historical Trajectories of Selected Water Bodies in the Seine River Basin

The application of the use of historical maps is illustrated here through three examples: (1) Versailles plateau headwaters (stream orders 1–3), (2) Bassée floodplain (order 6) and (3) the Poses sector (order 7), which marks the limit between the Lower Seine and the fluvial estuary. All case studies show a pronounced human impact on the natural river system in the last 200–300 years.

Table 2 Quantitative dynamics of the fluvial system and its uses retrieved from the comparison of historical maps

Compared elements	Physical dynamics	Ecological dynamics	Socio-economic issues
First-order stream capture (Fig. 5)	Basin size and shape		Water resources imbalance
Island modification (levelling, attachment to banks) (Fig. 9)	Local flow acceleration	Loss of habitat diversity	Navigation improvement
Man-made channel correction (Figs. 6, 7 and 9)	Disconnection of secondary arms (hydraulic decoupling)	Loss of linkages between rivers and floodplains; loss of rearing habitat	Modification of flood risks
River channel adjustment (Fig. 8)	Channel responses, lateral erosion or aggradation after river works	Changes in river habitat distribution	Changes in land property
Bathymetric profiles (reported on a altitudinal landmark)	Erosion zones (incision), changes in water level, dredging impacts	Dried riparian vegetation. Loss of habitat diversity	Modification of flood risks
Agricultural intensification (soil crustability enhancement)	Increased storm runoff, gullying		Competition between water users
Evolution of gravel pits (Fig. 7)	Exchanges between groundwater table and river		Groundwater exposure to pollution
Land use change in urban area (soil impermeabilisation)	Increase of rain storm runoff, riverbed erosion ("urban stream syndrome" [55]; increased flow variability)	Artificialisation of river habitat	

4.1 Stream Network Modification on the Versailles-Saclay Plateau (1670–1860)

The Versailles plateau, situated 50 km west of Paris, is one of the earliest and best documented examples of urban impacts on local streams (see location on Fig. 1). The Versailles château, its gardens and fountains, and its homonymous city were constructed in the late seventeenth century. At that time, the region was devoid of major water resources, and the Sun King mandated his engineers to supply his properties from two main water sources, (1) the famous Marly Machine, which elevated the Seine River water (Fig. 5), and (2) the derivation of all headwaters of the Saclay and Rambouillet plateaus to Versailles instead of flowing to the Yvette or Bièvre river, through a complex network of dugout ponds, ditches and aqueducts,

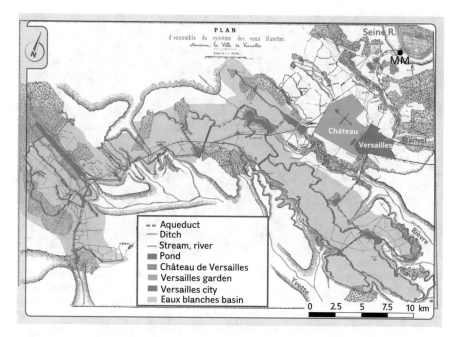

Fig. 5 Modifications of the headwater stream networks on the Versailles-Saclay plateaus to supply water to the Versailles chateau, fountains and city beginning in 1670. Digitalised networks of the 1861 map by La Vallée. MM, Marly Machine network of aqueducts and reservoirs using Seine River waters; blue and violet, Eaux Blanches networks on the plateau

the so-called Eaux Blanches (white waters), in operation for 290 years. The water supply map of 1861 by La Vallée illustrates these derivations, totalling 135 km in length and covering 192 km^2. The discharge of the Ru de Gally, the first-order stream which starts in the Versailles Gardens, was in turn multiplied by a factor of three given that it received all waste waters from the chateau and city, untreated until the 1950s [56].

4.2 Man-Made Heterogeneity of the Floodplain: Channelisation and Sandpits in the Bassée Floodplain

The Bassée alluvial plain is the major wetland of the Seine basin and stands in the Seine valley upstream of the Seine/Yonne confluence (see location in Fig. 1).

Archive sources show that sporadic river channel works were already undertaken in this sector before the 1600s to improve navigation and timber rafting [57]. In the second half of the nineteenth century, waterway transport was again promoted by the construction of dikes and then navigation canals, some 10 km long, excavated across the floodplain as the Beaulieu-Villiers canal (1885) and the Bray-La Tombe canal

Fig. 6 Aerial photograph of the Bassée floodplain, upstream of the Seine-Yonne confluence, showing the multiple sandpits, generally filled by groundwaters, and the straightened, deepened and dredged navigation channel with its levees and dikes. A former river meander part, now a disconnected water body, is visible on the right centre of the picture. Only patches of the original riparian forest remain between the sandpits and the agricultural areas [© La Pérouse 2005]

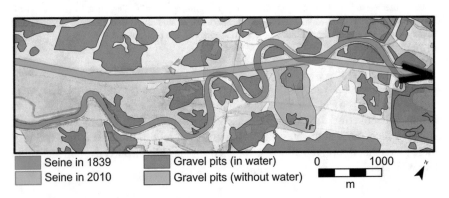

Fig. 7 Example of the Bassée floodplain transformations between 1839 and 2010, showing the dominance of man-made water bodies (pink = Grand Gabarit canal and red = sandpits) on a 6-km-long river sector (same sector as Fig. 6). Black arrow shows the point of view of the aerial photograph (Fig. 6)

(1899). The latter was abandoned in 1979, when the so-called Grand Gabarit canal (4.8 m draught) cut across the lower Bassée from La Grande Bosse to Montereau (Fig. 6).

In the 1970s, the hydrological dynamics of the Bassée were also impacted by the operation of two large reservoirs (Aube and Seine Reservoirs; see Fig. 1): the low flows are now artificially increased to sustain the summer water discharge in Paris, some 150 river km downstream, by a factor 3–4.

Figure 7 shows the evolution of the Seine River and the adjacent alluvial plain as depicted on a map from 1839 compared to the current situation, in the same area as that shown in Fig. 6. This 1/5000 1839 map covers a portion of the Bassée floodplain between Nogent-sur-Seine and Montereau-Fault-Yonne, 43 km long, 3–5 km wide, totalling 14,828 ha. It was drawn by the engineers of the Ponts et Chaussées in preparation for the first phase of the river works in this sector. The current situation is recorded on the 1/25,000 Scan 25© map of the IGN.

In this sector of the Bassée, one can observe the Grand Gabarit canal which has cut across the original meandering river course and created artificial and/or re-profiled bank. Today the Bassée is still a major wetland sector, according to the EU definition, but the riparian forest, its oxbows and other types of wetlands have been transformed by dozens of sandpits, some of which, still in operation, are not yet under water.

The accuracy of the 1839 map [58] quantifies slow changes of the Seine River as the secular natural meandering dynamics (Fig. 8). This map was drawn up as follows: (1) GIS geo-referencing; (2) identification and listing of the common spot, linear and surface elements; (3) digitalisation of these elements; (4) superimposition of these cartographic elements across time; and (5) determination of the many metrics (position, perimeter, length, area, etc.) and their temporal evolution [38, 58]. The rectification of the Seine in the whole Bassée led to a decrease in the length of the thalweg from 66.7 to 63.0 km. At the same time, the bed of the Seine was deepened and the average width of the Seine River decreased from 79.2 to 59.4 m. The meander shift rate of the natural Seine River was around 5 m/100 years

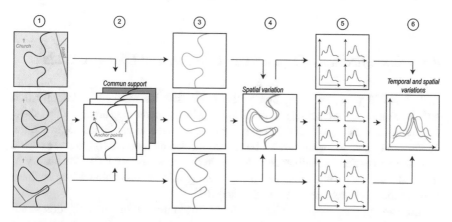

Fig. 8 Methodological approach for the processing and analysis of cartographic archives concerning river courses. Simplified diagram, adapted from the treatment chain developed by [59]. *1* Compilation of successive maps for the same sector; *2* scaling, geo-referencing; *3* digitalisation; *4* qualitative analysis (visualisation of lateral changes); *5* quantitative analysis (sinuosity, width, length, etc.); *6* spatio-temporal interpretation of fluvial changes

between 1839 and the middle of the twentieth century, thus illustrating the slow dynamic of the Seine River here [2].

From the 1950s onwards, the Bassée alluvial plain was exploited as gravel pits, with an average of 37 ha transformed into gravel pits every year since then (BRGM, 1995). They are typically between a few hectares and 100 ha (Fig. 6), a few metres deep, and most of them are filled with waters from the alluvial aquifer, which is found between 0.5 and 1.25 m from the ground level in most locations [21].

Between 1839 and 2010, the waterscape of the Bassée was completely transformed: in 1839, the natural Seine River channel covered 400 ha, and other bodies contributed 70 ha. In 2010, the artificial navigated channel covered 340 ha, the former river channels and associated natural waterbodies 360 ha and the water-filled sandpits 1,783 ha. As a whole, the open waterbody coverage of this sector has been increased five times. This tendency will continue in the near future with the construction of artificial water storage compartments, totalling 2,300 ha for the whole project, to store river water when Paris and its suburbs, some 100 river km downstream, are threatened by inundation. Ten compartments are planned, separated by 58 km of artificial dikes, up to 4.7 m high (http://www.seinegrandslacs.fr/le-projet-global-damenagement. Accessed 15 May 2019).

As a whole, this natural major wetland has been severely transformed for navigation (channel rectification), for construction (gravel pits) and in the future for flood control purposes (artificial water storage compartments): the mosaic of former natural water bodies, flooded forest and wetlands has been replaced by another mosaic of artificial water bodies, navigation channels and sandpits, with greater depths, often steep slopes, limited aquatic vegetation and direct exposure of the phreatic aquifer (which may result in a deterioration of the quality of ground-water used for drinking water supply). In the last 25 years, efforts have been made to "renaturalise" these artificial waterbodies, particularly as bird refuges.

4.3   Simplification of Seine River Channel and Loss of Islands

In the last 200 years, the navigation on the Seine River has resulted in simplification and deepening of the channel and regulation by locks. River map archives also show that simultaneously most islands have been reconnected to the banks. Between the 1830s and today, 405 islands have been reconnected along the 340 km river course from the Seine-Yonne confluence to Rouen in the Seine estuary [60]. The history of these islands, their settlements and uses, is rarely told in classical archives as these parts of the river area are seldom inhabited [61].

The cartographic archives, along with historical information on navigation structures, can be used to reconstruct their history. Figure 9 shows the modifications of the river channel in the Poses sector (for location, see Fig. 1), which marks today's upper limit of the tidal influence, between 1846 (Ponts et chaussées map) and 2015 (IGN map).

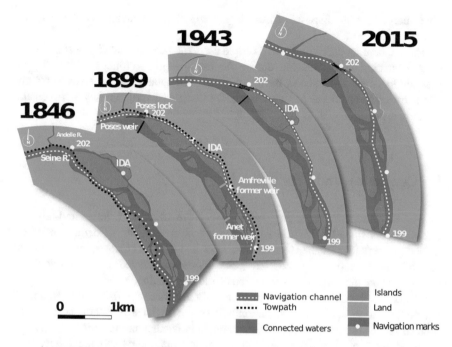

Fig. 9 Evolution of the Seine River course from geo-referenced and digitalised cartographic documents (1846–2015), showing the trajectories of river morphology and islands, towpath and navigation channel. Current navigation marks (I4), each river km downstream of Paris, are used as points of reference. *IDA* Ile des Deux Amants [37]

In 1846, Ile des Deux Amants Island (Fig. 9, IDA) was already reconnected to the right bank. Then its toponym rapidly disappeared from maps. The towpath in the Poses sector reveals, within a few kilometres, a complex shift between the banks and numerous islands, which implied complex manoeuvres of men and horses two or four times across the river. In 1852 the first Poses weirs and locks (Amfreville and Anet) were established. The navigation time between Rouen and Paris was accordingly shortened [42]. In 1886, the locks were moved downstream to Poses. In 1899 the Vuillaume map shows the fusion of several islands and the simplification of the navigated channel. The 1943 state is based on the navigation map published by IGN. The aerial photos (Geoportail, IGN) show that the islands upstream of the locks were separated in 1963–1964, to create a new navigation channel. Poses weir was again rebuilt in 1967, resulting in a drop of the water level between 5.4 and 8 m, depending on the tidal amplitude. This weir was considered as the essential obstacle to migratory fish circulation, such as salmon, lampreys, sea trout and shad (*Alosa alosa*) until the installation of two fish ladders in 1991 and 2018 to facilitate the upward circulation of these fishes.

These changes in island configuration were essentially anthropogenic and cannot be assumed to result from sediment supply increase – e.g. by enhanced erosion of agricultural areas – as featured in the Schumm diagram [2, 4]. Here, the islands were often merged by the disposal of materials, dredged in the navigation channel, in a non-navigated arm.

These island modifications had multiple consequences: the relatively still waters, used as resting areas, refuges during floods, spawning grounds and/or fishing grounds for some fish species have decreased. The total length of river banks per kilometre of river course has also markedly decreased (around 30% between the 1880s and 2010) as well as river habitat heterogeneity.

5 Conclusion and Perspectives

In Western European rivers, physical changes have affected the lower and medium stream orders since the Middle Ages, and sometimes earlier, as demonstrated by archeo-geology [3]. The heritage of these slow and gradual changes, and the related cultural elements (mills, ponds, etc.), has been represented with accuracy and precision since the early 1800s in various cartographic documents.

The selected cartographic documents mainly represent the visions that engineers had on these areas at each period, particularly on the fluvial area. But they can also be used to reconstruct, within our contemporary knowledge and technical means, the longue durée evolution of the river and stream network, the river channel in its three dimensions – longitudinal, lateral and vertical – and its related floodplain and the river-related natural and cultural elements represented. Our contemporary interpretations of these documents must be cautious: the reliability of these ancient documents must be assessed. The context and the objective of the cartographic work must be established, with its quality, accuracy, imprecision and relevance with regard to the contemporary research objectives [62].

These well-contextualised cartographic documents were used to perform quantified geo-historical analyses through two main approaches: (1) the analysis of the state of the river at a given period and (2) the comparison of two states, which provides the dynamics of the elements represented.

The applications of the analysis of historical cartographic documents are numerous: they can be used to investigate hydro-morphology, hydro-ecology, water resources and river use history, as well as river restoration and management. These natural and/or human-made changes lead to new hydraulic and hydrodynamic conditions which eventually reshape the river course and the aquatic environment, as evidenced by the evolution of the morphometric parameters. In turn the aquatic biota readapts to these new conditions. In the case of river works, the most frequent one, the related adjustments provide the response times of the aquatic environment to the pressures exerted on them. As these responses are often on the decadal or multi-decadal scale, analysis of cartographic archives is essential.

The physical modifications documented here have been found for every stream order, from the headwaters (e.g. Versailles plateau, Morvan forest) to the river-estuary limit (Poses sector, order 7). Several drivers are responsible for these existing and planned physical modifications of the fluvial system, thus shaping its trajectories:

Urban development, perfectly documented since the seventeenth century for Versailles chateau, gardens and city, results in major transformations of stream orders 1–3. In the nineteenth and twentieth centuries, the spatial and demographic expansion of Paris and its surroundings have also generated major changes in medium-sized urban streams (orders 2–4) [17].

The intensification of agriculture has largely modified first-order streams, converted into ditches to which underground tile drains are connected.

The intensification of navigation has been continuous since 1830, resulting in multiple transformation of the river channel in orders 5–8. It is largely driven by the urban development of Paris and by the export of agricultural products from the Seine basin. This activity is still progressing.

The demand for sand and gravels, extracted from the Seine, Marne and Oise floodplains, has been generated by urban and infrastructure development after 1950 and has reshaped the landscape of the alluvial plains.

River flow regulation and flood security have already resulted in the construction of large reservoirs able to double or triple the natural Seine River flow at Paris in summer and reduce the flood stage. New water storage facilities are now planned in the Bassée floodplain, finalising the physical transformation of the major wetland of the Seine Basin, already impacted by sandpits and river channelisation.

Unlike many pollutions caused by the Paris region, which have decreased or have been treated since 1960 or 1970 (see the following chapters), the physical modifications of the river network, channel and floodplain, and of its related habitats that have taken place in the last 200 years, are mostly irreversible, at least for order 5–7 rivers.

Today, many actors – river and water users, decision-makers and citizens – have a stake in the balance between river biodiversity and good ecological state, flood risk control, recreational uses and economic uses: some of these uses control the rivers' future, others react and adapt. For these reasons, the Seine River basin and its society is a typical example of a territory in the Anthropocene era, where society has gradually and permanently controlled a major part of the natural system [63]. Aside from physical control, other examples of biochemical, chemical or ecological controls are presented and illustrated in the following chapters.

Further applications of this work, which has required the identification of numerous archives over the past 8 years, during the previous phases of the PIREN-Seine programme, already include the evolution of obstacles to migratory fish and the identification of sediment archive storage sites to reconstruct trajectories of contaminants in the Seine basin [54]. One may also consider integrating the vertical evolution of the channel into the groundwater/surface water exchange models (see [64]). When focusing on the socio-economic and ecological consequences of river evolution, we can see that the entire extent of alluvial plains is in fact impacted. Specific studies, based on the same methodology of quantitative analysis of historical maps, will be carried out in the next phase of the PIREN-Seine project (2020–2023) to describe trajectories of the river corridor as a whole.

Acknowledgements This work is a contribution to the PIREN-Seine research programme (www.piren-seine.fr), which belongs to the Zone Atelier Seine part of the international Long-Term Socio-Ecological Research (LTSER) network. ArchiSeine is the result of a work carried out within a convention between the PIREN-Seine research programme and the French National Archives.

References

1. Bravard JP, Piegay H (2000) L'interface Nature – Sociétés dans les hydrosystèmes fluviaux. Géocarrefour 75(4):273–274
2. Amoros C, Petts GE (1996) Hydrosystèmes fluviaux. Chapman and Hall, London
3. Lespez L (2012) Paysages et gestion de l'eau: sept millénaires d'histoire des vallées et des plaines littorales en Basse-Normandie. Presses University, Caen
4. Downs PW, Gregory KJ (2004) River channel management: towards sustainable catchment hydrosystems. Arnold, London
5. Gurnell AM, Piery JL, Petts GE (2003) Using historical data in fluvial geomorphology. In: Kondolf M, Piégay H (eds) Tools in fluvial geomorphology. Wiley, Chichester
6. Jacob-Rousseau N (2009) Géohistoire/géo-histoire: quelles méthodes pour quel récit ? Géocarrefour 84(4):211–216
7. Bravard JP (2010) La valorisation de la cartographie historique des rivières d'Europe, de la recherche sur la dynamique des paysages à la gestion des eaux. In: Masotti L (ed) Il paesaggio dei tecnici, Attualità della cartographia storica per il governo delle acque. Marsiolio, Bolognia
8. Robert S (2011) Sources et techniques de l'archéogéographie planimétrique. Presses Universitaires de Franche-Comté, Besançon
9. Braga G, Gervasoni S (1989) Evolution of the Po river: an example of the application of historical maps. In: Petts GE, Möller H, Roux AL (eds) Historical change of large alluvial rivers: Western Europe. Wiley, Chichester, pp 113–126
10. Bravard JP, Gaydou P (2015) Historical development and integrated management of the Rhône River floodplain, from the Alps to the Camargue Delta, France. In: Hudson PF, Middlekoop H (eds) Geomorphic approaches to integrated floodplain management of lowland fluvial systems in North America and Europe. Springer, New York, pp 289–320
11. Hohensinner S, Lager B, Sonnlechner C et al (2013) Changes in water and land: the reconstructed Viennese riverscape from 1500 to the present. Water History 5(2):145–172
12. Arnaud F, Piégay H, Schmitt L et al (2015) Historical geomorphic analysis (1932–2011) of a by-passed river reach in process-based restoration perspectives: The Old Rhine downstream of the Kembs diversion dam (France, Germany). Geomorphology 236:163–177
13. Cioc M (2002) The Rhine, an eco-biography, 1815–2000. University of Washington Press, Seattle/London
14. Mauch C, Zeller T (eds) (2008) Rivers in history. Perspectives on waterways in Europe and North America. University of Pittsburgh Press, Pittsburgh
15. Castonguay S, Evenden M (2012) Urban rivers. Remaking rivers, cities, and space in Europe and North America. University of Pittsburgh Press, Pittsburgh
16. Knoll M, Lübken U, Schott D (2017) River lost, rivers regained. Rethinking city-river relations. University of Pittsburgh Press, Pittsburgh
17. Lestel L, Carré C (eds) (2017) Les rivières urbaines et leur pollution. Quae, Paris
18. Petts GE, Möller H, Roux AL (1989) Historical change of large alluvial rivers: Western Europe. Wiley, Chichester
19. Gregory KJ (2006) The human role in changing river channels. Geomorphology 79 (3–4):172–191

20. Mouchel JM, Boet P, Hubert P et al (1998) Un bassin et des hommes: une histoire tourmentée. In: Meybeck M, de Marsily G, Fustec E (eds) La Seine en son bassin. Elsevier, Paris, pp 77–125
21. Fustec E, Greiner I, Schanen O et al (1998) Les zones humides riveraines: des milieux divers aux multiples fonctions. In: Meybeck M, de Marsily G, Fustec E (eds) La Seine en son bassin. Elsevier, Paris, pp 211–262
22. Bendjoudi H, Weng P, Guérin R et al (2002) Riparian wetlands of the middle reach of the Seine river (France): historical development, investigation and present hydrologic functioning. A case study. J Hydrol 263(1):131–155
23. Guerrini MC, Mouchel JM, Meybeck M et al (1998) Le bassin de la Seine: la confrontation du rural et de l'urbain. In: Meybeck M, de Marsily G, Fustec E (eds) La Seine en son bassin. Elsevier, Paris, p 29
24. Poux AS, Gob F, Jacob-Rousseau N (2011) Reconstitution des débits des crues artificielles destinées au flottage du bois dans le massif du Morvan (centre de la France, XVIe-XIXe siècles) d'après les documents d'archive et la géomorphologie de terrain. Geomorphologie 17 (2):143–156
25. Rouillard J, Benoit P, Morera R (2011) L'eau dans les campagnes du bassin de la Seine avant l'ère industrielle, vol 10. PIREN-Seine, Paris. Available via PIREN Seine. https://www.piren-seine.fr/fr/content/l'eau-dans-les-campagnes-du-bassin-de-la-seine-avant-l'ère-industrielle. Accessed 15 May 2019
26. Passy P, Garnier J, Billen G et al (2012) Restoration of ponds in rural landscapes: modelling the effect on nitrate contamination of surface water (the Seine watershed, France). Sci Total Environ 430:280–290
27. Boët P, Belliard J, Berrebi D, Thomas R et al (1999) Multiple anthropogenic impacts induced by Paris on fish populations in the Seine Basin, France. Hydrobiologia 410:59–68
28. Le Sueur B (2015) Le domaine public des rivières et de canaux. Histoire culturelle et enjeux contemporains. L'Harmattan, Paris
29. Flipo N, Lestel L, Labadie P et al (2020) Trajectories of the Seine River basin. In: Flipo N, Labadie P, Lestel L (eds) The Seine River basin. Handbook of environmental chemistry. Springer, Cham. https://doi.org/10.1007/698_2019_437
30. Billen G, Barles S, Garnier J et al (2009) The food-print of Paris: long-term reconstruction of the nitrogen flows imported into the city from its rural hinterland. Reg Environ Chang 9(1):13–24
31. Guillerme A (1990) Le testament de la Seine/the legacy of the Seine. Géocarrefour 65 (4):240–250
32. Jugie M, Gob F, Virmoux C et al (2018) Characterizing and quantifying the discontinuous bank erosion of a small low energy river using structure-from-motion photogrammetry and erosion pins. J Hydrol 563:418–434
33. Cotte M (2002) L'amélioration de la navigation sur les rivières françaises au XIXe siècle: le cas de la Haute-Seine et de l'Yonne. In: Hilaire-Pérez L, Massounie D, Serna V (eds) Archives, objets et images des constructions de l'eau du Moyen Âge à l'ère industrielle. SFHST, ENS éditions, Paris, pp 265–279
34. Barles S (2015) The main characteristics of urban socio-ecological trajectories: Paris (France) from the 18th to the 20th century. Ecol Econ 118:177–185
35. Houbrechts G, Levecq Y, Vanderheyden V et al (2011) Long-term bedload mobility in gravel-bed rivers using iron slag as a tracer. Geomorphology 126:233–244
36. Harley JB, Gould P, Bailly A et al (1995) Le Pouvoir des cartes: Brian Harley et la cartographie. Anthropos, Paris
37. ArchiSEINE. Site d'archives et de données historiques sur le bassin versant de la Seine. http://archiseine.sisyphe.jussieu.fr/site/. Accessed 15 May 2019
38. Lestel L, Eschbach D, Steinmann R et al (2019) ArchiSEINE: une approche géohistorique du bassin de la Seine, vol 18. PIREN-Seine, Paris. Available via PIREN Seine. https://www.piren-seine.fr/fr/fascicules/archiseine-une-approche-géohistorique-du-bassin-de-la-seine. Accessed 15 May 2019

39. Verdier N (2009) Les plans et cartes du XIXe siècle. Introduction. In: Costa L, Robert S, Foucault M (eds) Guide de lecture des cartes anciennes. Illustrations dans le Val d'Oise et le Bassin parisien. Errance, Paris, pp 7–9

40. Buchotte (1722) Les règles du dessein et du lavis. Chez Claude Jombert, Paris

41. Archives Nationales (2009) Etat général des fonds. F/14. Travaux publics. http://www. archivesnationales.culture.gouv.fr/chan/chan/series/pdf/F14-2011.pdf. Accessed 15 May 2019

42. Merger M (1994) La canalisation de la Seine (1838-1939). La Seine et son histoire en Ile-de-France, Paris et Ile-de-France Mémoires, vol 45. Editions du CTHS, Paris, pp 107–124

43. Beyer A (2016) Les grands jalons de l'histoire des voies navigables françaises. Pour mémoire. halshs-01664447

44. Touzery M, Ladurie ELR (1995) Atlas de la généralité de Paris au XVIIIe siècle: un paysage retrouvé. Comité pour l'Histoire économique et financière, Paris

45. Costa L, Robert S, Foucault M (2009) Guide de lecture des cartes anciennes: illustrations dans le Val d'Oise et le Bassin parisien. Errance, Paris

46. Lagarde L (1987) Philippe Buache, ou le premier géographe français, 1700-1773. Mappemonde 2:26–30

47. Coic J, Duleau A (1830) Reconnaissances de la Seine, de Rouen à Saint-Denis, en 1829 et 1830, et travaux proposés pour rendre cette partie de la Seine facilement navigable. Paris

48. White GF (1945) Human adjustments to floods, Department of Geography Research, Paper no. 20, 29, University of Chicago

49. Di Baldassarre GD, Viglione A, Carr G et al (2013) Socio-hydrology: conceptualising human-flood interactions. Hydrol Earth Syst Sci 17(8):3295–3303

50. Belliard J, Beslagic S, Delaigue O et al (2018) Reconstructing long-term trajectories of fish assemblages using historical data: the Seine River basin (France) during the last two centuries. Environ Sci Pollut R 25(24):23430–23450

51. Belliard J, Beslagic S, Tales E (2020) Changes in fish communities of the Seine Basin over a long-term perspective. In: Flipo N, Labadie P, Lestel L (eds) The Seine River basin. Handbook of environmental chemistry. Springer, Cham. https://doi.org/10.1007/698_2019_380

52. Guerrin J, Bouleau G (2014) Remparts ou menaces? Trajectoires politiques de l'endiguement en France, aux Pays-Bas et aux Etats-Unis. Rev Int Polit Comp 21(1):89–109

53. Auerswald K, Moyle P, Seibert SP et al (2019) HESS opinions: socio-economic and ecological trade-offs of flood management–benefits of a transdisciplinary approach. Hydrol Earth Syst Sci 23(2):1035–1044

54. Ayrault S, Meybeck M, Mouchel JM et al (2020) Sedimentary archives reveal the concealed history of micropollutant contamination in the Seine River basin. In: Flipo N, Labadie P, Lestel L (eds) The Seine River basin. Handbook of environmental chemistry. Springer, Cham. https://doi.org/10.1007/698_2019_386

55. Walsh CJ, Roy AH, Feminella JW et al (2005) The urban stream syndrome: current knowledge and the search for a cure. J N Am Benthol Soc 24(3):706–723

56. Dmitrieva T, Lestel L, Meybeck M et al (2018) Versailles facing the degradation of its water supply from the Seine River: governance, water quality expertise and decision making, 1852–1894. Water History 10(2–3):183–205

57. Dzana JG (1997) Le lit de la Seine de Bar à Montereau: étude morphologique et rôle des aménagements. Thèse de doctorat en géographie, University of Paris 1, Paris

58. Eschbach D, Lestel L (2018) Dynamique hydromorphologique historique de la Seine dans le secteur de la Bassée aval. PIREN Seine report. https://www.piren-seine.fr/sites/default/files/PIREN_documents/phase_7/rapports_annuels/2016/a2b4_Eschbach_PIREN_2018.pdf. Accessed 6 August 2019

59. Miramont C, Jorda M, Pichard G (1998) Évolution historique de la morphogenèse et de la dynamique d'une rivière méditérannéenne: l'exemple de la moyenne Durance (France du sud-est). Geogr Phys Quatern 52(3):381–392

60. Lescure S, Arnaud-Fassetta G, Cordier S (2011) Sur quelques modifications hydromorphologiques dans le Val de Seine (Bassin Parisien, France) depuis 1830: quelle part accorder aux facteurs hydrologiques et anthropiques ? EchoGéo 18:15
61. Charbit M (2016) Îles de la Seine. Edition du Pavillon de l'Arsenal, Paris
62. Grosso E (2010) Integration of historical geographic data into current geo-referenced frameworks: a user-centered approach. Proceedings of the 5th international workshop on digital approaches in cartographic heritage (Vienna 2010), pp 107–117
63. Meybeck M, Lestel L (2017) A Western European River in the Anthropocene. The Seine, 1870-2010. In: Kelly JM, Scarpino P, Berry H et al (eds) Rivers of the anthropocene. University of California Press, Oakland, pp 84–100
64. Flipo N, Gallois N, Labarthe B et al (2020) Pluri-annual water budget on the Seine basin: past, current and future trends. In: Flipo N, Labadie P, Lestel L (eds) The Seine River basin. Handbook of environmental chemistry. Springer, Cham. https://doi.org/10.1007/698_2019_392

Pluri-annual Water Budget on the Seine Basin: Past, Current and Future Trends

Nicolas Flipo, Nicolas Gallois, Baptiste Labarthe, Fulvia Baratelli, Pascal Viennot, Jonathan Schuite, Agnès Rivière, Rémy Bonnet, and Julien Boé

Contents

1 Introduction 60
2 Historical Records of the Seine River Discharge in Paris 61
3 Development of the Seine Basin Model 65
   3.1 CaWaQS Model 65
   3.2 Implementation of the Seine Basin Model 66
4 Current State of the Seine Hydrosystem 66
   4.1 General Two-Step Calibration Strategy of Hydrosystem Models 67
   4.2 Average Water Budget 1993–2010 68
5 A Two-Century-Long Trajectory of the Seine Basin 71
   5.1 Estimating Land Cover Changes 72
   5.2 Estimating Anthropogenic Water Uptake 73
   5.3 Climate Scenarios 74
   5.4 Water Resources Trajectory from the 1900s to the 2100s 76
6 Conclusion 82
References 83

Abstract The trajectory of the Seine basin water resources is rebuilt from the early 1900s to the 2000s before being projected to the end of the twenty-first century. In the first part, the long-term hydrological data of the Paris gauging stations are analysed beginning in 1885, highlighting the effect of anthropogenic water management on the Seine River discharge. Then a detailed water budget of the Seine basin is proposed. It quantifies for the first time the water exchanges between aquifer units and the effect of water withdrawals on river–aquifer exchanges. Using this model,

---

The copyright year of the original version of this chapter was corrected from 2019 to 2020. A correction to this chapter can be found at https://doi.org/10.1007/698_2020_667

N. Flipo (✉), N. Gallois, B. Labarthe, F. Baratelli, P. Viennot, J. Schuite, and A. Rivière
Geosciences Department, MINES ParisTech, PSL University, Fontainebleau, France
e-mail: Nicolas.Flipo@mines-paristech.fr

R. Bonnet and J. Boé
CECI, Université de Toulouse, CNRS, CERFACS, Toulouse, France

Nicolas Flipo, Pierre Labadie, and Laurence Lestel (eds.), *The Seine River Basin*,
Hdb Env Chem (2021) 90: 59–90, https://doi.org/10.1007/698_2019_392,
© The Author(s) 2020, corrected publication 2020, Published online: 3 June 2020

the trajectory of the system is evaluated based on a downscaled climate reanalysis of the twentieth century and a reconstruction of the land use in the early 1900s, as well as the choice of a climate projection which favours the model that best reproduces the low frequency of precipitation. The trajectory is synthesised as average regimes, revealing a relative stability of the hydrosystem up to the present, and drastic changes in the discharge regime in the future, especially concerning the decreased amount of low flow and its increased duration. These expected changes will require the definition of an adaptation strategy even though they are rather limited in the Seine basin when compared to other French regions.

Keywords CaWaQS, Climate change, Groundwater, Hydrological distributed modelling, Land use, Past and future scenarios, Seine basin, Surface water, Water budget

1   Introduction

As mentioned by Flipo et al. [1] and Billen et al. [2], the Seine hydrosystem is unique due to the tremendous pressure exerted by the largest metropolis in Europe, Greater Paris, on water resources and the fact that it contains the largest groundwater reservoir in Europe, the Paris basin [3]. Today, the total water withdrawal reaches the enormous amount of 3 $Gm^3 a^{-1}$. Coupled with climate change, this pressure may hinder the sustainability of this unique hydrosystem.

Since the international effort to quantify the effect of climate change on global water circulation crystallised around the successive climate model intercomparison projects [4, 5], the expected effects of climate change on French water resources have been estimated [6, 7]. To evaluate water resources at the scale of a regional hydrosystem, it is now acknowledged that regional-scale models calibrated and validated against observed discharge should be used [8]. Following this effort, a first regional assessment at the Seine basin scale [9] led to the same conclusion as the nation-wide assessment [6, 7], meaning that heavier rainfall events are expected during winter and longer, more intense low-flow events may occur from May to late October.

Assessments of climate change impacts on regional hydrosystems are not sufficiently mature to be synergistically used with climate change adaptation decision-making [10, 11] designed to optimise what are called climate services [12]. These processes involve the full understanding of the regional system trajectory over decades or centuries. These approaches usually only consider the trajectory from now to tomorrow. At the Seine basin scale, the PIREN-Seine research programme promotes the study of trajectories from the past to today until tomorrow. This attempt to map the hydrosystem trajectory is the goal of this chapter. We believe that switching the cognitive reference from today to yesterday provides a broader view of the combined functioning of the anthro-eco-hydrosystem and it makes complete sense to evaluate climate services and possible adaptation strategies in a "safe operating space" [13].

Rebuilding a past trajectory involves the reconstruction of both past climate and past land use. Data assimilation in global circulation models was proposed in the 1990s and allowed the reanalysis of meteorological data over 40 years [14], which has led, in the 2010s, to multiple decade-long climate reanalysis [15–17] and also century-long reanalysis [18, 19]. Bonnet et al. [20] leveraged on those century-long reanalyses by downscaling them at the Seine basin scale. In the continuity of this work, a first reconstruction of the past hydrological trajectory of the Seine basin was proposed by Bonnet [21]. However, the Anthropocene era [22, 23] also introduced major land use changes over the past century [24–26], which we will account for in our attempt to rebuild the hydrosystem trajectory since the early 1900s.

This work is the accomplishment of 30 years of hydrogeological studies on the Seine basin within the PIREN-Seine programme. Starting from local studies at the river–aquifer interface [27, 28] and the knowledge on deep long-term water circulations in the Paris basin aquifer system [29–32], a first coupled hydrological–hydrogeological model of the Seine basin was proposed by Gomez et al. [33] and further enhanced for the simulation of surface–subsurface interactions [34, 35]. This modelling tool spread in the hydrometeorological community inspiriting global circulation models [36] and initiating combinations with soil-vegetation-atmosphere-transfer models [9, 37, 38]. It was also used for various combined applications to study the trajectory of nitrate due to agricultural practices [39–41] as well as the impact of climate change on water quality [42]. All these studies led to the distributed process-based hydrological–hydrogeological model CaWaQS [43–45] that is used in this anniversary chapter on the trajectory of the Seine basin water resources.

2   Historical Records of the Seine River Discharge in Paris

The Pont d'Austerlitz gauging station offers the longest and most viable mean daily discharge time series for the Seine River. The first estimates of discharge go back to 1885. At this station, the Seine River drains an area of 43,800 km^2, which corresponds to 60% of the total basin. Hence, this data set provides a valuable glimpse into the large-scale hydrological functioning of a significant part of the basin. It also bears remarkable traces of the long-term evolution of environmental factors that influence flow dynamics, whether natural or human-induced. Furthermore, the Austerlitz discharge data provide the opportunity to analyse the response of the basin to historical extreme events (floods and droughts).

Overall, the Seine reaches high-flow conditions in mid-February with a multi-annual average flow rate of 583 m^3 s^{-1} and smoothly arrives at low-flow periods in August with an average rate of 125 m^3 s^{-1} (Fig. 1a). Yet, the interannual variability of flow is remarkably high, especially between wet periods where the difference between the 0.95 and 0.05 quantiles is on the order of 1,000 m^3 s^{-1}. This is also reflected in historical extrema: discharge records range from a minimum 20 m^3 s^{-1} (historical drought of 1921) to over a maximum 2,600 m^3 s^{-1} (historical flood of

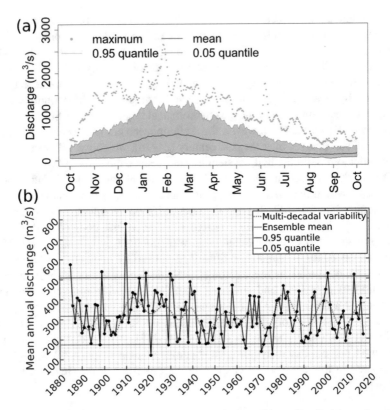

Fig. 1 Seine discharge estimates from water levels at the Pont d'Austerlitz (Paris) gauging station since 1885. (**a**) Mean and maximum daily discharge over the hydrological year. The 5th and 95th quantiles are also displayed. Calculation based on the R package FlowScreen [46]. (**b**) Mean annual discharge (black dots and black solid line) and multi-decadal variability (dashed red line) obtained by summation of the last 3 details out of 14 from the discrete wavelet transform of mean daily discharge, using an order 4 Symlet mother wavelet

1910) (Fig. 2a). Hence, the discharge rates at Pont d'Austerlitz vary within two orders of magnitude.

In addition to the strong interannual variability, the Seine is influenced by clear interdecadal variability (Fig. 1b). Indeed, the long-term average discharge measured at Pont d'Austerlitz between 1885 and 2018 is 319 m^3 s^{-1}, but mean annual flow rates typically range from 125 m^3 s^{-1} to over 750 m^3 s^{-1}. In general, above and below average years tend to follow each other, and the succession of drier and wetter decades follows a distinguishable low-frequency pseudo-cyclicity (Fig. 1b). Such multi-decadal variability stems from known global-scale climatic and oceanic fluctuations, mainly the North Atlantic Oscillation (NAO) and the Atlantic Multi-decadal Variability (AMV) [47–52]. The NAO index is defined by anomalies in the normalised atmospheric pressure difference between Reykjavik (Iceland) and Lisbon (Portugal) [48]. During positive NAO phases, strong anticyclonic conditions

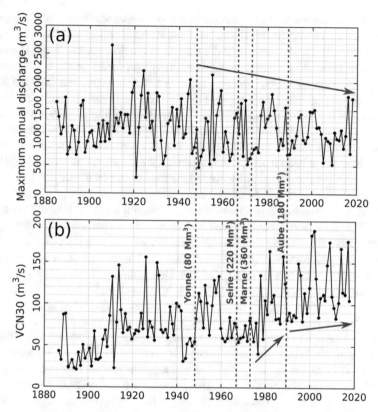

Fig. 2 Two statistical features of the Seine discharge data recorded at the Pont d'Austerlitz (Paris, France) gauging station since 1885. (**a**) Maximum annual discharge. (**b**) VCN30, defined as the annual minimum of a 30-day running average of daily mean discharge. The vertical dashed lines mark the commissioning of reservoir lakes in derivation of the Seine and its main tributaries upstream from Paris (names of the rivers are shown along with the reservoir volumes in millions of cubic metres, Mm^3)

prevail over the Azores, so that humid air masses are driven towards the north of Europe, where winters typically become wet and mild [53–55]. Hence, drier conditions over the Seine basin are encountered during negative NAO phases. The mechanisms driving the AMV are more controversial, since there is evidence for both natural and anthropogenic controls on its variability. Nevertheless, the role of Atlantic buoyancy-driven circulation on the AMV is generally recognised as predominant (see [47] and reference therein). In terms of a temporal signature, the NAO's periodicity mainly lies within two bands, 5–7 years and 16–19 years, whereas the AMV is characterised by timescales varying between 60 and 100 years [21, 47, 52, 56, 57], which makes it more difficult to detect given the relative shortness of the available records [52, 58]. These global-scale climatic processes were proven to be largely responsible for the Seine's observed variability in mean annual discharge [47, 50, 59, 60].

Despite the presence of clear interannual and interdecadal variability throughout the discharge records, the most extreme flood events all occurred before the 1960s. Only in 1910, 1924, 1945 and 1955 did the maximum flow rates at Pont d'Austerlitz reach over 2,000 m^3 s^{-1} (Fig. 2a). This is partly due to the creation and commissioning, since the mid-1960s, of large-capacity reservoirs upstream from the highly urbanised and populated Paris region [1]. The three largest reservoirs were built in derivation of the Seine in 1966 (220 Mm^3 capacity), the Marne in 1974 (360 Mm^3) and the Aube in 1990 (180 Mm^3). A smaller reservoir (80 Mm^3) was created in 1949, but since it is located far upstream of the Yonne watershed, its impact on extreme events observed in Paris is less pronounced than the impact of the others. Until now, these reservoirs have successfully limited the impact of potentially damaging flood events by storing up to 840 Mm^3 of water. Alternatively, in the summer and during particularly dry periods, they release water to maintain a minimal target flow rate of 60 m^3 s^{-1} in the Seine through Paris. This human-induced perturbation of the Seine's hydrodynamic functioning is noticeable in discharge data in at least two ways. First, the variance of mean and maximum annual discharges has decreased since the creation of the reservoirs (Figs. 1b and 2a). This demonstrates the ability to reduce the impact of climatic extremes on river flow by smoothing the flow signal, that is to say, by distributing rainfall inputs over time more efficiently than in natural (undisturbed) conditions. Second, the VCN30 values, the yearly minima of 30-day running average windows [61], have never dropped below 55 m^3 s^{-1} since the beginning of the 1960s. The only exception occurred in 1976 during an extreme drought (VCN30, 40 m^3 s^{-1}, Fig. 2b) concomitantly with an extreme dry year (annual discharge, 120 m^3 s^{-1}, Fig. 1b). Furthermore, the VCN30 has only rarely dropped below 80 m^3 s^{-1} since the beginning of the 1980s (Fig. 2b). This shows that the reservoirs considerably helped to mitigate the risk of extended low-flow periods, which may alter the integrity of river ecological habitats, as well as human activities.

The long discharge records at Pont d'Austerlitz hold valuable information on the overall dynamical response of the Seine hydrosystem to climatic inputs. Like any other hydrosystem on Earth, the Seine acts as a low-pass filter of effective precipitation, meaning that the highest frequencies in the climatic input signal are transformed by the various flow processes, whereas the lowest frequencies (typically, the interdecadal fluctuations) are generally transcribed exactly as they are into the discharge signal. From spectral analysis of effective precipitation and discharge data, it is possible to obtain a first-order estimate of the mean flow response time of the three fundamental compartments of a hydrosystem: the surface, the unsaturated layer and the aquifer [62]. When applied to the Austerlitz data since 1885, we estimate that the mean response times to climatic events are 1.7 ± 0.2 days, 5.0 ± 1.3 days and 2.5 ± 0.2 years for surface runoff, vadose zone flow and aquifer flow, respectively. Moreover, from the same analysis, we estimate that 81% (±2) of the long-term river flow is sustained by groundwater. Unsurprisingly, this preliminary analysis shows that the Seine is largely dependent on groundwater stocks, which are very capacitive and transmissive given the sedimentary nature of the basin. It also shows that the Seine is highly vulnerable to extreme events, especially floods, given that the response time of the runoff component is short. Nevertheless, as reservoir lakes have altered the natural functioning of the basin since the mid-1970s, the apparent aquifer response time increases to 7.3 ± 0.2 years, and the fraction of groundwater supply to fluvial systems appears to be about 90% when the spectral analysis of discharge [62] is applied to data between 1974 and 2018. It makes sense that, somehow, the capacitive component of the system is perceived by simple interpretation tools to have a higher dampening effect, and a slightly higher contribution to total fluxes, in the presence of reservoirs. However, it is not possible to attribute this effect to subsurface property changes at this time scale, as it is conceptualised by these tools [62]. This demonstrates once more the sensitivity of river flow dynamics to territory developments, but it also shows the necessity of adopting more complex theoretical and modelling schemes to precisely understand, quantify and predict flow paths and processes throughout the basin, which is the exact purpose of this chapter.

3 Development of the Seine Basin Model

3.1 CaWaQS Model

The functioning of the Seine hydrosystem is nowadays simulated with CaWaQS (CAtchment WAter Quality Simulator) [43–45, 62, 63]. It is a spatially distributed model which simulates the water balance and dynamics of water flow in all compartments of a hydrosystem based on the blueprint published by de Marsily et al. [64] and first implemented as the MODCOU model [65–67]. A first attempt of reprogramming in Fortran90 was achieved with the EauDyssée project [34, 35, 68–71], which mostly accounted for river stage fluctuations [34, 35, 69]. Following the extension by Flipo et al. [72] of the nested groundwater flow concept [73] in the stream–aquifer interface, CaWaQS 2.x was recoded [45] combining multiple C libraries. Running at a daily time step, calculations of surface, subsurface and groundwater flows are structured around five main components [45, 68]:

- A surface module, mainly conditioned by land use, climate and parent soil materials, which calculates the surface water balance via a conceptual reservoir-based approach [74, 75] using rainfall and potential evapotranspiration (PET) data to estimate actual evapotranspiration (AET), runoff and infiltration fluxes on each surface layer cell.
- An unsaturated module, which transfers water infiltration from the subsurface to the water table using a set of reservoirs so that the infiltration is diffused in time to form the aquifer recharge [76, 77].
- A saturated module that solves the pseudo 3D-diffusivity equation [78] in a multilayer aquifer system with a finite volume numerical scheme that uses water recharge as well as water withdrawals as forcings.
- A conductance model that represents surface–subsurface water exchanges [72, 79, 80].

- A hydraulic module that routes in-stream water flows using a Muskingum scheme [81, 82], from which water levels are derived using rating curves [34] or a manning approximation of a steady state [35, 69]. Computed discharges in each river cell of the hydrographic network result from both stream–aquifer fluxes and contributions from subsurface runoff.

3.2 Implementation of the Seine Basin Model

The surface layer of the model [45] covers an area of 81,200 km^2 including the entire 76,300-km^2 Seine basin. The grid is divided into elementary watersheds of an average area of 11 km^2, on which the surface water balance is calculated. The river network implemented in the model is directly provided by the Carthage national database[1]. In this application, both discharges and stream–aquifer exchange calculations are restricted to the main river network (Fig. 6) describing 4,520 km of rivers from Strahler orders 3–8 [83] at the mouth of the Seine River. While the first implementation of the model described three aquifer units [33, 39], the aquifer system was refined up to six aquifer units by Pryet et al. [35]. The current version, developed by Labarthe [45], discretises the multilayered structure in seven aquifer units, including alluvial plains, using a progressive multi-scale grid of nested square meshes, with cells varying from 200 m to 3,200 m in size. The seven aquifer units can be regrouped, from the oldest to the most recent, in three main ensembles [1]:

- A Cretaceous chalk aquifer that has by far the largest impluvium
- A five-layer Tertiary complex ensemble, located in the centre of the basin, covering aquifer units mainly made of limestones and sands, dating from Palaeocene to Miocene
- Quaternary alluvial deposits surrounding the basin's main rivers

Jurassic aquifers, which outcrop at the eastern border of the basin, are not represented as explicit aquifer layers and are treated by a subsurface procedure using a simplified reservoir model in order to route infiltration fluxes towards the hydraulic network.

4 Current State of the Seine Hydrosystem

A 17-year periodicity associated with the NAO was previously identified [50] on observed climatic and stream discharge data over the Seine basin. Both groundwater and river water storage have also been proved to be stationary over a 17-year period

---

[1]http://www.sandre.eaufrance.fr.

[68]. All further results are therefore averaged over the 1993–2010 period to ensure their significance [84].

The model uses daily rainfall and PET provided by SAFRAN [85, 86], a mesoscale atmospheric analysis system for surface variables, producing data at a daily time step on an 8 × 8-km grid. SAFRAN uses the Météo-France observation network and data from a number of well-instrumented stations. The anthropogenic pressure on water resources is taken into account by means of water withdrawals in each aquifer unit. The yearly amount of each uptake is provided by the Seine-Normandie Water Agency.

4.1 General Two-Step Calibration Strategy of Hydrosystem Models

Flipo et al. [68] proposed a stepwise calibration procedure for hydrosystem models. This method is further developed here in the form of an innovative two-step calibration methodology (Fig. 3) applied to the Seine basin [45]. The first step consists in the calibration of surface parameters with a process-based multi-objective function, which accounts for cumulative total river discharge and runoff dynamics. The estimation of the runoff dynamics is based on a proper hydrograph separation. Further details on the first step are given below. In the second step, subsurface

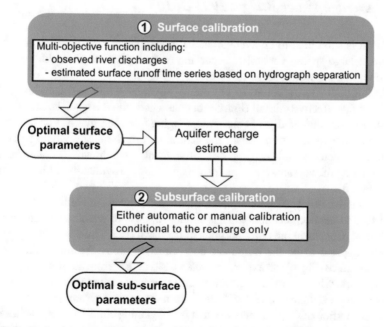

Fig. 3 Innovative two-step calibration procedure of the Seine basin with the CaWaQS model. The outcome of each step is an optimal parameter set (bold case)

parameters are calibrated given the infiltration rate calculated at the end of the first step. For this step, a classical trial-and-error method is used [35]. The conductance coefficient is automatically calibrated [35] using the horizontal permeability of the aquifer near the river to calculate the conductance at each river cell [79].

The first step is crucial in our approach, since the second step is conditional to the aquifer recharge estimated by the first step. Hydrosystem internal water fluxes estimated by surface–subsurface-coupled models are highly sensitive to recharge estimation [87], and subsurface parameters are highly sensitive to baseflow estimates [88, 89], which are correlated with aquifer recharge [90, 91]. Taking into account these crucial considerations, the core idea of our method is to add a baseflow estimate to the classical measurements (total river discharge and groundwater levels) that compose the multi-objective function to be minimised. Baseflow estimations were calculated directly from observed discharge time series, distributed in 30 gauging stations across the Seine basin [45], with the one parameter recursive digital filter first proposed by Lyne and Hollick [92] and later improved by Chapman [93]. Hydrograph separation is based on the estimation of the recession parameter, which is achieved by fitting the slope of logQ(t) after rainfall events [94].

River discharges are simulated properly by the CaWaQS model over 1993–2010 at various stations of the basin (Table 3). For instance, the Nash Efficiency [95] at the Paris Austerlitz station reaches 0.9 [45].

4.2 Average Water Budget 1993–2010

The basin is submitted to an average annual precipitation rate of 812 mm a^{-1} and exhibits a large spatial variability according to both distance from the ocean and elevation, ranging from 595 to 1,370 mm a^{-1} (Fig. 4a).

The infiltration rate amounts to 111 mm a^{-1} over the whole basin, corresponding to 56% of the effective rainfall (Fig. 5). In coherence with the spatial distribution of rainfall, a centripetal gradient is observed, ranging from 181 mm a^{-1} on the eastern Jurassic border to 82 mm a^{-1} over the central part of the basin, where an aquifer system is explicitly simulated (Fig. 4c). Groundwater withdrawals account for 14% of the total aquifer system recharge, of which 21% is provided by infiltration from rivers (Fig. 5).

Within the multilayer aquifer system, a dominant downward vertical flux from the subsurface to the deep chalk aquifer is simulated, 44% of which is redistributed to the alluvial deposits in the outer portions of the basin (Fig. 5). Supply from aquifers to rivers appears to be the dominant flow direction along the modelled stream–aquifer interface (Fig. 6); an exfiltration rate of 140 $m^3\ s^{-1}$ from aquifers supplies the river network, while 47 $m^3\ s^{-1}$ infiltrates the other way around (Fig. 5). On average the river network drains 10 $L^{-1}\ s^{-1}\ km^{-1}$ from the aquifer system [35, 45].

The proportion of the river network that is in a gaining configuration reaches 82% and would rise to 97% if all water withdrawals in the aquifer system were stopped. The surface–subsurface functioning is thus significantly altered by the

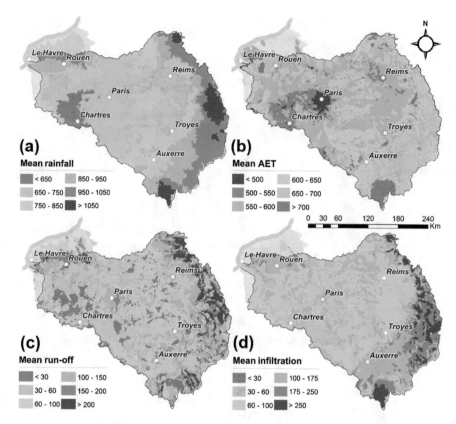

Fig. 4 Averaged spatial distributions over the 1993–2010 period of (**a**) SAFRAN annual rainfall and simulated, (**b**) annual AET, (**c**) annual infiltration, (**d**) annual runoff. Units in mm a^{-1}

anthropogenic pressure on the groundwater resources, which in turn modifies the biogeochemical functioning of the interface and greenhouse gas emissions [96]. The consequences of such alterations on river ecology are still poorly understood and are the focus of ongoing research [97, 98]. Heavy pumping removes 10 m^3 s^{-1} from alluvial plains connected to the downstream part of the river network (Strahler orders >3). Half of it is taken up from the alluvial aquifers, and the other half is taken directly from rivers through the natural upward motion of the water from the aquifer to the river [35].

At the river–aquifer interface, water is exchanged upward and downward. The absolute summation of these two is called "gross flux", while the difference is called "net flux". At the Seine basin scale, the gross flux is 50% larger than the net flux. During dry years, those proportions remain unchanged, while they increase by 50% during wet years [35].

As conceptualised by Flipo et al. [72], most of the water exchanges along the main river network take place in the stream–aquifer interface represented at this scale

N. Flipo et al.

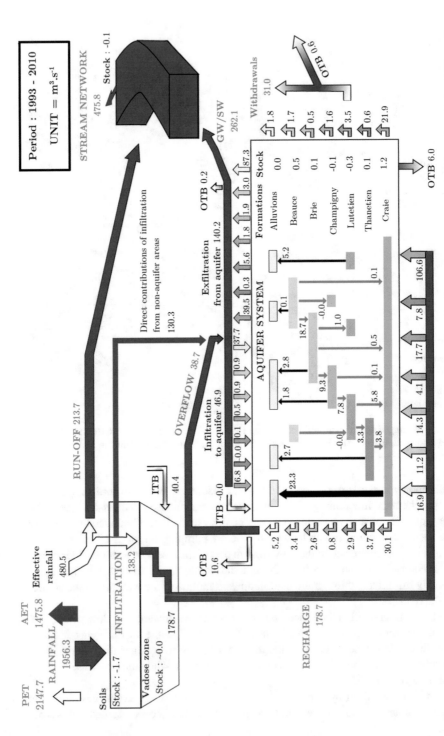

Fig. 5 The Seine basin water balance calculated over the 1993–2010 period. All fluxes in m³ s⁻¹. ITB, inflow through surface basin boundaries; OTB, outflows through surface basin boundaries. Red and black arrows distinguish downward and upward vertical fluxes, respectively

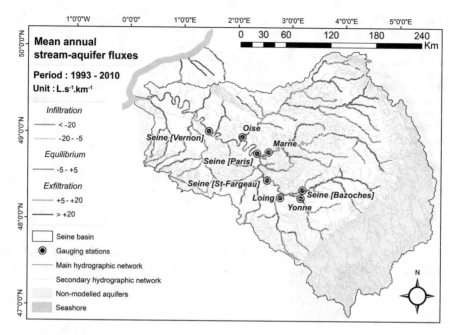

Fig. 6 Spatial distribution of stream–aquifer exchanges averaged over the 1993–2010 period. The main gauging stations of the basin are also displayed

by the alluvial plains. Overflows as well as direct exchanges at the stream–aquifer interface from aquifers contribute 55% of the river network discharge at the basin outlet (Fig. 5).

5   A Two-Century-Long Trajectory of the Seine Basin

The model provides the opportunity to characterise modifications in behaviour of the Seine hydrosystem under different scenarios combining preselected constraints on climate, land use and anthropogenic pressure data. It is therefore possible to reconstruct the trajectory of the system from the early 1900s to the end of the current century. Two scenarios are proposed to re-evaluate water fluxes: (1) a simulation associated with the first part of the twentieth century (1917–1934), which integrates past daily climate data as well as modifications in the surface distribution of land use, and (2) a second one (2083–2100) using climate data derived from a general circulation model (GCM) forced by the high emission scenario RCP8.5 [5] to evaluate the response of the Seine system in a context of climate change. Radiative forcings in RCP8.5 increase throughout the twenty-first century before reaching a level of 8.5 W m^{-2} at the end of the century. Modifications in groundwater withdrawal rates are also implemented. For the sake of clarity, results from both

Table 1 Key figures on distributions of land cover and withdrawal types in groundwater as well as population rates at the scale of the Seine basin for the 2000s simulation. Other values identify respective relative variations (%) for the 1900s and 2100s scenarios when compared to the 2000s reference

Scenario	1900s	2000s	2100s
Withdrawal types (in Mm^3 a^{-1})			
Drinking water	−100%	731	−12.9%
Industrial processes	−100%	169	−52.6%
Irrigation	−100%	96	−2.5%
Total	−100%	996	−16.7%
Land use (in % of the surface domain)			
Urban areas	−3.3%	7.9	+0.1%
Forested lands	−2.7%	23.2	0%
Agricultural lands	+6.0%	68.9	−0.1%
Demography (in 10^6 inhabitants)			
Population rate	−47%	16.7	+26%

simulations will be referred to as "1900s" and "2100s" in the remainder of the paper. Results from the SAFRAN reference simulation will be referred to as "2000s" or "reference situation".

5.1 Estimating Land Cover Changes

The land cover and soil texture define the hydrological response unit on which the water balance is calculated. A crucial step for the simulation of scenarios is therefore to estimate the changes in the land cover for the three main types of land use: urban areas, agricultural lands and forested lands (Table 1).

At the scale of all 24 administrative districts of the basin, 1901 and 2009 population census data provided by the National Institute of Statistics and Economic Studies (INSEE[2]) have allowed evaluating urban areas for the 1900s scenario, using equation (1), which relates the surface of artificialised area S_a (in hectares for 10^3 inhabitants) and the population density p (in inhabitants/km^2) [99]:

$$S_a = 1,475 \, p^{-0.6}, \tag{1}$$

In the case of the 2100s run, a similar method was implemented using results from the Explore 2070 project [100], which provides projection data on how the population is expected to evolve by the year 2070. This was considered steady until the year 2100.

Regarding forested areas, data between 1950 and 2010, also provided at the administrative district scale, were extracted from the annual agricultural statistics

---

[2]https://www.insee.fr/en/accueil.

data sets [101] for the 1900s scenario. However, no raw data are available before 1950. Thus, these spatial distributions were considered to be steady over the 1900–1950 period, given its negligible variation during the first half of the twentieth century [102]. Forested areas were also considered steady in the 2100s scenario when compared to the present day.

Then, in the 1900s scenario only, land use was refined over the Paris area [1] using data made available by the Urbanism and Land Use Institute for the Île-de-France area (IAU[3]). This additional step tends to integrate, as precisely as possible, the significant extension of the urban area over time and its consequences on local water budget calculations. In both cases, forested and urban area adjustments were made at the expense or benefit of the proportion of agricultural lands over the entire domain. These modifications were made assuming no significant changes in wetlands areas.

5.2   Estimating Anthropogenic Water Uptake

Projections from the Explore 2070 project [100] allowed integrating variation coefficients on present pumping rates in groundwater into the prospective 2100s simulation. These coefficients have been differentiated according to pumping types and geographical location (data not shown). These projections, which were elaborated in conjunction with demographic growth hypotheses and population migration processes, are based on extension of INSEE forecasts, i.e. the overall population rate of the basin to over 21 million inhabitants in 2070 (Table 1). Regarding drinking water withdrawals, forecasts expect a 40% increase in the number of households over the 2006–2070 period along with a decrease in water consumption, which has been differentiated according to the habitat type (-0.3% per year for buildings, -0.6% per year for private housing). A slight improvement in supply system efficiency is also integrated ($+0.2\%$ per year). Three main variables are involved in hypotheses on industrial-related withdrawals: production rate, past observed trends in water savings and types of cooling circuits. These forecasts are based on a general 4.0% decrease in water withdrawals per year, partly compensated with an increase in industrial production. They also emphasise the pursuit of cooling circuit closures along continuous improvements in water use efficiency of production processes. Finally, irrigation requirements are calculated according to crop types, integrating plant water demand, rainfall over agricultural lands as well as water efficiency of irrigation techniques (gravity-fed or drip irrigation). It remains important to bear in mind that these projections do not account for any impact of climate change, especially in the case of seasonal pumpings.

To summarise, the implementation of the Explore 2070 projections resulted in a decrease of the total volume withdrawn of 16.7% with ranges of variation of -2.5,

---

[3]https://en.iau-idf.fr/.

−12.9 and −52.6% for irrigation, drinking water and industrial processes, respectively (Table 1). Current groundwater total uptake is 996 Mm^3 a^{-1}, divided into 73%, 17% and 10% for drinking water, industrial processes and irrigation, respectively. Regarding the 1900s simulation, the assumption is made that most of the withdrawals were taken from surface water in the early twentieth century. For this simulation, there are no groundwater withdrawals.

5.3   Climate Scenarios

5.3.1   Reanalysis of the Past

The 1900s scenario climate variables are described using newly materialised reconstructions over long-term atmospheric reanalysis elaborated by Bonnet [21]. The reanalysis is based upon the twentieth-century NOAA 20CRv2c reanalysis [18], which was downscaled [20] following a statistical downscaling strategy [103, 104] that mobilises the ISBA-MODCOU chain [38, 105]. These data, downscaled at the SAFRAN grid scale, integrate the use of homogenised observations in the process of statistical downscaling in order to ensure a correct reproduction of the spatio-temporal variability of precipitation, temperature and river flows.

5.3.2   Selecting an Appropriate Climate Product for the Projection

Many worldwide climatic reanalysis and prediction products exist. Owing to differences in model structure, parametrisation and regionalisation, these products generate dissimilar results for precipitation and potential evapotranspiration, which eventually lead to distinct hydrological predictions [104, 106–108]. More precisely, while hydrological parametrisation and regionalisation are of the utmost importance for the evaluation of hydrological functioning at the seasonal scale, climate modellers agree on the fact that climate models are the dominant source of uncertainty in future climate projection [106–108]. To evaluate the impact of climate change on regional hydrosystems, various approaches are used [109], mostly based on statistical downscaling [6, 104, 109] or the use of a regional climate model such as in the EURO-CORDEX initiative [110]. To analyse the Seine hydrosystem trajectory in the future, we therefore decided to pay careful attention to the ability of the method to reproduce the current state of the system, as recommended by Radanovics et al. [111]. Projection data from four GCMs and one regional climate model from the Fifth Coupled Model Intercomparison Project (CMIP5) [5] were therefore disaggregated at the scale of SAFRAN grid and made available for the 1850–2100 period. Data from the following models were analysed: CanESM2 (Canada), MIROC5 (Japan), BCC-CSM-1-1-m (China), CSIRO-Mk3-6-0 (Australia) and Aladin-Climat (France).

As shown in Sect. 1.2, the hydrology of the Seine is responsive to long periods of climatic fluctuations due to large-scale climatic phenomena, such as the NAO.

Consequently, it is important to establish the presence and consistency of such fluctuations in the climatic product that is used for hydrological simulations, especially given that low-frequency fluctuations are difficult to simulate [112]. Therefore, we applied wavelet transformation techniques to compare the five climatic precipitation products to the SAFRAN precipitation reanalysis in four different sectors of the Seine basin. The four sectors chosen, thought to be representative of different local hydroclimatic conditions, are Pays de Caux (NW), Beauce (S), Champagne (NE) and the Morvan plateau (SE). For each sector, we considered the daily precipitation average over three-by-three grid cells, hence covering a sample area of 576 km^2. Time series of daily precipitation were transformed using a continuous wavelet transformation with a Morlet mother wavelet, in order to disentangle the various scales of variability in the signal (Fig. 7). Only the results for the Beauce

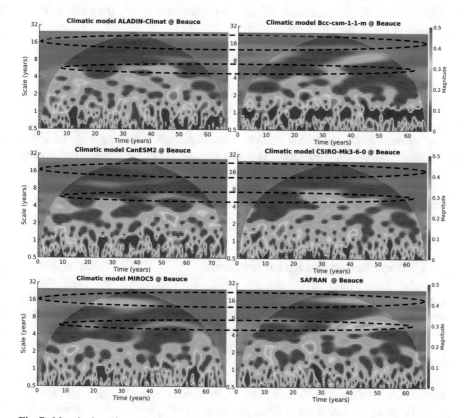

Fig. 7 Magnitude scalograms of daily precipitation time series in the Beauce sector of five climatic model products and for the SAFRAN data, used as the reference. The magnitude is set to a threshold of 0.5 mm in all scalograms in order to enhance visibility of low-frequency (high-scale) components. The MIROC5 model was chosen for hydrological simulation because the structure of its scalogram is the closest to SAFRAN's scalogram, especially concerning the 5–7- and 16–19-year scales. The white dashed line delimits the cone of influence, beyond which estimates of magnitude cannot be trusted

sector are displayed; the other sectors did not display any fundamentally different result. From this analysis, it appears that all climatic models differ significantly from the SAFRAN reanalysis, which is not surprising given the well-known difficulty of climatic modelling and downscaling [104]. In particular, the BCC-CSM, CanESM2, CSIRO-Mk3 and Aladin-Climat models do not bear the appropriate scales of low-frequency variability. For instance, the low-frequency variability found in BCC-CSM does not have the same scales as in the SAFRAN reanalysis (i.e. the 5–7- and 16–19-year scales), and the magnitudes associated with the annual cycle are overestimated. The CSIRO-Mk3 and CanESM2 products fail to show any significant variability above the 10-year scale, which is inconsistent with observations [50, 59] (Fig. 7). In contrast, MIROC5 precipitation outputs generally entail the closest-matching scales of variability, even if a lack of energy can be noted for the 5–7-year scale, which is the case for all models. Therefore, MIROC5 was selected as input for hydrological forecasts of the 2100s scenario to illustrate the potential changes in the hydrosystem functioning.

5.4   Water Resources Trajectory from the 1900s to the 2100s

How the Seine basin evolved from the 1900s to the 2100s was evaluated. For each case evaluated, the variations in fluxes are expressed according to a reference simulation under the assumption that the bias on climate forcings is stationary between the simulated and the reference periods. The reference simulations are either (1) the average state of the system previously simulated (2000s, Fig. 5) for the 1900s scenario[4] or (2) the average simulated state using MIROC5 climate data over the same period, for the 2100s scenario.

5.4.1   Water Budget and Recharge Modes

Like many other basins across western and northern Europe, the Seine discharge displays strong seasonality that is mainly driven by the quasi-sinusoidal fluctuation of the actual evapotranspiration throughout the hydrological year, whereas the precipitation input is much more stable over time in the reference simulation (Fig. 8). This growing seasonal gap in rainfall has an impact on the infiltration/runoff partition rate, which can especially be observed on the 2100s scenario. Indeed, despite a similar effective rainfall rate when compared to the 2000s simulation (Table 2), a relative variation in runoff of +14.0% is simulated, along with a

---

[4]A proper comparison between the 1900s and the 2000s should consider the average hydrosystem state forced by Bonnet's meteorological forcings [21] rather than the SAFRAN reanalysis. However, a visual comparison between the two simulations shows that they are very close (data not shown). For readability the SAFRAN reanalysis is used in the remainder of the paper.

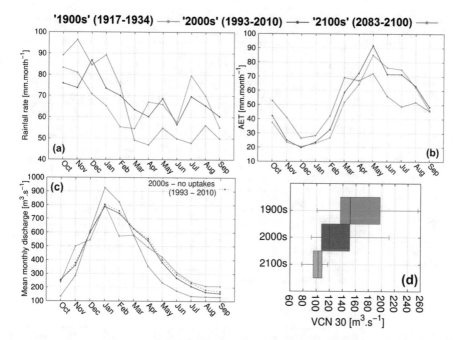

Fig. 8 Distribution of average monthly (**a**) rainfall rate (mm) and (**b**) actual evapotranspiration rates (mm) over the Seine basin; (**c**) monthly distribution of mean discharge (m³ s⁻¹) of the Seine River at the basin outlet (Seine at Vernon); (**d**) distribution of simulated annual minimum moving average over a 30-day period at the basin outlet. Blue, 2000s simulation forced with SAFRAN; green, 1900s simulation forced with Bonnet's reanalysis [21]; red, 2100s simulation forced with MIROC5

decrease of infiltration of −6.0%, which is due to spatio-temporal variations in the meteorological forcings (Fig. 9). To a lesser extent, a contrary observation could have been formulated for the 1900s, which is the result of both climate forcings and land use changes, particularly in the lower extent of the Paris urban area in the 1900s. In this case, it is not possible to disentangle the two origins based solely on these simulations.

Although the 2100s scenario is slightly drier in terms of recharge over the Seine basin (Table 2), recharge variation is not homogeneous spatially (Fig. 10b, d). Indeed, the lower Eocene and Palaeocene aquifer layers undergo a slight increase of recharge, +3.0% and +8.7%, respectively. Negative variations are more pronounced further west over the Cenomanian aquifer (−9.5%) and the eastern Jurassic border (−8.6%).

Table 2 Characterisation of the impact of the 1900s and 2100s scenarios on the behaviour of the Seine hydrosystem

	Variable	1900s	2000s	RV	CGM ref.	2100s	RV
Surface domain variables	Rainfall	1,870.5	1,956.3	−4.4	1,817.6	1,850.7	+1.8
	PET	1,916.1	2,147.7	−10.8	1,943.1	2,455.9	+26.4
	AET	1,408.6	1,475.8	−4.6	1,415.4	1,437.6	+1.6
	Effective rainfall	461.9	480.5	−3.9	402.2	413	+2.7
	Runoff	178.7	213.7	−16.4	170.8	194.6	+14.0
	Infiltration on whole domain	287	268.5	+6.9	231.6	217.8	−6.0
	Infiltration on non-aquifer areas	144.4	130.3	+10.8	121.3	110.8	−8.7
Aquifer system	GW recharge	189.1	178.7	+5.8	142.3	135.4	−4.9
	GW uptakes	0.00	31.6	–	31.6	26.3	−16.8
SW-GW exchanges	Infiltration from SW to GW	39.8	46.8	−15.1	46.9	47.3	+0.9
	Overflows from GW	51.4	38.7	+32.9	27.6	28.1	+1.7
	Exfiltration from GW to SW	160.9	140.2	+14.8	123.5	124.9	+1.1
	Contribution from GW to SW	316.7	262.1	+20.9	225.5	216.9	−4.1
	Discharge at outlet	495.6	475.8	+4.2	396.3	411.1	+3.7
	Infiltration length[*]	615.9	794.7	−22.5	935.4	907.3	−3.0
	Exfiltration length[*]	3,475.3	3,314	+4.9	3,146.9	3,171.3	+0.8

All relative variations (RV) are a % expressed according to their respective reference. Main water fluxes calculated over 17-year periods in $m^3\ s^{-1}$ except where noted [*] in kilometres. *GW* groundwater, *SW* surface water

5.4.2 Surface Water–Groundwater Exchanges

As no groundwater uptakes were integrated into the 1900s scenario, a comparison between the 1900s and 2000s simulations clarifies the impact of groundwater uptakes on stream–aquifer exchanges (Fig. 11), showing juxtaposed locations of (1) portions of the hydrographic network for which the average stream–aquifer direction flow switches from an exfiltration mode (i.e. aquifer feeding the river) to an infiltration mode (i.e. aquifer being fed from the river) and (2) locations of the current main pumping sites, thus identifying 179 km of rivers where stream-aquifer exchanges are modified. At the scale of each river cell, additional simulations (not shown) distinguished the origin of the stream reversal, either climatic (shown in black in Fig. 11) or directly related to local pumping (in red). A significant perturbation of stream–aquifer relations (Table 2) by proximal pumping sites was thus

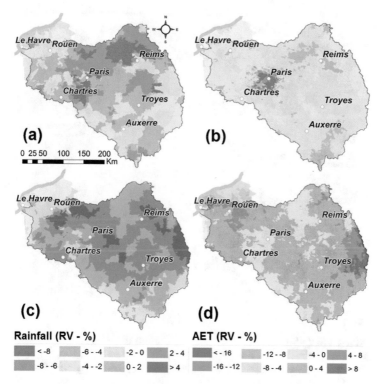

Fig. 9 Relative variations (RV) of average rainfall (**a, c**) and AET (**b, d**) for the 1900s (**a, b**) and 2100s scenarios (**c, d**). RVs are calculated with regard to their respective reference (2000s for the 1900s scenario and MIROC5 in the current days for the 2100s scenario)

identified since a 158-km proportion is directly linked to groundwater uptakes. Therefore, at the scale of the Seine basin, simulated conditions in the 1900s resulted in a significant 20.9% increase of the contribution from the groundwater compartment to the surface waters. On the other hand, in the 2100s scenario, slight variations only in fluxes from underground to surface waters are simulated when compared to the MIROC5 reference (Table 2), highlighting the relative stability of the stream–aquifer exchanges.

5.4.3 Hydrological Regimes

A comparison approach for all three cases on simulated discharge rates was carried out to identify any significant changes in hydraulic behaviour of the main rivers. Therefore, an analysis based on the usual characteristic flow rates was conducted on eight gauging stations located either at the downstream limits of the Seine's main tributaries or along the Seine River itself (Table 3). The mean annual flow rates

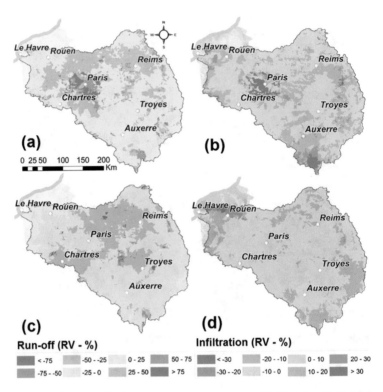

Fig. 10 Relative variations (RV) of average runoff (**a**, **c**) and infiltration rates (**b**, **d**) for the 1900s (**a**, **b**) and 2100s scenarios (**c**, **d**). RVs are calculated with regard to their respective reference (2000s for the 1900s scenario and MIROC5 in the current days for the 2100s scenario)

(module) all fit around the 2000s values within a range of ±10%. This first-order relative stability of the average discharge has to be weighted, however, by the temporal distribution of discharges, also called regime, and the analysis of extremes, especially low-flow values. Two usual flow rates were therefore considered: (1) the annual minimum flow rate calculated over a consecutive 30-day period (VCN30) and (2) the mean monthly annual minimum discharge associated with a 5-year return period (QMNA5).

The evolution of monthly discharges is in line with previous observations on precipitation dynamics, meaning an evolution towards a winter situation characterised by an increase of mean discharge rates as opposed to a decrease in summer (Fig. 8). The hydrological regime of the Seine basin appears stable between the 1900s and the 2000s, even though (1) natural discharges decrease significantly during low flow everywhere in the basin (see QMNA5 in Table 3) and (2) the spread of the low-flow discharge also significantly decreases, meaning more recurrent low-flow periods (Fig. 8d). In September, one-third of the discharge decrease is

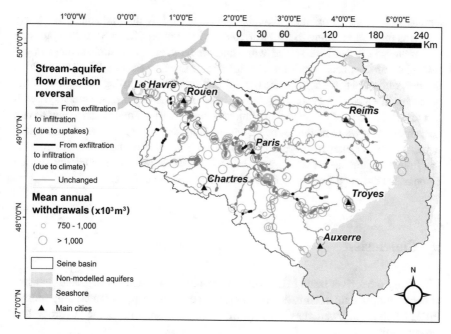

Fig. 11 Modifications of river–aquifer exchanges from the 1900s to the current scenario (2000s)

Table 3 Relative variations (RV, %) calculated on mean monthly annual minimum discharges and mean annual discharges when compared to respective reference simulations. In the latter case, values in brackets indicate the associated relative variation of standard deviation σ_{RV}

Gauging station	Efficiency[a]	QMNA5		Mean annual discharges	
		RV_{1900s}	RV_{2100s}	RV_{1900s} (σ_{RV})	RV_{2100s} (σ_{RV})
Yonne (Courlon)	0.81	55.5	−25.2	15.3 (29.3)	−2.3 (19.5)
Loing (Episy)	0.57	32.7	−7.2	16.1 (2.5)	3.7 (15.9)
Marne (Gournay)	0.90	18.5	−3.1	4.4 (−10.3)	3.3 (−6.7)
Oise (Pontoise)	0.69	17.0	2.5	−3.5 (−31.1)	12.1 (1.1)
Seine (Bazoches)	0.63	17.3	−1.3	7.5 (2.5)	1.8 (2.9)
Seine (St-Fargeau)	0.78	29.5	−15.7	12.6 (13.6)	0.5 (12.2)
Seine (Paris)	0.87	25.0	−8.6	8.8 (1.0)	2.2 (5.8)
Seine (Vernon)	0.77	22.9	−5.9	5.6 (−11.7)	5.1 (5.8)

[a]Nash efficiency [95] of simulated discharges for the 2000s simulation. Location of gauging stations is displayed in Fig. 6

due to pumping, while the remaining is due to the climate itself (Fig. 8c). This trend may continue and be reinforced in the future. The overall evolution from the 2000s to the 2100s may indeed lead to more abundant winter precipitation rates while being significantly reduced in the summer period (Fig. 8c, d). Although, in the case of the 2100s simulation, no change appears to be significant enough to be notable at the annual scale in terms of water availability (Table 3), low-flow rate analysis clearly confirms the trend leading to lower discharges in summer (Table 3), with a progressive reduction in VCN30 spread with time, underlying a potential increase in the frequency of the occurrence of severe low-water discharges (Fig. 8d). This trend is not new and rather in line with the transformation of the hydrological regime of the Seine basin since the 1900s (Fig. 8c). These observations are in agreement with former simulations of the impact of climate change [6, 9]. Climate change will have a large impact on water resources management in the future, with higher discharges in winter and lower ones from April to October. The increase of the low-flow period in the future should also lead to very critical low-flow periods in October, which emphasises the need for water management adaptation strategies.

6 Conclusion

Like every other river basin in the world, the Seine River basin faces global changes, either anthropogenic or climatic. The long-term environmental research programme, PIREN-Seine, allowed us to perform research on the Seine hydrosystem sustainability.

First, the analysis of long-term discharge data over 130 years at Paris displays the past trends, meaning long-term control of the discharge by climate and also the development of large water reservoirs that store 840 Mm^3 of water during winter and its release for sustaining low-flow discharge over 60 $m^3 s^{-1}$ at Paris. The latter has significantly transformed the hydrological regime of the Seine River by reducing the variability of monthly discharges since its implementation in the last quarter of the twentieth century.

In the last few decades, a distributed physically based coupled model of surface and subsurface flows, CaWaQS, was also developed and progressively refined in terms of aquifer units and the river network description, as well as processes such as river–aquifer exchanges. It establishes the first published water budget of the whole Seine basin hydrosystem over a 17-year NAO oscillation period, including average exchange fluxes between aquifer layers, as well as between the river network and its underlying free aquifer units, mostly composed of an alluvial plain for Strahler orders higher than 3.

Coupled with significant progress in hydrometeorology and climate research, this model was used to assess the Seine basin hydrological trajectory from the 1900s to the 2100s. It reveals relative stability in the average annual discharges over the entire basin over time coupled with substantial changes in the hydrological regime since the 1900s, starting one century ago with the decrease of low-flow discharges during low flow in September and the decrease of its annual variability. This first change led to a 50% decrease of the average August discharge at the basin's outlet, one-third of it due to the gradual implementation of groundwater uptakes. The analysis of the climate projection under the RCP 8.5 scenario simulated with the MIROC5 model reveals a strengthening of the tendency, with another 50% reduction of the actual average August discharge, but a more worrisome increase of the low-flow period, which may extend until the end of October in the future and start 1 month earlier. This drastic mutation of the hydrological regime may be manageable considering the projected significant increase of the January and February discharges but reveals the crucial need for the elaboration of adaptation strategies for water resources management.

This work is a first analysis based on a careful selection of only one climate projection that best mimics the low frequency of the rainfall signal. It will need to be strengthened in the future by a more thorough analysis of more climate projections with the definition of a pertinent and not redundant ensemble of climate projections. The more progress made in climate projections, the more sensitive the results will be to uncertainties due to model errors and especially to processes controlling surface and subsurface exchanges, i.e. aquifer recharge processes as well as river–aquifer exchanges. The PIREN-Seine will therefore continue to dedicate a significant part of its research resources to those key scientific questions.

Acknowledgements This study is a contribution to the PIREN-Seine research programme (www. piren-seine.fr), which belongs to the Zone Atelier Seine part of the international Long-Term Socio-Ecological Research (LTSER) network.

References

1. Flipo N, Lestel L, Labadie P et al (2020) Trajectories of the Seine River basin. In: Flipo N, Labadie P, Lestel L (eds) The Seine River basin. Handbook of environmental chemistry. Springer, Cham. https://doi.org/10.1007/698_2019_437
2. Billen G, Garnier J, Mouchel J-M, Silvestre M (2007) The Seine system: introduction to a multidisciplinary approach of the functioning of a regional river system. Sci Total Environ 375:1–12
3. Guillocheau F, Robin C, Allemand P et al (2000) Meso-Cenozoic geodynamic evolution of the Paris Basin: 3D stratigraphic constraints. Geodin Acta 13:189–245. https://doi.org/10.1080/09853111.2000.11105372
4. Meehl GA, Covey C, Delworth T et al (2007) The WCRP CMIP3 multimodel dataset: a new era in climate change research. Bull Am Meteorol Soc 88:1383–1394. https://doi.org/10.1175/BAMS-88-9-1383
5. Taylor KE, Stouffer RJ, Meehl GA (2012) An overview of CMIP5 and the experiment design. Bull Am Meteorol Soc 93:485–498. https://doi.org/10.1175/BAMS-D-11-00094.1
6. Dayon G, Boé J, Martin E, Gailhard J (2018) Impacts of climate change on the hydrological cycle over France and associated uncertainties. C R Geosci 350:141–153. https://doi.org/10.1016/j.crte.2018.03.001
7. Déqué M, Somot S, Sanchez-Gomez E et al (2012) The spread amongst ENSEMBLES regional scenarios: regional climate models, driving general circulation models and interannual variability. Climate Dynam 38:951–964. https://doi.org/10.1007/s00382-011-1053-x
8. Hattermann FF, Krysanova V, Gosling SN et al (2017) Cross-scale intercomparison of climate change impacts simulated by regional and global hydrological models in eleven large river basins. Clim Change 141:561–576. https://doi.org/10.1007/s10584-016-1829-4

9. Habets F, Boé J, Déqué M et al (2013) Impact of climate change on the hydrogeology of two basins in northern France. Clim Change 121:771–785. https://doi.org/10.1007/s10584-013-0934-x

10. Webber S, Donner SD (2017) Climate service warnings: cautions about commercializing climate science for adaptation in the developing world. Wiley Interdiscip Rev Clim Chang 8:e424. https://doi.org/10.1002/wcc.424

11. Donnelly C, Ernst K, Arheimer B (2018) A comparison of hydrological climate services at different scales by users and scientists. Clim Serv 11:24–35. https://doi.org/10.1016/j.cliser.2018.06.002

12. Hewitt C, Mason S, Walland D (2012) The global framework for climate services. Nat Clim Chang 2:831–832. https://doi.org/10.1038/nclimate1745

13. Rockström J, Steffen W, Noone K et al (2009) A safe operating space for humanity. Nature 461:472–475. https://doi.org/10.1038/461472a

14. Kalnay E, Kanamitsu M, Kistler R et al (1996) The NCEP/NCAR 40-year reanalysis project. Bull Am Meteorol Soc 77:437–472. https://doi.org/10.1175/1520-0477(1996)077<0437:TNYRP>2.0.CO;2

15. Saha S, Moorthi S, Pan H-L et al (2010) The NCEP climate forecast system reanalysis. Bull Am Meteorol Soc 91:1015–1058. https://doi.org/10.1175/2010BAMS3001.1

16. Dee DP, Uppala SM, Simmons AJ et al (2011) The ERA-Interim reanalysis: configuration and performance of the data assimilation system. Q J Roy Meteorol Soc 137:553–597. https://doi.org/10.1002/qj.828

17. Kobayashi S, Ota Y, Harada Y et al (2015) The JRA-55 reanalysis: general specifications and basic characteristics. J Meteorol Soc Japan Ser II 93:5–48. https://doi.org/10.2151/jmsj.2015-001

18. Compo GP, Whitaker JS, Sardeshmukh PD et al (2011) The twentieth century reanalysis project. Q J Roy Meteorol Soc 137:1–28. https://doi.org/10.1002/qj.776

19. Poli P, Hersbach H, Dee DP et al (2016) ERA-20C: an atmospheric reanalysis of the twentieth century. J Climate 29:4083–4097. https://doi.org/10.1175/JCLI-D-15-0556.1

20. Bonnet R, Boé J, Dayon G, Martin E (2017) Twentieth-century hydrometeorological reconstructions to study the multidecadal variations of the water cycle over France. Water Resour Res 53:8366–8382. https://doi.org/10.1002/2017WR020596

21. Bonnet R (2018) Variations du cycle hydrologique continental en France des années 1850 à aujourd'hui. Université de Toulouse 3 Paul Sabatier

22. Crutzen PJ (2002) Geology of mankind. Nature 415:23. https://doi.org/10.1038/415023a

23. Crutzen PJ, Steffen W (2003) How long have we been in the Anthropocene era? An editorial comment. Clim Change 61:251–257. https://doi.org/10.1023/B:CLIM.0000004708.74871.62

24. Ellis EC, Ramankutty N (2008) Putting people in the map: anthropogenic biomes of the world. Front Ecol Environ 6:439–447. https://doi.org/10.1890/070062

25. Ellis EC, Goldewijk KK, Siebert S et al (2010) Anthropogenic transformation of the biomes, 1700 to 2000. Glob Ecol Biogeogr 19:589–606. https://doi.org/10.1111/j.1466-8238.2010.00540.x

26. Ramankutty N, Foley JA (1999) Estimating historical changes in global land cover: croplands from 1700 to 1992. Global Biogeochem Cycles 13:997–1027. https://doi.org/10.1029/1999GB900046

27. Doussan C, Toma A, Paris B et al (1994) Coupled use of thermal and hydraulic head data to characterize river-groundwater exchanges. J Hydrol 153:215–229. https://doi.org/10.1016/0022-1694(94)90192-9

28. Doussan C, Poitevin G, Ledoux E, Detay M (1997) River bank filtration: modelling of the changes in water chemistry with emphasis on nitrogen species. J Contam Hydrol 25:129–156

29. Wei HF, Ledoux E, De Marsily G (1990) Regional modelling of groundwater flow and salt and environmental tracer transport in deep aquifers in the Paris Basin. J Hydrol 120:341–358. https://doi.org/10.1016/0022-1694(90)90158-T

30. Gonçalvès J, Violette S, Robin C et al (2004) Combining a compaction model with a facies model to reproduce permeability fields at the regional scale. Phys Chem Earth 29:17–24. https://doi.org/10.1016/j.pce.2003.11.009

31. Gonçalvès J, Violette S, Guillocheau F et al (2004) Contribution of a three-dimensional regional scale basin model to the study of the past fluid flow evolution and the present hydrology of the Paris basin, France. Basin Res 16:569–586. https://doi.org/10.1111/j.1365-2117.2004.00243.x

32. Jost A, Violette S, Gonçalvès J et al (2007) Long-term hydrodynamic response induced by past climatic and geomorphologic forcing: the case of the Paris basin, France. Phys Chem Earth Parts A/B/C 32:368–378. https://doi.org/10.1016/j.pce.2006.02.053

33. Gomez E, Ledoux E, Viennot P et al (2003) Un outil de modélisation intégrée du transfert des nitrates sur un système hydrologique: application au bassin de la Seine. La Houille Blanche 3–2003:38–45

34. Saleh F, Flipo N, Habets F et al (2011) Modeling the impact of in-stream water level fluctuations on stream-aquifer interactions at the regional scale. J Hydrol 400:490–500. https://doi.org/10.1016/j.jhydrol.2011.02.001

35. Pryet A, Labarthe B, Saleh F et al (2015) Reporting of stream-aquifer flow distribution at the regional scale with a distributed process-based model. Water Resour Manag 29:139–159. https://doi.org/10.1007/s11269-014-0832-7

36. Ducharne A, Golaz C, Leblois E et al (2003) Development of a high resolution runoff routing model, calibration and application to assess runoff from the LMD GCM. J Hydrol 280:207–228. https://doi.org/10.1016/S0022-1694(03)00230-0

37. Habets F, Noilhan J, Golaz C et al (1999) The ISBA surface scheme in a macroscale hydrological model applied to the Hapex-Mobilhy area part I: model and database. J Hydrol 217:75–96

38. Rousset F, Habets F, Gomez E et al (2004) Hydrometeorological modeling of the Seine basin using the SAFRAN-ISBA-MODCOU system. J Geophys Res 109:D14105. https://doi.org/10.1029/2003JD004403

39. Ledoux E, Gomez E, Monget JM et al (2007) Agriculture and groundwater nitrate contamination in the Seine basin. The STICS-MODCOU modelling chain. Sci Total Environ 375:33–47

40. Flipo N, Even S, Poulin M et al (2007) Modelling nitrate fluxes at the catchment scale using the integrated tool CaWaQS. Sci Total Environ 375:69–79. https://doi.org/10.1016/j.scitotenv.2006.12.016

41. Flipo N, Jeannée N, Poulin M et al (2007) Assessment of nitrate pollution in the Grand Morin aquifers (France): combined use of geostatistics and physically-based modeling. Environ Pollut 146:241–256. https://doi.org/10.1016/j.envpol.2006.03.056

42. Ducharne A (2007) Importance of stream temperature to climate change impact on water quality. Hydrol Earth Syst Sci Discuss 4:2425–2460

43. Flipo N (2005) Modélisation intégrée des transferts d'azote dans les aquifères et les rivières: application au bassin du Grand Morin. Centre d'Informatique Géologique, Ecole Nationale Supérieure des Mines de Paris

44. Flipo N, Even S, Poulin M, Ledoux E (2005) Hydrological part of CaWaQS (catchment water quality simulator): fitting on a small sedimentary basin. Verh Internat Verein Limnol 29:768–772

45. Labarthe B (2016) Quantification des échanges nappe-rivière au sein de l'hydrosystème Seine par modélisation multi-échelle. MINES ParisTech, PSL Research University

46. Dierauer JR, Whitfield PH, Allen DM (2017) Assessing the suitability of hydrometric data for trend analysis: the FlowScreen package for R. Can Water Res J 42:269–275. https://doi.org/10.1080/07011784.2017.1290553

47. Boé J, Habets F (2014) Multi-decadal river flow variations in France. Hydrol Earth Syst Sci 18:691–708. https://doi.org/10.5194/hess-18-691-2014

48. Hurrell JW (1995) Decadal trends in the North Atlantic oscillation: regional temperatures and precipitation. Science 269:676–679. https://doi.org/10.1126/science.269.5224.676

49. Hurrell JW, Deser C (2014) Northern hemisphere climate variability during winter: looking back on the work of Felix Exner. Meteorol Z 24:113–118. https://doi.org/10.1127/metz/2015/0578

50. Massei N, Laignel B, Deloffre J et al (2010) Long-term hydrological changes of the Seine River flow (France) and their relation to the North Atlantic Oscillation over the period 1950-2008. Int J Climatol 30:2146–2154. https://doi.org/10.1002/joc.2022

51. Kerr RA (2000) A North Atlantic climate pacemaker for the centuries. Science 288:1984–1985. https://doi.org/10.1126/science.288.5473.1984

52. Schlesinger ME, Ramankutty N (1994) An oscillation in the global climate system of period 65-70 years. Nature 367:723–726. https://doi.org/10.1038/367723a0

53. Hurrell JW, Kushnir Y, Visbeck M (2001) The North Atlantic oscillation. Science 291:603–605. https://doi.org/10.1126/science.1058761

54. Visbeck MH, Hurrell JW, Polvani L, Cullen HM (2001) The North Atlantic oscillation: past, present, and future. Proc Natl Acad Sci U S A 98:12876–12877. https://doi.org/10.1073/pnas.231391598

55. Hurrell JW, Kushnir Y, Ottersen G, Visbeck M (2003) An overview of the north atlantic oscillation. Geophys Monogr Ser 134:1–35. https://doi.org/10.1029/134GM01

56. Dieppois B, Lawler DM, Slonosky V et al (2016) Multidecadal climate variability over northern France during the past 500 years and its relation to large-scale atmospheric circulation. Int J Climatol 36:4679–4696. https://doi.org/10.1002/joc.4660

57. Wang J, Yang B, Ljungqvist FC et al (2017) Internal and external forcing of multidecadal Atlantic climate variability over the past 1200 years. Nat Geosci 10:512–517. https://doi.org/10.1038/ngeo2962

58. Sutton RT, Dong B (2012) Atlantic Ocean influence on a shift in European climate in the 1990s. Nat Geosci 5:788–792. https://doi.org/10.1038/ngeo1595

59. Massei N, Dieppois B, Hannah DM et al (2017) Multi-time-scale hydroclimate dynamics of a regional watershed and links to large-scale atmospheric circulation: application to the Seine river catchment, France. J Hydrol 546:262–275. https://doi.org/10.1016/j.jhydrol.2017.01.008

60. Massei N, Fournier M (2012) Assessing the expression of large-scale climatic fluctuations in the hydrological variability of daily Seine river flow (France) between 1950 and 2008 using Hilbert-Huang Transform. J Hydrol 448–449:119–128. https://doi.org/10.1016/j.jhydrol.2012.04.052

61. Galéa G, Mercier G, Adler MJ (1999) Low flow-duration-frequency models. Concept and use for a regional approach to watershed low-flow regimes in the Loire (France) and Crisu-Alb (Romania) regions. Rev des Sci l'Eau 12:93–122

62. Schuite J, Flipo N, Massei N et al Improving the spectral analysis of hydrological signals to efficiently constrain watershed properties. Water Resour Res. https://doi.org/10.1029/2018WR024579

63. Baratelli F, Flipo N, Rivière A, Biancamaria S (2018) Retrieving river baseflow from SWOT spaceborne mission. Remote Sens Environ 218:44–54. https://doi.org/10.1016/j.rse.2018.09.013

64. de Marsily G, Ledoux E, Levassor A et al (1978) Modelling of large multilayered aquifer systems: theory and applications. J Hydrol 36:1–34

65. Ledoux E (1980) Modélisation intégrée des écoulements de surface et des écoulements souterrains sur un bassin hydrologique. ENSMP, UPMC

66. Ledoux E, Girard G, Villeneuve JP (1984) Proposition d'un modèle couplé pour la simulation conjointe des écoulements de surface et des écoulements souterrains sur un bassin hydrologique. La Houille Blanche (1–2):101–110

67. Ledoux E, Girard G, de Marsily G et al (1989) Unsaturated flow in hydrologic modeling – theory and practice. In: Morel-Seytoux HJ (ed) NATO ASI series. Kluwer, Norwell, pp 435–454

68. Flipo N, Monteil C, Poulin M et al (2012) Hybrid fitting of a hydrosystem model: long term insight into the Beauce aquifer functioning (France). Water Resour Res 48:W05509. https://doi.org/10.1029/2011WR011092

69. Baratelli F, Flipo N, Moatar F (2016) Estimation of distributed stream-aquifer exchanges at the regional scale using a distributed model: sensitivity to in-stream water level fluctuations, riverbed elevation and roughness. J Hydrol 542:686–703. https://doi.org/10.1016/j.jhydrol.2016.09.041

70. Labarthe B, Pryet A, Saleh F et al (2015) Distributed simulation of daily stream-aquifer exchanged fluxes in the Seine River basin at regional scale. In: Lollino G, Arrattano M, Rinaldi M et al (eds) Engineering geology for society and territory, vol 3. Springer, Berlin, pp 261–265

71. Vergnes J-P, Habets F (2018) Impact of river water levels on the simulation of stream–aquifer exchanges over the Upper Rhine alluvial aquifer (France/Germany). Hydrgeol J 26:2443–2457. https://doi.org/10.1007/s10040-018-1788-0

72. Flipo N, Mouhri A, Labarthe B et al (2014) Continental hydrosystem modelling: the concept of nested stream-aquifer interfaces. Hydrol Earth Syst Sci 18:3121–3149. https://doi.org/10.5194/hess-18-3121-2014

73. Tóth J (1963) A theoretical analysis of groundwater flow in small drainage basins. J Geophys Res 68:4795–4812

74. Girard G, Ledoux E, Villeneuve J-P (1980) An integrated rainfall, surface and underground runoff model. La Houille Blanche 4/5:315–320

75. Deschesnes J, Villeneuve J-P, Ledoux E, Girard G (1985) Modeling the hydrologic cycle: the MC model. Part I – principles and description. Hydrol Res 16:257–272

76. Nash JE (1959) Systematic determination of unit hydrograph parameters. J Geophys Res 64:111–115. https://doi.org/10.1029/JZ064i001p00111

77. Besbes M, De Marsily G (1984) From infiltration to recharge: use of a parametric transfer function. J Hydrol 74:271–293. https://doi.org/10.1016/0022-1694(84)90019-2

78. de Marsily G (1986) Quantitative hydrogeology – groundwater hydrology for engineers. Academic Press, London

79. Rushton K (2007) Representation in regional models of saturated river-aquifer interaction for gaining/losing rivers. J Hydrol 334:262–281. https://doi.org/10.1016/j.jhydrol.2006.10.008

80. Ebel BA, Mirus BB, Heppner CS et al (2009) First-order exchange coefficient coupling for simulating surface water-groundwater interactions: parameter sensitivity and consistency with a physics-based approach. Hydrol Process 23:1949–1959. https://doi.org/10.1002/hyp.7279

81. David CH, Habets F, Maidment DR, Yang Z-L (2011) RAPID applied to the SIM-France model. Hydrol Process 25:3412–3425. https://doi.org/10.1002/hyp.8070

82. David CH, Yang Z-L, Famiglietti JS (2013) Quantification of the upstream-to-downstream influence in the Muskingum method and implications for speedup in parallel computations of river flow. Water Resour Res 49:2783–2800

83. Strahler AN (1957) Quantitative analysis of watershed geomorphology. Geophys Union Trans 38:913–920

84. de Fouquet C (2012) Environmental statistics revisited: is the mean reliable? Environ Sci Technol 46:1964–1970

85. Quintana-Seguí P, Le MP, Durand Y et al (2008) Analysis of near-surface atmospheric variables: validation of the SAFRAN analysis over France. J Appl Meteorol Climatol 47:92–107

86. Vidal J-P, Martin E, Franchistéguy L et al (2010) A 50-year high-resolution atmospheric reanalysis over France with the Safran system. Int J Climatol 30:1627–1644. https://doi.org/10.1002/joc.2003

87. Wu B, Zheng Y, Tian Y et al (2014) Systematic assessment of the uncertainty in integrated surface water-groundwater modeling based on the probabilistic collocation method. Water Resour Res 50:5848–5865. https://doi.org/10.1002/2014WR015366

88. Hunt R, Strand M, Walker J (2006) Measuring groundwater-surface water interaction and its effect on wetland stream benthic productivity, Trout Lake watershed, northern Wisconsin, USA. J Hydrol 320:370–384. https://doi.org/10.1016/j.jhydrol.2005.07.029

89. Yager RM (1998) Deflecting influential observations in nonlinear regression modelling of groundwater flow. Water Resour Res 34:1623–1633

90. Brutsaert W, Nieber JL (1977) Regionalized drought flow hydrographs from a mature glaciated plateau. Water Resour Res 13:637–643. https://doi.org/10.1029/WR013i003p00637

91. Arnold JG, Muttiah RS, Srinivasan R, Allen PM (2000) Regional estimation of base flow and groundwater recharge in the Upper Mississippi river basin. J Hydrol 227:21–40. https://doi.org/10.1016/S0022-1694(99)00139-0

92. Lyne V, Hollick M (1979) Stochastic time variable rainfall-runoff modelling. In: Proceedings of the hydrology and water resources symposium, Perth, 10–12 September,1979. Institution of Engineers National Conference Publication, pp 89–92

93. Chapman T (1991) Comment on "Evaluation of automated techniques for base flow and recession analyses" by R.J. Nathan and T.A. McMahon. Water Resour Res 27:1783–1784

94. Chapman T (1999) A comparison of algorithms for stream flow recession and baseflow separation. Hydrol Process 13:701–714

95. Nash JE, Sutcliffe JV (1970) River flow forecasting through conceptual models. {P}art {I}, a discussion of principles. J Hydrol 10:282–290

96. Newcomer ME, Hubbard SS, Fleckenstein JH et al (2018) Influence of hydrological perturbations and riverbed sediment characteristics on hyporheic zone respiration of CO2 and N2. Eur J Vasc Endovasc Surg 123:902–922. https://doi.org/10.1002/2017JG004090

97. Marmonier P, Archambaud G, Belaidi N et al (2012) The role of organisms in hyporheic processes: gaps in current knowledge, needs for future research and applications. Int J Limnol 48:253–266

98. Boano F, Harvey JW, Marion A et al (2014) Hyporheic flow and transport processes: mechanisms, models, and biogeochemical implications. Rev Geophys 52:603–679. https://doi.org/10.1002/2012RG000417

99. Couturier C, Charru M, Doublet S, Pointereau P (2017) Le scénario Afterres 2050. Solagro

100. Ministère de l'Ecologie du Développement Durable et de l'Energie (2012) Explore 2070: prospective socio-économique et démographique – pressions anthropiques

101. AGRESTE (2009) La statistique agricole annuelle: Présentation générale

102. Koerner M, Cinotti B, Jussy J-H, Benoit M (2000) Evolution des surfaces boisées en France depuis le début du XIXème siècle: identification et localisation des boisements des territoires agricoles abandonnés. Rev For Fr 3:249–269. https://doi.org/10.4267/2042/5359

103. Boé J, Terray L, Habets F, Martin E (2007) Statistical and dynamical downscaling of the Seine basin climate for hydro-meteorological studies. Int J Climatol 27:1643–1655. https://doi.org/10.1002/joc.1602

104. Dayon G, Boé J, Martin E (2015) Transferability in the future climate of a statistical downscaling method for precipitation in France. J Geophys Res 120:1023–1043. https://doi.org/10.1002/2014JD022236

105. Habets F, Boone A, Champeaux J et al (2008) The SAFRAN-ISBA-MODCOU hydrometeorological model applied over France. J Geophys Res 113:D06113. https://doi.org/10.1029/2007JDOO8548

106. Hattermann FF, Vetter T, Breuer L et al (2018) Sources of uncertainty in hydrological climate impact assessment: a cross-scale study. Environ Res Lett 13:15006. https://doi.org/10.1088/1748-9326/aa9938

107. Her Y, Yoo S-H, Cho J et al (2019) Uncertainty in hydrological analysis of climate change: multi-parameter vs. multi-GCM ensemble predictions. Sci Rep 9:4974. https://doi.org/10.1038/s41598-019-41334-7

108. Ashraf Vaghefi S, Iravani M, Sauchyn D et al Regionalization and parameterization of a hydrologic model significantly affect the cascade of uncertainty in climate-impact projections. Climate Dynam. https://doi.org/10.1007/s00382-019-04664-w

109. Maraun D, Wetterhall F, Ireson AM et al (2010) Precipitation downscaling under climate change: recent developments to bridge the gap between dynamical models and the end user. Rev Geophys 48:RG3003. https://doi.org/10.1029/2009RG000314

110. Jacob D, Petersen J, Eggert B et al (2014) EURO-CORDEX: new high-resolution climate change projections for European impact research. Reg Environ Chang 14:563–578. https://doi.org/10.1007/s10113-013-0499-2

111. Radanovics S, Vidal J-P, Sauquet E et al (2013) Optimising predictor domains for spatially coherent precipitation downscaling. Hydrol Earth Syst Sci 17:4189–4208. https://doi.org/10.5194/hess-17-4189-2013

112. Martin ER, Thorncroft C, Booth BBB (2014) The multidecadal Atlantic SST – sahel rainfall teleconnection in CMIP5 simulations. J Climate 27:784–806. https://doi.org/10.1175/JCLI-D-13-00242.1

The Seine Watershed Water-Agro-Food System: Long-Term Trajectories of C, N and P Metabolism

Gilles Billen, Josette Garnier, Julia Le Noë, Pascal Viennot, Nicolas Gallois,
Thomas Puech, Celine Schott, Juliette Anglade, Bruno Mary,
Nicolas Beaudoin, Joël Léonard, Catherine Mignolet, Sylvain Théry,
Vincent Thieu, Marie Silvestre, and Paul Passy

Contents

1   Introduction 93
2   Material and Methods 93
3   Trajectory and Biogeochemical Functioning of the Agricultural System 95
   3.1   Long-Term Changes in the Structure of the Northern France Agricultural System .   95
   3.2   Changes in Land Use and Crop Rotations 98
   3.3   Yield-Fertilisation Relationship 98
4   Soil Biogeochemistry Reflects This Trajectory 100
   4.1   Soil Organic Carbon Storage 100
   4.2   Agricultural Greenhouse Gas Emissions 100
   4.3   Nitrogen Soil Storage and Leaching 102
   4.4   Phosphorus Dynamics and Erosion 103

The copyright year of the original version of this chapter was corrected from 2019 to 2020.
A correction to this chapter can be found at https://doi.org/10.1007/698_2020_667

G. Billen (✉), J. Garnier, and V. Thieu
UMR 7619 Metis, Sorbonne-Université CNRS EPHE, Paris, France
e-mail: gilles.billen@sorbonne-universite.fr

J. Le Noë
Institut für Soziale Ökologie, BOKU Universität, Vienna, Austria

P. Viennot and N. Gallois
Geosciences Department, MINES ParisTech, PSL University, Fontainebleau, France

T. Puech, C. Schott, J. Anglade, and C. Mignolet
ASTER, INRA, Centre Grand Est, Nancy, France

B. Mary, N. Beaudoin, and J. Léonard
AgroImpact, INRA, Laon, France

S. Théry, M. Silvestre, and P. Passy
CNRS SU FR 3020 FIRE, Paris, France

Nicolas Flipo, Pierre Labadie, and Laurence Lestel (eds.), *The Seine River Basin*,
Hdb Env Chem (2021) 90: 91–116, https://doi.org/10.1007/698_2019_393,
© The Author(s) 2020, corrected publication 2020, Published online: 3 June 2020

5   Hydrosystem Response to Agricultural Trajectories 104
     5.1   Aquifer Storage of Nitrogen 105
     5.2   Riparian Processes 105
     5.3   Point and Diffuse Sources of Nutrients to the River System 107
     5.4   N and P Budget of the Water-Agro-Food System 108
6   Conclusion and Scenarios for the Future 109
     6.1   The Importance of Long-Term Storage Processes 109
     6.2   The Importance of the Structural Pattern of Agro-Food Systems
           on the Environmental Imprint 110
References 111

Abstract   Based on the GRAFS method of biogeochemical accounting for nitrogen (N), phosphorus (P) and carbon (C) fluxes through crop, grassland, livestock and human consumption, a full description of the structure and main functioning features of the French agro-food system was obtained from 1850 to the present at the scale of 33 agricultural regions. For the period since 1970, this description was compared with the results of an agronomic reconstitution of the cropping systems of the Seine watershed based on agricultural census and detailed enquiries about farming practices at the scale of small agricultural regions (the ARSeine database), which were then used as input to an agronomical model (STICS) calculating yields, and the dynamics of N and C. STICS was then coupled with a hydrogeological model (MODCOU), so that the entire modelling chain can thus highlight the high temporal inertia of both soil organic matter pool and aquifers. GRAFS and ARSeine revealed that the agriculture of the North of France is currently characterised by a high degree of territorial openness, specialisation and disconnection between crop and livestock farming, food consumption and production. This situation is the result of a historical trajectory starting in the middle of the nineteenth century, when agricultural systems based on mixed crop and livestock farming with a high level of autonomy were dominant. The major transition occurred only after World War II and the implementation of the Common Agricultural Policy and led, within only a few decades, to a situation where industrial fertilisers largely replaced manure and where livestock farming activities were concentrated either in the Eastern margins of the watershed in residual mixed farming areas or in specialised animal production zones of the Great West. A second turning point occurred around the 1990s when regulatory measures were taken to partly correct the environmental damage caused by the preceding regime, yet without in-depth change of its logic of specialisation and intensification. Agricultural soil biogeochemistry (C sequestration, nitrate losses, P accumulation, etc.) responds, with a long delay, to these long-term structural changes. The same is true for the hydrosystem and most of its different compartments (vadose zone, aquifers, riparian zones), so that the relationship between the diffuse sources of nutrients (or pesticides) and the agricultural practices is not immediate and is strongly influenced by legacies from the past structure and practices of the agricultural system. This has strong implications regarding the possible futures of the Seine basin agriculture.

Keywords   Agriculture, Aquifers, Carbon, Denitrification, Fertilisers, Greenhouse gases, Leaching, Nitrogen, Nutrients, Phosphorus, Riparian wetlands, Soil

1 Introduction

Given that it deeply affects the functioning of terrestrial ecosystems, agriculture is not only the major determinant of landscape structure, biodiversity and soil biogeochemistry but also an essential factor in determining the hydrology and water quality of river systems and their receiving marine coastal waters. In particular, the nutrient (C, N, P, Si) composition of ground- and surface water is largely dependent on diffuse sources from the watershed which respond to land use and agricultural practices. This response, however, is far from being simple and direct, due to the complex cascade of processes, including storage and elimination, that nutrients, emitted from the root zone of cropping systems, have to move across, with temporalities ranging from sub-hourly to multi-decadal.

The Seine watershed, with a catchment area of about 70,000 km^2, is entirely located within one of the most fertile areas of Western Europe, the Paris Basin. This geological unit consists of concentric tertiary sedimentary formations (alternating clay, sandstone and limestone), covered by loess in its central part and lying on a basement of ancient crystalline rock formations outcropping at the extreme South-East and North-East (Fig. 1a). Paris developed in the middle of the drainage network, at the convergence area of large tributaries draining this basin, which historically was a favourable factor in terms of the city's food, feed and fuel supply. Currently, the central zone of the Seine watershed, around the huge Paris conurbation, is oriented towards mass production of cereals and industrial crops, while animal breeding is restricted to the peripheral areas of the basin where pedoclimatic conditions are less suitable to stockless cropping systems (Fig. 1b).

The Seine River basin has been subject to intensive research for 30 years within the PIREN-Seine programme [2, 3]. Here we present a synthesis of this work, addressing the interrelated issues of agricultural dynamics, soil biogeochemistry as well as ground- and surface water nutrient contamination. The purpose of this chapter is to describe, over a 150-year period, the long-term dynamics through which the current state of the agricultural system has gradually been constructed, in order to understand both the drivers of change and the inertia of the different environmental compartments of the water-agro-food system of the Seine watershed. Based on this long-term view of the role of legacies on the current system functioning, the issue of its possible future evolution will be shortly addressed.

2 Material and Methods

This chapter is mainly based on the results of two complementary integrated research efforts developed in the PIREN-Seine programme (www.piren-seine.fr), namely, the GRAFS-Riverstrahler and the ARSeine-STICS-MODCOU approaches, which are here compared and merged for the very first time. These two approaches differ in their level of detail, time and space resolution and the duration of the historical period they are able to encompass.

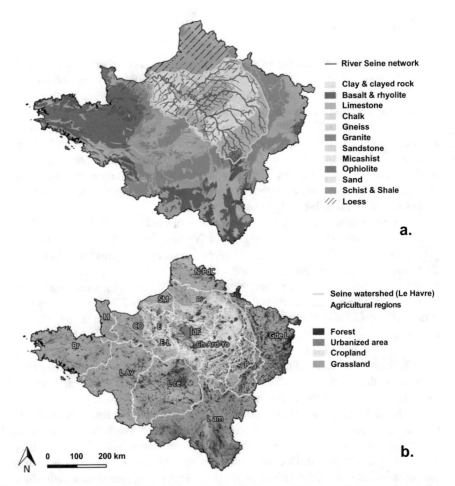

Fig. 1 The Seine basin within in its wider geographical context. (**a**) The Seine drainage network and the lithological zones within and around the Paris Basin (source: BRGM www.brgm.fr). (**b**) Land use (source: Corine Land Cover 2012, www.data.gouv.fr/fr/datasets/corine-land-cover-occupation-des-sols-en-france) and agricultural regions defined by Le Noë et al. [1]. *NPdC* Nord-Pas-de-Calais, *Pic* Picardy, *SM* Seine Maritime, *M* Manche, *CO* Calvados-Orne, *E* Eure, *E-L* Eure-et-Loir, *IdF* Île-de-France, *Ch-Ard-Yo* Champagne-Ardenne-Yonne, *GdeL* Grande Lorraine, *Brg* Burgundy, *Br* Brittany, *L Av* Loire Aval, *Lce* Loire centrale, *L am* Loire Amont

GRAFS (for Generalized Representation of the Agro-Food System) is a biogeochemical accounting tool for describing the N, P and C fluxes across the crop- and grassland, livestock and human population of a given territory [4]. It is conceived as a framework for analysing the functioning of agricultural systems, their requirements in terms of resources and their environmental losses, as well as their long-term trajectories, since 1850 [5], based on data mostly derived from the compilation of official agricultural statistics available at the *département* scale (typically 6,000 km^2). It provides the required data for running the Riverstrahler model [6, 7] (www.fire.upmc.fr/rive), which calculates the nutrient transfers and the ecological functioning of each tributary of the river system, given the diffuse and point sources of nutrient and organic matter from the watershed. The calculated nutrient fluxes at the outlet of the river system can then be used by a coastal marine model such as ECO-MARS 3D to assess the eutrophication generated by these fluxes [8–11].

The ARSeine database [12] offers a spatially detailed and distributed description of the Seine-Normandie cropping systems over the 1970–2015 period, including land use, crop rotations and detailed management techniques at the *Petites Régions Agricoles* scale (typically 1,000 km^2). It has been designed to provide the inputs to a 2D-distributed version [13] of the STICS model [14–18]. STICS is an agronomical crop model simulating crop production and the components of the N cycle at the same space and time resolution. Input soil parameters have been defined for each soil unit of the Soil Geographic Database of France at the 1:1,000,000 scale [19], using local pedotransfer functions [20]. Daily values of nitrate leaching predicted by STICS are used as an input to the hydrogeological MODCOU model [21], which calculates the recharge and nitrate contamination of the basin's main aquifer formations [13, 22].

Evaluating the uncertainty on the results of such long-term reconstruction of environmental data is a critical task. As far as modelling approaches are concerned, two types of uncertainty can be distinguished: structural uncertainties related to the adequacy of the model's representation of the system and operational uncertainties related to the accuracy in the data and parameters used [23]. The latter can be evaluated using Monte Carlo methods to assess how uncertainty on the raw data propagates to final model results; this approach shows typical uncertainties of approximately 25% for the GRAFS approach [4]. Structural uncertainties are by essence much more difficult to assess. They have been roughly estimated at 15% for the STICS model [20].

3 Trajectory and Biogeochemical Functioning of the Agricultural System

3.1 Long-Term Changes in the Structure of the Northern France Agricultural System

Until the beginning of the twentieth century, mixed crop and livestock farming systems dominated everywhere in France (Fig. 2a). Manure and symbiotic N fixation by grassland and legume crops inserted in rotations were the only sources of cropland fertilisation. Specialisation into stockless cropping systems, relying on

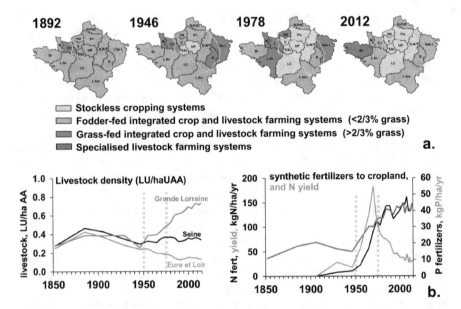

Fig. 2 (**a**) Gradual specialisation of agricultural systems in the North of France from 1850 to 2015. (**b**) Long-term variation of number of livestock density (in livestock units per ha of agricultural land) (right), N (black curve) and P (blue curve) fertilisation and crop yield (right) in the Seine basin (after [5]) (Three periods from 1850 to 1950, from 1950 to 1975 and from 1975 to 2015 can be distinguished)

industrial N and P fertilisers, is developed first in the Île-de-France and Eure-et-Loir regions in the first half of the twentieth century, owing to the proximity of Paris and transport infrastructures. After World War II, a voluntarist state policy of agriculture modernisation led to increased farm size, the rural exodus, the rapid increase of industrial fertiliser use and regional specialisation [24, 25] into either stockless cropping systems (dominating in the middle of the Paris Basin) or intensive livestock farming systems (in the Great West), often highly dependent on the import of feed (Fig. 2a). This resulted in an unprecedented opening of the nutrient cycles, with increasing environmental losses and growing insertion into international markets. After the 1980s, public policies shifted from interventionist support in favour of increasing production to give way to greater liberalism. However, since the 1990s, the rise of fertiliser prices together with the implementation of agro-environmental measures to limit nutrient losses resulted in an inversion in the trends of mineral fertilisation. N and P soil balances decreased, even becoming negative in the case of P in arable soils (Figs. 2b and 10a). These changes are clearly reflected in the patterns of N fluxes between arable land, permanent grassland, livestock and human nutrition (Fig. 3); similar trends are also apparent in terms of P and C fluxes [4, 5].

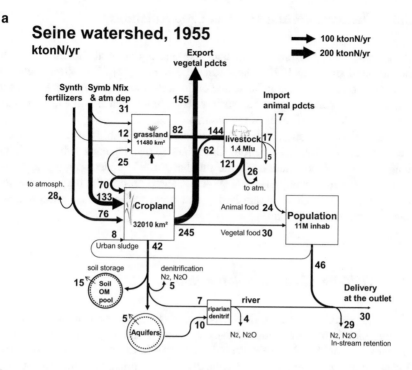

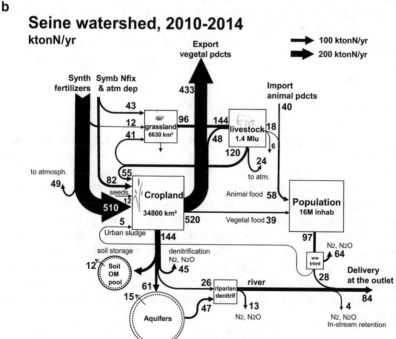

Fig. 3 GRAFS representation of N fluxes between cropland, grassland, livestock, human population and the hydrosystem in the Seine basin around (**a**) 1955 and (**b**) 2010–2014. (approximated fluxes in thousand metric tons of N)

3.2 Changes in Land Use and Crop Rotations

At a finer scale, the changes in land use and agricultural practices have been documented since the 1970s (the ARSeine database [12]). The specialisation in a stockless cropping system in the centre of the watershed went together with a strong reduction of permanent grassland surfaces (Fig. 4a), which are now restricted to the Eastern and Western fringes of the basin.

A significant reduction of the length and diversity of arable crop rotations has also occurred during the same period. Grain and forage legumes, which were basic components of crop rotations in the middle of the twentieth-century agriculture, were abandoned in many places (Fig. 4a). A sharp drop in the frequency of spring crops (Fig. 4b), such as spring barley and grain maize, is also observed, while rapeseed has gained ground.

3.3 Yield-Fertilisation Relationship

While the variations of crop productivity during the second half of the nineteenth century closely followed those of livestock density and the resulting availability of manure, the rapid yield rise observed after 1950 is the direct consequence of the increased use of mineral fertilisers (Fig. 2b). The historical trajectory followed until 1980 by agriculture in terms of crop yield (Y, in kgN/ha/year) and total N inputs to the soil (F, in kgN/ha/year) (through manure, synthetic fertilisers, symbiotic N fixation and atmospheric deposition) followed a hyperbolic curve reflecting the non-linear agronomical relationship between yield and fertilisation [26] (Fig. 5a) expressed as

$$Y = Y\max \cdot F/(F + Y\max)$$

where Ymax is a parameter representing the maximum yield at saturating fertilisation.

After 1980, owing to improvements in agronomic practices, a shift occurred towards another trajectory with higher yields, in spite of lower fertilisation rates in the most recent period. The new trajectory is coherent with the yield-fertilisation relationship observed, although with considerable variability, for individual crop rotation systems, in both conventional and organic farming systems (Fig. 5b). It is remarkable that no significant difference in the yield-fertilisation relationship, expressed in total protein production over the whole crop rotation, is apparent between organic and conventional systems of the same pedoclimatic contexts, contrary to the common opinion that organic systems would be intrinsically less productive.

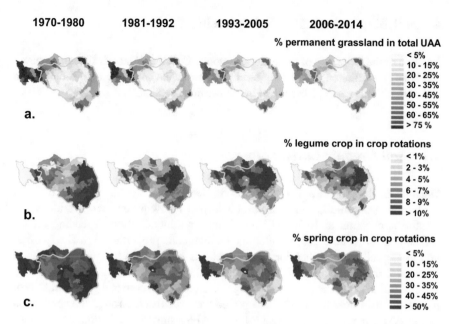

Fig. 4 Long-term changes in the frequency of (**a**) permanent grassland (in % of usable agricultural area), (**b**) legume crops, (**c**) spring crops (in % of crop rotation), during the 1970–2014 period. The limits of the Seine watershed are shown as a fine yellow line

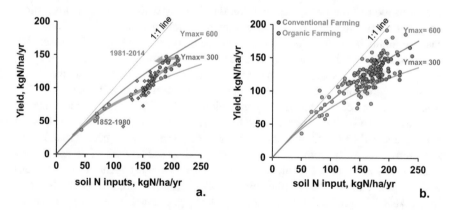

Fig. 5 (**a**) Long-term trajectory of cropland yield and soil N inputs averaged over the Seine watershed, as revealed by the GRAFS analysis for 1852–2014 (circles) [5] and by the application of the STICS model to the crop rotation and technical itinerary (diamonds) [13, 22] reconstituted in the ARSeine database. Until 1980 (blue symbols), the trajectory follows a single hyperbolic curve with the Ymax parameter close to 300 kgN/ha/year, while a shift to a higher curve is observed for the more recent period (red symbols). (**b**) Yield vs soil N input relationships for current conventional (red points) and organic (green points) farming crop rotation. Data from documented crop rotations gathered by Anglade et al. [27, 28] and Rakotovololona et al. [29]

4   Soil Biogeochemistry Reflects This Trajectory

Because of the large size of these element pools, C, N and P metabolism in cropland soil is largely affected by the long-term structural changes in the agro-food system and agricultural practices described in the previous section.

4.1   Soil Organic Carbon Storage

Organic C storage in agricultural soil is determined by the balance between (1) humified organic C inputs from above- and below-ground crop residues and manure application and (2) soil organic matter mineralisation depending on the soil content of labile organic matter and pedoclimatic properties [30]. C sequestration in agricultural soil therefore always reflects a long-term temporary imbalance between C inputs to soil, which are determined by agricultural practices and soil C mineralisation [31]. The dynamics of organic C storage in crop and grassland soils of the Seine basin over the 1850–2015 period was calculated through the application of the AMG model [32, 33], using the GRAFS estimation of humified C inputs [30] (Fig. 6a, b). A low rate of C sequestration occurred during the second half of the nineteenth century, followed by a destocking period until 1950 (Fig. 6c). Then a period of enhanced C sequestration occurred during the phase of rapid modernisation of the basin's agriculture, due to increased net primary production, and associated underground residue inputs, counterbalancing the reduction in manure inputs in the regions adopting stockless cropping systems. The mean rate of C sequestration then gradually levelled off and is now close to 40 kgC/ha/year, i.e. about 1‰ of the Corg stock in the top 30-cm top of soil. However, in the North-Western regions of France, destocking of organic C in cropland soil occurs (Fig. 6d). These results are consistent with the predictions of the STICS model concerning the variations of soil organic nitrogen (SON), which are strongly linked with those of soil organic carbon (SOC) in arable crops (Fig. 6e). The current agriculture in France is therefore far away from the objective assigned by the COP21 of an annual 4‰ increase in soil organic C content to counterbalance anthropogenic CO_2 emissions and mitigate climate change (the so-called 4‰ initiative; http://www.4p1000.org/ [34]). This is reinforced by the fact that total agricultural area has significantly decreased in the same period, at a rate comparable to that of organic C storage.

4.2   Agricultural Greenhouse Gas Emissions

Besides C emissions related to a possible negative C soil balance, greenhouse gas (GHG) emissions by the agricultural sector include N_2O emissions linked to nitrification and denitrification in cropland and grassland soils, CH_4 emissions mostly related

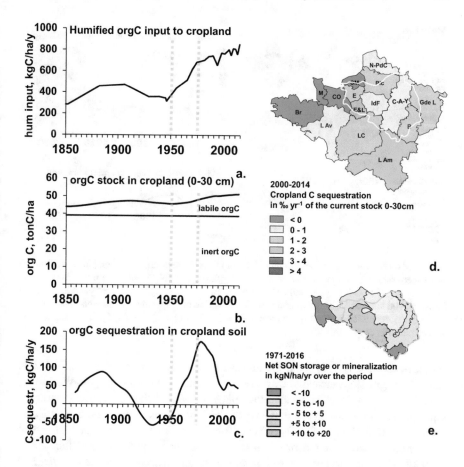

Fig. 6 (**a**) Long-term trends of humified organic carbon inputs to cropland soil of the Seine basin. (**b**) Calculated evolution of the labile and inert C stock in the 30-cm top soil of cropland. (**c**) Rate of C sequestration in kgC/ha/year and (**d**) in ‰ year^{-1} of the total stock in the 30-cm top soil of cropland (data from [30]). (**e**) Distribution of SON storage per agricultural district simulated by STICS in arable land from 1971 to 2013 [20]

to enteric fermentation of livestock and manure management as well as CO_2 emissions linked to fossil fuel consumption for mechanisation, heating, transport of feed and fertiliser manufacture. These components of the GHG balance were estimated by Garnier et al. [35] for the Seine basin agriculture over the 1852–2015 period. Soil N_2O emissions were estimated from an empirical relationship with exogenous N inputs (as synthetic fertilisers and manure), rainfall and temperature (Fig. 7a). They fit well with the predictions of the STICS model in agricultural zones [13]. CH_4 emissions from ruminants and monogastrics were estimated using livestock numbers and time-dependent emissions factors. CO_2 emissions were calculated following the ClimAgri methodology [36]. While CH_4 emissions did not change much over the period under study in the Seine basin, owing to the gradual reduction of

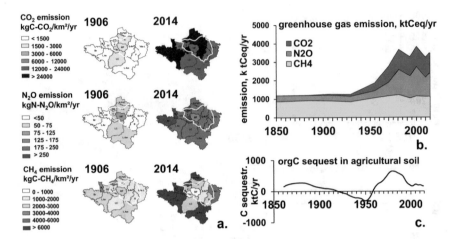

Fig. 7 GHG emissions from the agricultural sector of the Seine basin. (**a**) Geographical distribution of CO_2 emissions from fuel combustion (kg $C-CO_2/km^2/year$), N_2O emissions from cropland and grassland (kg $N-N_2O/km^2/year$) and CH_4 emissions by livestock (kg $C-CH_4/km^2/year$) in the Seine basin in 1906 and 2014. (**b**) Long-term variation of agricultural greenhouse gas emissions from the Seine basin expressed in $C-CO_2$ equivalent (ktonC-CO2equ/year) (after [35]). (**c**) Long-term C sequestration in crop soils [30]

livestock farming activity from the greatest part of the territory (Fig. 2b), N_2O and direct CO_2 emissions increased by more than a factor of 4 during the post-World War II period and then levelled off after the 1980s (Fig. 7b). When expressed in terms of equivalent C emissions, the current level of GHG emissions by agriculture in the Seine basin is about 3,400 ktonC-CO2 eq/year. This is one order of magnitude higher than the current C sequestration rate into the organic matter pool of agricultural soils (180 ktonC-CO2/year, [30]), as well as the maximum sequestration rate ever reached over the 1850–2015 period (Fig. 7c), showing that the 4‰ initiative, although desirable in terms of improvement of the soil quality, cannot be considered as a very significant climate change mitigation strategy, at least for France.

4.3 Nitrogen Soil Storage and Leaching

The balance of N inputs to cropland soils (as manure, fertilisers, symbiotic fixation and atmospheric deposition) and N export through harvest represents the potential N losses to the atmosphere (mostly as denitrification and ammonia volatilisation) or the hydrosphere (as nitrate leaching) (Fig. 3). Part of this balance is retained, however, within the organic N pool of the soil, depending on both the nature on the N inputs and the pedoclimatic conditions. As the C:N ratio of the soil organic matter does not deviate much from a mean value of 10 gC/gN, the above estimate of the C sequestration rate (Fig. 7c) can be used to calculate the long-term storage of N in

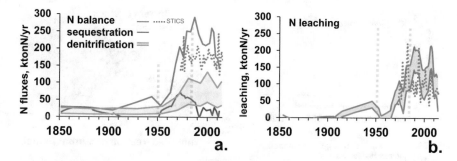

Fig. 8 (**a**) N balance of agricultural soils estimated from the GRAFS approach over the 1852–2015 period and its breakdown in terms of N storage and denitrification. The dotted red line is the estimation of N balance according to the coupled ARSeine database/STICS model. (**b**) N leaching calculated as the difference between N balance and N storage and denitrification. The dotted red line represents the N leaching calculated by the STICS model during the 1970–2015 period

organic form (Fig. 8a). We have no direct estimate of N loss through soil denitrification, which is very difficult to measure and to model. However, the estimate of N_2O emissions (Fig. 7) can be used to calculate a range of denitrification rates (Fig. 8a), assuming that the average N_2O/N_2 ratio lies between 10 and 30% [37–40]. Leaching is the remaining part, as shown in Fig. 8b.

The application of the STICS model at the scale of the Seine basin since 1970 allows a direct estimation of N leaching (Fig. 8b). These values match reasonably well with the estimation by difference between N balance, soil N storage and denitrification (Fig. 8b). The distribution of N surplus between N storage, denitrification and leaching during the last two decades (8–10%, 15–55%, 35–75%, respectively) is consistent with similar budgets experimentally established in long-term agronomical experiments in the Paris Basin [29, 31, 41].

4.4 Phosphorus Dynamics and Erosion

Contrasting with the high environmental mobility of N, P, once applied to soils in excess over the requirements of crop growth, accumulates within the soil where it remains strongly adsorbed. The only significant loss mechanism is net erosion, which mostly affects cropland. It has been estimated at 0.6 t soil/ha/year for the Seine basin based on the data calculated by Borelli et al. [42]. This represents a net erosion loss rate of about 0.00015 $year^{-1}$ for the cropland soils of the Seine basin when expressed relative to the soil mass in the 0 to 30-cm layer.

Using this estimate, the long-term P balance of cropland (Fig. 9a) can be used to calculate the storage of this element in the soil pool (Fig. 9b). While P stocks decreased during the 1850–1950 period, due to a low fertilisation rate, a sharp increase is observed during the 1950–1980 period, characterised by considerable overfertilisation. For the past 30 years, P fertilisation levels have considerably

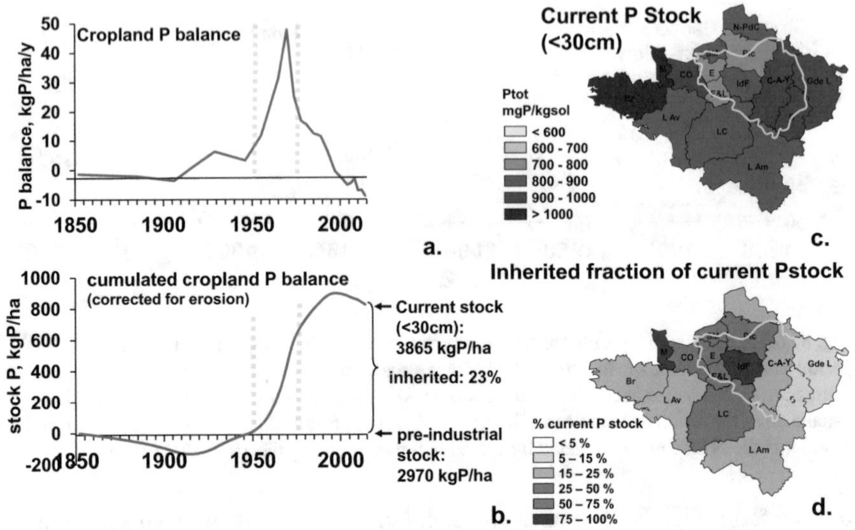

Fig. 9 (**a**) P balance of cropland soils of the Seine watershed estimated from the GRAFS approach over the 1852–2015 period. (**b**) Accumulation of P in cropland soil estimated from the cumulated P balance corrected for net erosion losses. (**c**) Current total P concentration in cropland soils of the Seine basin, according to Delmas et al. [43]. (**d**) Inherited fraction of total P in cropland soils

reduced (Fig. 2b), with P balance becoming even negative in recent periods (Fig. 9a), but the legacy of accumulated P is large enough to keep sustaining high crop productivity for one decade or more [44].

Comparing the regional estimates of cumulated P storage with the data reported by Delmas et al. [43], providing the distribution of measured total P concentration in agricultural soil at the scale of France (Fig. 9c), reveals that the inherited amount of P in cropland accounts for 7–80% of the total stock, with an average value of 23% over the Seine basin (Fig. 9d).

5   Hydrosystem Response to Agricultural Trajectories

Groundwater quality closely reflects the trends of agriculture changes, particularly regarding nitrate concentration, but also pesticide contamination which is dealt with in detail in chapter "How Should Agricultural Practices Be Integrated to Understand and Simulate Long-Term Pesticide Contamination in the Seine River Basin?" As far as surface water quality is concerned, both diffuse sources from agriculture and point sources from urban wastewater together determine their level of contamination. All along the continuum from land to river and to sea, a cascade of transfer, retention and elimination processes affects the budget of nutrients and their ultimate delivery at the outlet of the watershed.

5.1 Aquifer Storage of Nitrogen

The central area of the Seine basin is characterised by the presence of large aquifers within sedimentary rock formations (Fig. 1), with decadal groundwater residence time. Nitrate concentrations of these aquifers monitored since the beginning of the twentieth century in several locations (Fig. 11a) show a significant increase from the beginning of the 1960s. The MODCOU model coupled with STICS [20, 21] simulates this evolution in the main aquifer formations (Fig. 10b) and provides a picture of the current level of N contamination in several aquifers at a rather fine resolution (Fig. 10c). The drinking water standard of 11 mgN/l is exceeded in many places.

The model also calculates the recharge of the aquifer formations (infiltration from agricultural, forested and urbanised soils of the basin) and its N concentration and the exfiltration from the aquifer to the river network for the period from 1970 to 2015. As a long-term average, about 56% of the total water runoff of the Seine watershed flows through aquifers, forming the base flow of the river network (with water ages about 10 years), while the rest forms the surface or sub-surface flow rapidly (weeks) reaching rivers. Although no denitrification process is taken into account within the aquifers, the model calculations show that the N flux associated with the base flow is 55% lower than the N flux contributing to the recharge of aquifers. This large budget default can be explained by two processes: (1) water extraction both for irrigation and drinking water provision, currently accounting for about 1.2 Gm^3/year, i.e. 13% of aquifer recharge, and (2) long-term storage of nitrate in the groundwater and the non-saturated zone. Both processes together reduce by more than half the amount of N transferred from watershed soils to the hydrosystem.

5.2 Riparian Processes

Before they reach the river bed, flows of superficial and phreatic water coming from the watershed, with their nutrient concentration determined by land use and agricultural practices as discussed above, have to cross a more or less extended riparian area where biogeochemically active superficial soils, often rich in organic matter, are in contact with the river water table. These soils have a significant denitrification capacity, as well as a propensity to reduce iron oxides, thus possibly releasing adsorbed phosphates. Unless the watershed area is equipped with tile drains, by-passing the riparian zone (as is the case in some areas), the flow of nitrate effectively reaching the river is therefore reduced by the denitrification capacity of the riparian wetland. Billen et al. [7] estimated the extent of riparian denitrification in the Seine watershed at 150 kgN/km^2/year. A more recent study, based on the coupling of Riverstrahler with STICS-MODCOU, yields a significantly higher figure of 270 kgN/km^2/year, i.e. 18% of the flux of nitrate coming from base and sub-surface runoff. As expected, this riparian retention mostly occurs in large

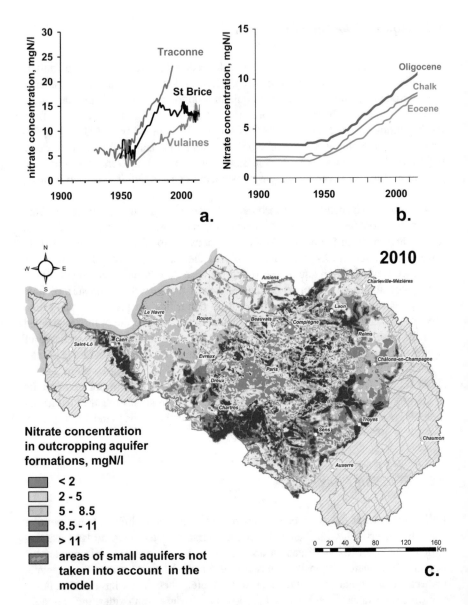

Fig. 10 (**a**) Long-term record of nitrate contamination in some springs in the Seine basin (springs of the Petite Traconne, Saint Brice and Vulaines, Brie limestone formation). (**b**) Long-term simulation of mean nitrate contamination in the major aquifer units of the Seine basin. (**c**) Map of 2010 level of nitrate contamination of sub-surface aquifer systems around the Seine watershed as calculated by the STICS-MODCOU modelling chain [13]

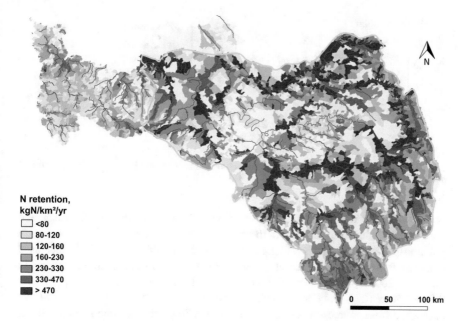

Fig. 11 Distribution of the mean annual flux of riparian denitrification among elementary watersheds of the Seine river system (calculation by the Riverstrahler model coupled with STICS-MODCOU for the 2010–2016 period)

alluvial valleys (Fig. 11), which both have largely developed riparian wetlands and receive substantial nitrate fluxes from the watershed.

Contrary to nitrate, particulate P accumulating downslope in riparian wetlands is prone to being remobilised as dissolved phosphate when anoxic conditions occur, as shown by Gu et al. [45] for the case of Brittany. This process has not been considered in the Seine and could cause higher diffuse P transfer from the watershed to the river system than our estimation based on net erosion fluxes.

5.3 Point and Diffuse Sources of Nutrients to the River System

Because of their different response to discharge variations and the different strategy to be implemented for their mitigation, diffuse and point sources of nutrients to the river network have to be distinguished. As far as N is concerned, diffuse sources are dominated by nitrate fluxes from groundwater and surface runoff, after transit through the riparian filter. For P, the diffuse sources are mostly made of the net erosion flux of cropland soils. Point sources are caused by the release of urban and industrial wastewater, eventually after treatment in wastewater purification plants. Their long-term evolution thus results from the combined effects of an increasing

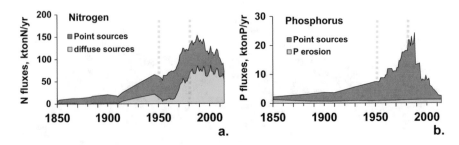

Fig. 12 Long-term variations of point and diffuse sources of N (**a**) and P (**b**) to the river network of the Seine watershed

population in the watershed, the progress of wastewater collection and treatment and the decline of industrial activity. In the case of P, the substitution period of soap with polyphosphate-containing synthetic detergents in the 1970s and 1980s increased domestic P loading by a factor of 3, before P was banned from washing powders in the early 2000s [46]. Figure 12 compares the long-term variations of point and diffuse sources of N and P to the river network of the Seine basin. During the last five decades, diffuse sources of N dominated river loading, and this trend is reinforced in the current period due to the progress in wastewater N treatment [47]. In contrast, for P point sources have always been the dominant source of surface water contamination. However, the spectacular reduction of point sources during the last two decades makes diffuse sources relatively more significant; in the current situation, diffuse and point sources have a nearly equal share in the total P loading of the river system.

5.4   N and P Budget of the Water-Agro-Food System

The concept of the water-agro-food system integrates water quality and agricultural issues, food and feed trade and human diet within a single perspective [46]. The system considered consists of the soils of the watershed, receiving rain and agricultural inputs, the underlying vadose zone and aquifers, the riparian wetlands, the discharging sewers collecting urban wastewater and the river network. Nutrient inputs to this system are the inputs to agricultural soils in excess over export by harvest, as well as point sources from urban wastewater. Only a limited part of these inputs are ultimately delivered by the river flow at the outlet of the basin; a large proportion is transiently retained (for P) or permanently eliminated (for N) along the entire soil-water continuum. Figure 13 gathers the available estimates of the relative value of these different storage or elimination processes and their long-term variations. The soil storage is particularly significant in the case of P, given that it accumulates most of the P brought to agricultural soil in excess of harvest export. Only soil erosion, a process of rather low intensity in the Seine watershed, responds to the increase of soil P by increasing inputs to the hydrosystem.

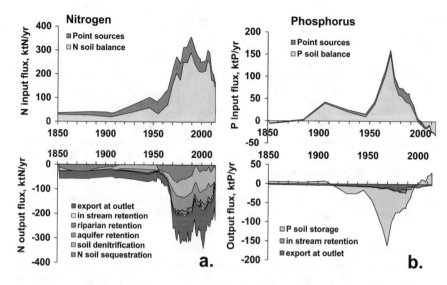

Fig. 13 Inputs and outputs of N (**a**) and P (**b**) fluxes from agricultural soils to the river network over the 1850–2015 period. The upper panels distinguish point sources and soil balance (excess inputs to harvest export); the lower panels show the fate of these inputs as export to the outlet of the watershed or retention/elimination processes in the soils, aquifers, riparian zones and streams

By comparison, N storage in soils is of lower significance and under the control of the soil C cycle. Although periods of increasing agricultural productivity, such as 1955–1980, were characterised by considerable C and organic N storage in cropland soil, most of the N brought to soils in surplus of harvest export is denitrified (in cropland soil itself, in riparian wetlands and to a much lesser extent in the river bed), is stored in the vadose zone and aquifer (the concentration of which takes decades to reach equilibrium) or is exported by the river flow to the outlet of the watershed. These differences in behaviour between N and P in the water-agro system explains the unbalanced nutrient loading delivered by the Seine River to the marine coastal waters of the Seine Bight, which is the source of severe eutrophication problems [9].

6 Conclusion and Scenarios for the Future

6.1 The Importance of Long-Term Storage Processes

Previous attempts at reconstructing the past chemical state of the Seine River [9, 48] were based on the implicit hypothesis of a direct and short-term relationship between land use (and agricultural practices) and diffuse sources of nutrients to the river network. This approach, however, did not account for delays linked to the storage of nutrients in the soil, the vadose zone and the aquifer compartment of the watershed,

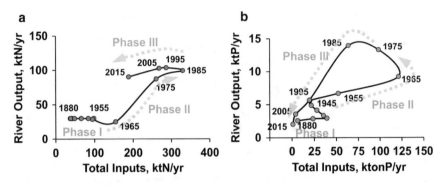

Fig. 14 Trajectory of the Seine River N (**a**) and P (**b**) delivery in response to total inputs to the water-agro-food system from 1850 to 2015 (the data shown represent an average over a 10-year period)

as already discussed by Thieu et al. [49]. In view of the length of these delays, considering long-term historical variations of agriculture is required for correctly understanding soil and water quality: many characteristics of these systems are inherited from past trajectories of agricultural systems. This is particularly true for pools of C and nutrients accumulated in the soil, as well as for nitrate (and pesticides) contaminating groundwater. The role of storage and elimination mechanisms of nutrients along the entire soil-river continuum also explains the non-linear response of the flows of nutrients delivered at the outlet with respect to the long-term changes in nutrient inputs to the water-agrosystem, with distinct hysteresis (Fig. 14). Three phases can be considered in the long-term trajectory of the Seine river system: during phase I, from the mid-nineteenth to the mid-twentieth century moderate increase of inputs are absorbed by retention processes; the short phase II from 1950 to 1975 is a time of rapid increase of nutrient inputs, with a visible response in terms of outputs at the outlet of the river; and phase III is the period of reduction of the inputs, with virtually no response of the outputs in the case of nitrogen, because of the dominance of diffuse sources buffered by aquifers, and a delayed response in the case of phosphorus in so far as point sources are reduced as well as fertilisers inputs. From a management point of view, these mechanisms prevent a rapid improvement of eutrophication conditions, particularly regarding measures taken to reduce diffuse sources of nutrient contamination, as their response to changes in agricultural practices and other environmental management measures may be delayed by several decades.

6.2 The Importance of the Structural Pattern of Agro-Food Systems on the Environmental Imprint

Another important conclusion from the studies summarised in this chapter is the link between the structure of the agro-food system flux pattern and the nutrient environmental losses or accumulation. Indeed, the major trends observed of a gradual intensification and specialisation of agricultural systems are associated with increased opening of the nutrient cycles and growing environmental losses, even though a significant reduction in fertiliser over-use has been observed since the 1980s as a result of agro-environmental regulations (Figs. 1, 8 and 9).

This link is suitably illustrated by two contrasted scenarios for the French agriculture at the horizon 2040, established by Billen et al. [50]. One of these scenarios assumes the pursuit of the trends towards opening to the global market and specialisation of territorial agricultural systems observed over the last 50 years, with, however, compliance to current agro-environmental regulations. The second scenario depicts an alternative option where generalisation of agroecological practices, reconnection of crop and livestock farming and of local food production and consumption, and a change in the human diet towards a Mediterranean diet with a much lower contribution of meat and milk allow a high level of autonomy of the agro-food system of the Seine watershed with respect to industrial and long-distance trade inputs. It was shown that both scenarios can feed the French population at the 2040 horizon and still export a significant amount of cereals, with, however, quite different environmental impacts. Only the latter scenario is able to halve GHG emissions [35] and to improve nitrate (and pesticide) contamination of groundwater and surface water [50]. It has also been shown that solving problems caused by noxious algal blooms in coastal marine waters at the outlet of large, human-impacted river systems would require this type of paradigmatic change in the structure of the agro-food systems [10, 51].

Acknowledgements This work was carried out in the scope of the PIREN-Seine programme, supported by the Seine-Normandie Water Agency and several other partners, and could benefit from the scientific context of the LTSER Zone Atelier Seine managed by the French National Center for Scientific Research (CNRS).

References

1. Le Noë J, Billen G, Lassaletta L, Silvestre M, Garnier J (2016) La place du transport de denrées agricoles dans le cycle biogéochimique de l'azote en France: un aspect de la spécialisation des territoires. Cah Agric 25:15004. https://doi.org/10.1051/cagri/2016002
2. Meybeck M, de Marsily G, Fustec É (1998) La Seine en son bassin, Fonctionnement écologique d'un système fluvial anthropisé. Elsevier, Paris
3. Billen G, Garnier J, Mouchel J-M, Silvestre M (2007) The Seine system: introduction to a multidisciplinary approach of the functioning of a regional river system. Sci Total Environ 375:1–12
4. Le Noë J, Billen G, Garnier J (2017) How the structure of agro-food systems shapes nitrogen, phosphorus, and carbon fluxes: the generalized representation of agro-food system applied at the regional scale in France. Sci Total Environ 586:42–55. https://doi.org/10.1016/j.scitotenv.2017.02.040

5. Le Noë J, Billen G, Esculier F, Garnier J (2018) Long-term socioecological trajectories of agro-food systems revealed by N and P flows in French regions from 1852 to 2014. Agric Ecosyst Environ 265:132–143. https://doi.org/10.1016/j.jenvman.2017.09.039

6. Ruelland D, Billen G, Brunstein D, Garnier J (2007) SENEQUE: a multi-scaled GIS interface to the RIVERSTRAHLER model of the biogeochemical functioning of river systems. Sci Total Environ 375:257–273

7. Billen G, Ramarson A, Thieu V, Théry S, Silvestre M, Pasquier C, Hénault C, Garnier J (2018) Nitrate retention at the river–watershed interface: a new conceptual modeling approach. Biogeochemistry 139:31–51. https://doi.org/10.1007/s10533-018-0455-9

8. Cugier P, Billen G, Guillaud JF, Garnier J, Ménesguen A (2005) Modelling eutrophication of the Seine Bight under present, historical and future Seine river nutrient loads. J Hydrol 304:381–396

9. Passy P, Le Gendre R, Garnier J, Cugier P, Callens J, Paris F, Billen G, Riou P, Romero E (2016) Eutrophication modelling chain for improved management strategies to prevent algal blooms in the Seine Bight. Mar Ecol Prog Ser 543:107–125. https://doi.org/10.3354/meps11533

10. Desmit X, Thieu V, Dulière V, Ménesguen A, Campuzano F, Lassaletta L, Sobrinho JL, Silvestre M, Garnier J, Neves R, Billen G, Lacroix G (2018) Reducing marine eutrophication may require a paradigmatic change. Sci Total Environ 635:1444–1466. https://doi.org/10.1016/j.scitotenv.2018.04.1

11. Romero E, Garnier J, Billen G, Ramarson A, Riou P, Le Gendre R (2018) The biogeochemical functioning of the Seine estuary and the nearby coastal zone: export, retention and transformations. A modelling approach. Limnol Oceanogr 64:895–912. https://doi.org/10.1002/lno.11082

12. Puech T, Schott C, Mignolet C (2018) Evolution des bases de données pour caractériser les dynamiques des systèmes de culture sur le bassin Seine-Normandie. Rapport d'étude, INRA SAD-Aster

13. Gallois N, Viennot P (2018) Modélisation de la pollution diffuse d'origine agricole des grands aquifères du bassin Seine-Normandie: Actualisation des modélisations couplées STICS-MODCOU – Modélisation de scénarios agricoles sous changement climatique. Rapport d'étude, Centre de Géosciences ARMINES/MINES ParisTech, PSL Université

14. Brisson N et al (2003) An overview of the STICS crop model. Eur J Agron 18:309–332

15. Brisson N, Launay M, Mary B, Beaudoin N (eds) (2009) Conceptual basis, formalisations and parametrization of the STICS crop model. INRA Science Update, 297 pp

16. Beaudoin N, Launay M, Sauboua E, Ponsardin G, Mary B (2008) Evaluation of the soil crop model STICS over 8 years against the "on farm" database of Bruyères catchment. Eur J Agron 29(1):46–57

17. Bergez JE, Debaeke P, Raynal H, Launay M, Beaudoin N, Casellas E, Caubel J, Chabrier P, Coucheney E, Dury J, de Cortazar-Atauri IG, Justes E, Mary B, Ripoche D, Ruget F (2014) Evolution of the STICS crop model to tackle new environmental issues: new formalisms and integration in the modelling and simulation platform RECORD. Environ Model Softw 62:370–384

18. Coucheney E, Buis S, Launay M, Constantin J, Mary B, Garcia de Cortazar-Atauri I, Ripoche D, Beaudoin N, Ruget F, Andrianorisoa S, Le Bas C, Justes E, Léonard J (2015) Accuracy, robustness and behavior of the STICS 8.2.2 soil-crop model for plant, water and nitrogen outputs: evaluation over a wide range of agro-environmental conditions in France. Environ Model Softw 64:177–190

19. INRA (1998) Base de données Géographique des Sols de France à l'échelle du 1/1.000.000. INRA, US Infosol, Orléans

20. Beaudoin N, Gallois N, Viennot P, Le Bas C, Puech T, Schott C, Mary B (2016) Evaluation of a spatialized agronomic model in predicting yield and N leaching at the scale of the Seine-Normandie basin. Environ Sci Pollut Res 25(24):23529–23558. https://doi.org/10.1007/s11356-016-7478-3

21. Ledoux E, Gomez E, Monget JM, Viavatene C, Viennot P, Ducharne A, Benoit M, Mignolet C, Schott C, Mary B (2007) Agriculture and groundwater nitrate contamination in the Seine basin. The STICS-MODCOU modeling chain. Sci Total Environ 375:33–47

22. Gallois N, Viennot P, Beaudoin N, Mary B, Le Bas C, Puech T (2015) Modélisation de la pollution nitrique des grands aquifères du bassin Seine Rapport final PIREN phase 6. https://www.piren-seine.fr/sites/default/files/PIREN_documents/phase_6/rapports_de_syntheses/Synthese_Ph6_Vol1.pdf

23. Oenema O, Kros H, de Vries W (2003) Approaches and uncertainties in nutrient budgets: implications for nutrient management and environmental policies. Eur J Agron 20:3–16

24. Mignolet C, Schott C, Benoit M (2007) Spatial dynamics of farming practices in the Seine basin: methods for agronomic approaches on a regional scale. Sci Total Environ 375(1–3):13–32. https://doi.org/10.1016/j.scitotenv.2006.12.004

25. Schott C, Puech T, Mignolet C (2018) Dynamiques passées des systèmes agricoles en France: une spécialisation des exploitations et des territoires depuis les années 1970. Fourrages 235:153–161

26. Lassaletta L, Billen G, Grizzetti B, Anglade J, Garnier J (2014) 50 year trends in nitrogen use efficiency of world cropping systems: the relationship between yield and nitrogen input to cropland. Environ Res Lett 9:105011. https://doi.org/10.1088/1748-9326/9/10/105011

27. Anglade J, Billen G, Garnier J (2015) Relationships for estimating N2 fixation in legumes: incidence for N balance of legume-based cropping systems in Europe. Ecosphere 6(3):37. https://doi.org/10.1890/ES14-00353.1

28. Anglade J, Billen G, Makridis T, Garnier J, Puech T, Tittel C (2015) Nitrogen soil surface balance of organic vs conventional cash crop farming in the Seine watershed. Agric Syst 139:82–92

29. Rakotovololona L, Beaudoin N, Ronceux A, Venet E, Mary B (2018) Driving factors of nitrate leaching in arable organic cropping systems in Northern France. Agric Ecosyst Environ 272:38–51

30. Le Noë J, Billen G, Mary B, Garnier J (2019) Drivers of long-term carbon dynamics in cropland: a bio-political history (France, 1852–2014). Environ Sci Policy 93:53–65. https://doi.org/10.1016/j.envsci.2018.12.027

31. Autret B, Mary B, Chenu C, Balabane M, Girardin C, Bertrand M, Grandeau G, Beaudoin N (2016) Alternative arable cropping systems: a key to increase soil organic carbon storage? Results from a 16 year field experiment. Agric Ecosyst Environ 232:150–164. https://doi.org/10.1016/j.agee.2016.07.008

32. Clivot H, Mouny JC, Duparque A, Dinh JL, Denoroy P, Houot S, Vertès F, Trochard R, Bouthier A, Sagot S, Mary B (2019) Modeling soil organic carbon evolution in long-term arable experiments with AMG model. Environ Model Softw 118:99–113

33. Saffih-Hdadi K, Mary B (2008) Modeling consequences of straw residues export on soil organic carbon. Soil Biol Biochem 40:594–607

34. Minasny B, Malone BP, McBratney AB, Angers DA, Arrouays D, Chambers A, Chaplot V, Zueng-Sang C, Cheng K, Das BS, Field DJ, Gimona A, Hedley CB, Young Hong S, Mandal B, Marchant BP, Martin M, McConkey BG, Leatitia Mulder V, O'Rourke S, Richer-de-Forges AC, Odeh I, Padarian J, Paustian K, Pan G, Poggio L, Savin I, Stolbovoy V, Stockmann U, Sulaemen TC-C, Vagen T-G, van Wesemael B, Winowiecki L (2017) Soil carbon 4 per mille. Geoderma 292:59–86. https://doi.org/10.1016/j.geoderma.2017.01.002

35. Garnier J, Le Noë J, Marescaux A, Sanz-Cobena A, Lassaletta L, Silvestre M, Thieu V, Billen G (2019) Long term changes in greenhouse gas emissions of French agriculture (1852–2014): from traditional agriculture to conventional intensive systems. Sci Total Environ 660:1486–1501. https://doi.org/10.1016/j.scitotenv.2019.01.048

36. Doublet S (2011) CLIMAGRI: bilan énergies et GES des territoires ruraux, la ferme France en 2006 et 4 scénarios pour 2030. Rapport ADEME. http://www.ademe.fr/sites/default/files/assets/documents/climagri-la-ferme-france-en-2006-et-4-scenarios-pour-2030.pdf

37. Wang JY, Jia JX, Xiong ZQ, Khalil MAK, Xing GK (2011) Water regime–nitrogen fertilizer–straw incorporation interaction: field study on nitrous oxide emissions from a rice agro-ecosystem in Nanjing, China. Agric Ecosyst Environ 141:437–446. https://doi.org/10.1016/j.agee.2011.04.009

38. Benoit M, Garnier J, Billen G (2014) Nitrous oxide production from nitrification and denitrification in agricultural soils: determination of temperature relationships in batch experiments. Process Biochem 50(1):79–85. https://doi.org/10.1016/j.procbio.2014.10.013

39. Vilain G, Garnier J, Decuq C, Lugnot M (2014) Nitrous oxide production from soil experiments: denitrification prevails over nitrification. Nutr Cycl Agroecosyst 98:169–186. https://doi.org/10.1007/s10705-014-9604-2

40. Autret B, Mary B, Gréhan E, Ferchaud F, Grandeau G, Rakotovololona L, Bertrand M, Beaudoin N (2019) Can alternative cropping systems mitigate nitrogen losses and improve GHG balance? Results from a 19-yr experiment in Northern France. Geoderma 342:20–33. https://doi.org/10.1016/j.geoderma.2019.01.039

41. Constantin J, Mary B, Laurent F, Aubrion G, Fontaine A, Kerveillant P, Beaudoin N (2010) Effects of catch crops, no till and reduced nitrogen fertilization on nitrogen leaching and balance in three long-term experiments. Agric Ecosyst Environ 135:268–278

42. Borrelli P, Van Oost K, Meusburger K, Alewell B, Lugato E, Panagos P (2018) A step towards a holistic assessment of soil degradation in Europe: coupling on-site erosion with sediment transfer and carbon fluxes. Environ Res 161:291–298

43. Delmas M, Saby N, Arrouays D, Dupas R, Lemercier B, Pellerin S, Gascuel-Odoux C (2015) Explaining and mapping total phosphorus content in French topsoils. Soil Use Manag 31:259–269

44. Le Noë J, Billen G, Garnier J (2018) Phosphorus management in cropping systems of the Paris Basin: from farm to regional scale. J Environ Manag 205:18–28. https://doi.org/10.1016/j.jenvman.2017.09.039

45. Gu S, Gruau G, Dupas R, Rumpel C, Crème A, Fovet O, Gascuel-Odoux C, Jeanneau L, Humbert G, Petitjean P (2017) Release of dissolved phosphorus from riparian wetlands: evidence for complex interactions among hydroclimate variability, topography and soil properties. Sci Total Environ 598:421–431

46. Garnier J, Lassaletta L, Billen G, Romero E, Grizzetti B, Némery J, Le QLP, Pistocchi C, Aissa-Grouz N, Luu MTN, Vilmin L, Dorioz J-M (2015) Phosphorus budget in the water-agro-food system at nested scales in two contrasted regions of the world (ASEAN-8 and EU-27). Global Biogeochem Cycles 29(9):1348–1368. https://doi.org/10.1002/2015GB005147

47. Meybeck M, Lestel L, Carré C, Bouleau G, Garnier J, Mouchel JM (2018) Trajectories of river chemical quality issues over the Longue Durée: the Seine River (1900s-2010). Environ Sci Pollut Res 25(24):23468–23484. https://doi.org/10.1007/s11356-016-7124-0

48. Billen G, Garnier J, Nemery J, Sebilo M, Sferratore A, Benoit P, Barles S, Benoit M (2007b) A long term view of nutrient transfers through the Seine river continuum. Sci Total Environ 275:80–97

49. Thieu V, Mayorga E, Billen G, Garnier J (2010) Sub-regional and downscaled-global scenarios of nutrient transfer in river basins: the Seine-Scheldt-Somme case study. Special issue "Past and future trends in nutrient export from global watersheds and impacts on water quality and eutrophication". Global Biogeochem Cycles 24:4. https://doi.org/10.1029/2009GB003561

50. Billen G, Le Noë J, Garnier J (2018) Two contrasted future scenarios for the French agro-food system. Sci Total Environ 637–638:695–705. https://doi.org/10.1016/j.scitotenv.2018.05.043
51. Garnier J, Ramarson A, Billen G, Théry S, Thiéry D, Thieu V, Minaudo C, Moatar F (2018) Nutrient inputs and hydrology together determine biogeochemical status of the Loire River (France): current situation and possible future scenarios. Sci Total Environ 637–638:609–624. https://doi.org/10.1016/j.scitotenv.2018.05.045

Past and Future Trajectories of Human Excreta Management Systems: Paris in the Nineteenth to Twenty-First Centuries

Fabien Esculier and Sabine Barles

Contents

1   Introduction 118
   1.1   Context 118
   1.2   Temporal and Geographical Frame 119
   1.3   Methodology 121
2   Resource-Oriented Management Leading to Circularity (1800–1905) 123
   2.1   Intention Without Achievement (1800–1868) 123
   2.2   Successful Mutualism (1868–1905) 126
3   Waste-Oriented Management Leading to Linearity (1905 to Today) 129
   3.1   The Sacrifice of the Seine (1905–1968) 129
   3.2   Pollution Treatment by Resource Destruction (1968 to Today) 132
4   Future Human Excreta Management Scenarios 133
   4.1   Lock-In and Opportunities 133
   4.2   Recovering Circularity Through Source Separation 135
5   Conclusion 137
References 138

The copyright year of the original version of this chapter was corrected from 2019 to 2020. A correction to this chapter can be found at https://doi.org/10.1007/698_2020_667

F. Esculier (✉)
Laboratoire Eau Environnement et Systèmes Urbains (LEESU), École des Ponts ParisTech, Université Paris-Est Créteil, AgroParisTech, Champs-sur-Marne, France

Milieux Environnementaux, Transferts et Interactions dans les hydrosystèmes et les Sols (METIS), Sorbonne Université, Centre National de la Recherche Scientifique, École Pratique des Hautes Études, Paris, France
e-mail: fabien.esculier@enpc.fr

S. Barles
UMR Géographie-Cités, Université Paris 1 Panthéon-Sorbonne, Université de Paris, École des Hautes Études en Sciences Sociales, Centre National de la Recherche Scientifique, Paris, France

Nicolas Flipo, Pierre Labadie, and Laurence Lestel (eds.), *The Seine River Basin*, Hdb Env Chem (2021) 90: 117–140, https://doi.org/10.1007/698_2019_407, © The Author(s) 2020, corrected publication 2020, Published online: 3 June 2020

Abstract This chapter addresses the fate of nutrients in agro-food systems after their ingestion by humans. Depending on how human urine and faeces are managed, they can become a source of pollution to the environment, or they can be used as a resource, notably as fertilisers, thus contributing to closing the loop of nutrients. Taking the city of Paris as a case study from the nineteenth to the twenty-first century, we analyse the fate of human excreta through the evaluation of corresponding nitrogen and phosphorus mass flows. We put forward two major phases concerning the management of human excreta:

1. The circularisation phase (1800s to 1900s): human excreta management is characterised by increasing circularity which peaks in the 1900s with around 50% of human excreta nutrients being recycled.
2. The linearisation phase (1900s–today): human excreta management is characterised by increasing linearity, i.e. a decrease in recycling rates of nutrients. Generalisation and improvement of wastewater treatment have led to decreasing pollution but also confirm the linearisation process (e.g. 5% recycling of human excreta nitrogen).

This increase in linearity came together with increased dependency of agro-systems on fossil resources. Ongoing climate change is also putting the current system under pressure since the dilution capacity of the Seine River is decreasing, while the population of Paris is increasing. We therefore analyse three scenarios of future human excreta management (incineration, end-of-pipe recycling and source separation) and show that source separation of human excreta may offer the per-spective of a sustainable human excreta management system.

Keywords Biogeochemical cycles, Circularity, Human excreta, Nitrogen, Phosphorus, Pollution, Socioecological trajectories, Source separation, Urban metabolism, Urine, Wastewater

1 Introduction

1.1 Context

"Flush and forget" toilets are so deeply rooted in the Western way of life that little attention is paid to the fate of our daily excreta. However, raising awareness of the disastrous ecological state to which the development model of Western countries has led, be it eutrophication, climate change or biodiversity loss – to quote only three major transgressions of planetary boundaries [1] – prompts us to question all aspects of the Western way of life. Conventional human excreta management is therefore currently being reconsidered by numerous research projects and innovative implementations, especially in Northern Europe [2] and also more recently in France [3], notably in the OCAPI programme (www.leesu.fr/ocapi) and the PIREN-Seine. On the other hand, past human excreta management has been carefully analysed for Paris in previous studies conducted within the PIREN-Seine programme. It has been shown that waste and wastewater are recent "inventions" of the twentieth century [4] and that this current linear economy comes after a period of mutualism where urban waste was considered valuable matter for industry and agriculture.

In this chapter, we aim at giving a long-term perspective of human excreta management by bridging historical studies and the current "reinvention" of alternative management methods. This case study is based on the city of Paris, particularly interesting in the sense that its human excreta management has been well documented since the beginning of the nineteenth century. In addition, its main river, the Seine River, has an increasingly low flow rate in relation to the population of Paris, making water and human excreta management critical issues for this city. We analysed the socioecological trajectory of this territory [5], based on substance flow analysis. As recommended in previous studies [6, 7], we focus on nitrogen (N) and phosphorus (P) mass flows to characterise the human excreta management of this area, and we mainly analyse two aspects of this management: circularity vs linearity and pollution. Circularity is defined by the proportion of human excreta, as reflected in N and P mass flows, that goes back to agricultural land. Linearity is the opposite of circularity: it is defined as the proportion of human excreta that does not return to agricultural land. Such fractions of human excreta are defined herein as pollution if it is released in a reactive form into the environment.

Our bibliographical review shows that such long-term analyses of human excreta management at the urban level are scarce. Only one study has been identified, on the city of Linköping in Sweden. It shows an abrupt decline of circularity since the beginning of the twentieth century, leading to nearly total linearity in the 1960s [8]. Previous studies by Barles provide a solid basis regarding the management of human excreta in Paris. Numerous historical data and figures concerning human excreta management have already been compiled. For example, Paris N mass flows were specifically analysed for the years 1817, 1852, 1869, 1888, 1913 and 1931 [6, 9]. More recently, supplementary data have been gathered to give an overview of N and P mass flows related to human excreta management for the 1850–2010 period in the Paris conurbation [7]. In this chapter, we aim at consolidating, updating and extending the scope of these previous studies with a focus on long-term analysis of circularity and pollution.

1.2   Temporal and Geographical Frame

Our temporal frame for past management of human excreta is the period 1800–2019. This makes it possible to understand the changing trajectory of N and P mass flows, from the search for circularity (nineteenth century) to the advent of linearity. Concerning the prospective approach, the temporal frame is 2020–2100. Given the characteristic rate at which management of human excreta has changed in the past centuries, this temporal frame enables one to project in new socioecological regimes concerning the management of human excreta.

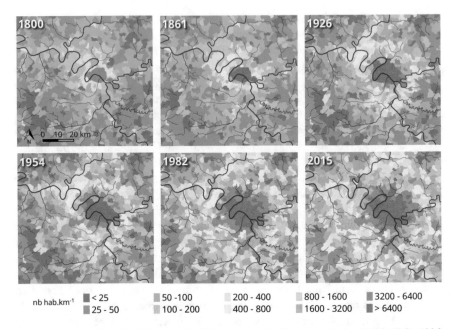

Fig. 1 Population density of the municipalities forming the Paris conurbation (1800, 1861, 1926, 1954, 1982 and 2015 general population censuses. Municipal perimeters for the six dates are based on the municipal perimeters of the year 2000. Credit: Sylvain Théry & Michel Meybeck)

Our geographical frame is the Paris conurbation. We follow the current INSEE[1] definition of an urban unit: its main characteristic is that the distance between two buildings does not exceed 200 m. During the last two centuries, the perimeter of the Paris conurbation has thus greatly varied (Fig. 1).

The Paris conurbation was included in the administrative limits of Paris at the beginning of the nineteenth century, with a population of approximately 575,000 people in the 1800s.[2] In 1860, the Paris administrative limits were extended to the nearby villages. The population of the city was multiplied by three and reached approximately 1,750,000 in the 1860s. In the 1920s, the conurbation had extended to the surrounding cities of the Seine department, and one-third of the department's population was located outside the city of Paris: the population was multiplied by 2.5 and totalled approximately 4,550,000 inhabitants. In the 1950s, the population had grown to approximately 5,600,000, one-half inside Paris, one-half in the 123 surrounding cities that formed three new administrative units in 1968 (Hauts-de-Seine, Val-de-Marne and Seine-Saint-Denis). In 2015, the INSEE considered that the

---

[1]Institut National de la Statistique et des Études Économiques: the French National Institute of Statistics and Economic Studies (www.insee.fr).

[2]Population data were taken from general censuses of the population, either directly or through the compilation made by Claude Motte of the Laboratoire de Démographie et d'Histoire Sociale (EHESS).

Paris conurbation comprised 432 municipalities. The city of Paris accounts for approximately one-fifth of this conurbation of approximately 10,700,000 inhabitants.

1.3 Methodology

For each decade between 1800 and 2010, we sought to estimate the degree of circularity of human excreta management, i.e. the proportion of N and P in human excreta that was returned to agricultural land. For this purpose, we estimated four main parameters:

1. N-DEC is the recycling rate, calculated in terms of N, of human excreta collected through a decentralised device, usually a cesspool, and returned to agricultural land.
2. P-DEC is the same as N-DEC calculated in terms of P.
3. N-CENT is the recycling rate, calculated in terms of N, of human excreta collected through a centralised device, i.e. a sewer.
4. P-CENT is the same as N-CENT calculated in terms of P.

These recycling rates were multiplied by the proportion of people connected to a sewer (parameter SEWER) or to a decentralised device (deduced from the parameter SEWER) to obtain the global proportion of N and P recycled in the Paris conurbation. Except for denitrification in wastewater treatment plants (WWTPs) and incineration of sludge, all N and P that is not recycled in agriculture is considered to be transferred in a reactive form to soil, underground, air or surface water and is thus considered as a pollution.

The data are selected to represent at best each decade: they correspond to the average of the decade when all annual data are available, otherwise the value at the middle of the decade. The parameter SEWER was calculated for the city of Paris on the basis of municipal statistical data [10]. Between the 1870s and the 1910s, the number of Parisians connected to the sewer via a filtering device called a *tinette filtrante* was estimated [11], considered as a combined centralised and decentralised system.

Outside the city of Paris, the parameter SEWER was linearly increased from 0% in the 1890s to 33% in the 1940s [12] and extrapolated, by taking into account the historical evolution of sewage management [13], to a total connection rate of 98% estimated in the 2010s [14]. Until the 1940s, it was considered that the connected population of the suburbs was equally distributed between connection to the main sewer network of the Paris conurbation and to an independent suburban sewer network [15].

Data from [7, 10, 13, 15, 16] enabled us to estimate the proportion of four different fates of N and P when collected in a sewer:

- Direct discharge in a river. In this case it was considered that N/P-CENT = 0% and that all N and P resulted in surface water pollution.
- Irrigation fields. In this case it was considered that N/P-CENT = 100% since all N and P was spread on fields. Not all N and P was actually taken up by crops, but the agronomic efficiency of plant uptake is not taken into account in the calculation of circularity.
- WWTPs with sludge spreading on fields. N/P-CENT is calculated by the proportion of N and P contained in sludge. The remaining P is released to a river as pollution. The remaining N is either released to a river as pollution or denitrified to the atmosphere in the non-reactive form N_2. Despite their importance in terms of the greenhouse gas effect and their high level of emission in Paris WWTPs [17, 18], N_2O emissions were not specifically estimated in this study since the different forms of reactive N are not distinguished.
- WWTPs with sludge incineration. In this case, N/P-CENT = 0%. Depending on the treatment efficiency, a proportion of N and P is released to a river as pollution; the rest of N is considered to go to the atmosphere in the non-reactive form N_2 and P not released to the river is contained in ashes and usually used as construction material with a loss of the fertilisation potential of P.

For households not connected to a sewer, we based our calculations on the fact that 90% of N and 65% of P is excreted via urine and the remaining via faeces [7]. N-DEC and P-DEC were calculated as follows:

- The N excretion of people was taken from N diet data [19] to which a 20% decrease was applied to take into account non-ingested food. The multiplication by the total population gave the total amount of N contained in excretions. N-DEC was calculated on the basis of total quantity and N content of collected night soil and final products obtained from cesspool management for the 1780s [20] and the years 1817, 1852, 1869, 1911 and 1926 [6, 9, 21, 22]. Data were linearly extrapolated for the missing decades. N-DEC was estimated at 0% for the period since the 1980s.
- The P content of products obtained from cesspool management was not available in previous studies. The P recycling rate is expected to be higher than the N recycling rate since N does not precipitate and is usually lost in the form of NH_3 emissions in cesspool treatment processes, whereas P remains in the solid and liquid phases [23]. Therefore, the best possible estimate for P-DEC was chosen as the mean value between N-DEC and the recycling rate of night soil calculated in volumes.

The combination of the five main parameters, N/P-DEC, N/P-CENT and SEWER, allowed us to estimate the circularity of human excreta management as the total amount of N and P returned to agricultural land via night soil processing, field irrigation or sewage sludge spread onto agricultural lands and the total amount of N and P pollution to soil, underground and surface water or air. Uncertainty on most data is difficult to quantify. It is considered high for data concerning decentralised management and concerning wastewater management in the suburbs of Paris until the 1960s. Concerning the prospective for 2020–2100, the methodology is described in Sect. 4.

2 Resource-Oriented Management Leading to Circularity (1800–1905)

2.1 Intention Without Achievement (1800–1868)

At the beginning of the nineteenth century, our figures show extremely low rates of circularity in human excreta management: 4% for N and 8% for P in the 1800s. The rest of N and P is mainly lost at three different stages: (1) just after excretion if excreta are not stored for collection, (2) during storage and (3) in the treatment process.

Some human excreta are abandoned in public and private spaces, even though this practice had been forbidden for centuries – unsuccessfully – and construction of cesspools for each house was mandatory before the sixteenth century [23]. When human excreta are stored in cesspools, only very little of it is eventually collected since leakages lead to transfer to soil and underground water. Cesspool watertightness is for the first time enforced by a decree in 1809 [23]. The proportion of leaking night soil has probably decreased since then, but in 1858, tremendous concentrations, between 29 and 300 mgN/L, were still reported in underground water [24].

Human excreta treatment processes are the third major cause of inefficient recycling. In the eighteenth century, the Paris authorities required that night soil be stored for 3 years before being applied to agricultural land, in order to guarantee salubrity. There is evidence, however, of many farmers being sued because they spread night soil directly on their land [23].[3] In 1781, night soil had to be transported to a single facility called the *voirie de Montfaucon* (Fig. 2). The drying process implemented in 1787 by Jacques Bridet created a greatly appreciated fertiliser called *poudrette* where nutrients are highly concentrated. But the overall efficiency of uptake of nutrients from the night soil content to the final *poudrette* product was low: before 1844, most liquids were evaporated, infiltrated in the soil or discharged to surface water, together with their N and P content, with specific volatilisation of N in the form of ammonia. Therefore, 90% of N is estimated to be lost in the process [23].

Although regulations tended to favour circularity, the actual materialisation of human excreta management in Paris led to major pollution of soil, underground and surface water and air and a very low recycling rate. Between the 1800s and the 1860s, however, there were major changes and numerous innovations in human

---

[3]This practice seems common in some places in the countryside but also in urban areas such as Grenoble and Lille [25]. In these areas, the circularity of the process of night soil treatment is probably very high through direct application of liquid night soil and all its N and P content.

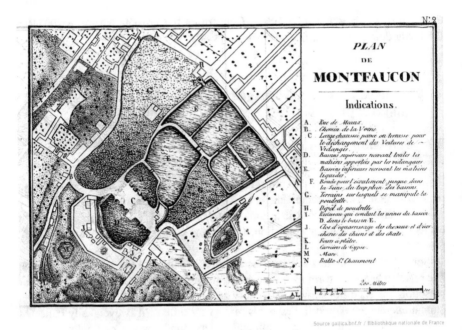

Fig. 2 Map of the *voirie de Montfaucon* in the beginning of the nineteenth century (Source: Perrot, M. Impressions de voyage – Montfaucon. Paris, chez l'éditeur – 1840. Credit: Gallica)

excreta management, in a context of the industrial revolution and a threefold increase in the urban population. Cesspool watertightness increased the efficiency of collection, but, at the same time, development of water distribution led to higher and more diluted volumes of night soil to collect. Per capita night soil collection was multiplied by five between the 1800s and the 1860s, and total collected night soil volumes were multiplied by 15. New storage techniques appeared, such as the *fosse mobile* where night soil was not emptied from a cesspool but stored in a barrel, which was collected together with its content (Fig. 3).

Improvements also appeared at the treatment facility. Numerous inventors, scholars or companies worked on processes that could extract or concentrate the valuable nutrients of human excreta. In 1820, for instance, Joseph Donat received a medal from the Royal and Central Society of Agriculture (known today as the Académie d'Agriculture de France) for his proposal of mixing the liquid part of night soil with an absorbent material to obtain a concentrated fertiliser [26]. Some of these technical improvements were implemented in Paris night soil treatment facilities. In 1852, most night soil treatment had been transferred to Bondy (10 km upstream of Paris along the Ourcq canal) where, alongside 10,000 m^3 of the traditional *poudrette*, three other products were obtained by the extraction of N from the liquid phase: 8 t of ammonium muriate (NH_4Cl), 40 t of volatile alkali (NH_4OH) and 835 t of ammonium sulphate (($NH_4)_2SO_4$), which was mostly shipped to England (these figures are for 1852) [27].

Fig. 3 Collection of night soil in a *fosse mobile* in 1820 (Detail. Musée d'hygiène de la ville de Paris. Credit: Jacques Boyer/Roger-Viollet)

The combination of these diverse improvements did not lead to radical change in the global circularity of human excreta management. Although between the 1800s and the 1850s N recycling had been multiplied by two, the total N circularity of Parisians' excreta still did not exceed 10%.

A major change appeared with the extension of sewers. The total length of Paris's sewers timidly rose from 20 km at the end of the eighteenth century to 40 km in 1831 and 168 km in 1858. Their main initial purpose was the collection of rainwater and cleaning water from street fountains. In the first half of the nineteenth century, houses were not connected to the sewers: few houses had a domestic water supply, and dumping human excreta in sewers was considered both unhygienic and a waste of a valuable resource, so it was forbidden. But this changed radically in the middle of the century. As the domestic water supply increased, direct dumping of domestic water on streets became less and less acceptable. In 1852, disposal of domestic water (what is now called grey water) in the sewer became compulsory and general servicing of Paris with the sewer was undertaken. In the 1860s, the total length of the sewers was close to 600 km [28].

The combination of a steep increase in night soil volumes, extended sewers and the declining quality of the Seine River because of sewage water discharge led engineers to consider another technique to manage human excreta. Flush toilets could be favourably combined with transport of human excreta in the sewers and spreading of the resulting fertiliser-rich water in sewage farms, a technique already implemented abroad [29]. In 1868, engineers Adolphe Mille and Alfred Durand-Claye received the authorisation to test sewage spreading on fields in Gennevilliers. Although dry management of human excreta was conducted for the purposes of circularity, its practical results led to low circularity, which started to be challenged by wet management of human excreta.

2.2   Successful Mutualism (1868–1905)

According to its supporters, the first experiments of sewage spreading on agricultural land were successful. The community of sewage spreading enthusiasts grew larger but still had many opponents: doctors who feared that diseases would be widely spread, fertiliser-makers subjected to competition, owners of buildings who feared the costs of connection to the sewer, the municipalities who were to receive sewage on their lands and of course cesspool emptiers who could see their activity disappear [11]. At the international hygiene congress of 1882 in Geneva, French doctors and hygienists were among the only people to express fears about mixing human excreta with sewage water. London had already launched combined sewers in 1858, and Berlin had made it mandatory in 1874 [12].

The challenge of managing night soil by dry means was also becoming increasingly acute. Between 1868 and 1885, the population had increased by 30% and collected night soil volumes by 50%. Night soil was becoming more and more diluted and less and less convenient to convert to fertiliser. Whereas London had implemented flush toilets and combined sewers without spreading sewage on agricultural land, circularity was considered a sine qua non condition for adopting flush toilets connected to a sewer in Paris because the Seine had to be protected and the human excreta recycled.

An intermediary solution adopted in Paris is worth mentioning: the *tinette filtrante* (Fig. 4). It consisted of a filtering device installed on a downpipe. It provided the convenience of having a flush toilet at home with the possibility of recycling organic matter. The *tinettes filtrantes* had two major flaws: they were subject to clogging, leading to overflows; they retained mostly solid matter, such as faeces or paper, but probably let most of the N, contained in the urine, flow to the sewer and then to the river. Nevertheless, in the 1880s, *tinettes filtrantes* peaked at 18% of buildings being equipped with in Paris.

Collection of human excreta in sewers together with spreading on fields finally came into practice in the 1880s. The connection of houses to sewers was made optional in 1885 and mandatory after the 1894 *tout-à-l'égout* law. This law stated that all houses that were served by the sewer network were compelled to discharge their excreta in the sewer within 3 years (Fig. 5). The connection rate rose but not as fast as expected due to various oppositions. There was less than 1% connection in the 1880s, 10% connection in 1895 and already 55% connection in 1905 (Fig. 8).

In a few decades, Paris had shifted from dry collection of human excreta to a prevailing collection in sewers. Sewage farms increased in size, covering 5,000 ha in the 1900s (Fig. 6). During this decade, the mean proportion of sewage water spread on land was 74%. Sewers were collecting more water due to the rise in the population and water consumption but also more organic matter and nutrients due

Fig. 4 Scheme of a *tinette filtrante*. Solids are kept in the barrel (C) while liquids flow to the sewer (I) [30] (Credit: Gallica)

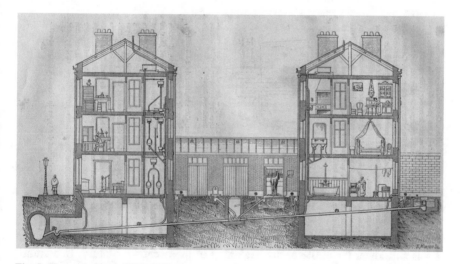

Fig. 5 House drainage, 1880s [31]

to the connection of toilets to the sewer. Much less of Paris sewage water was directly discharged into the river.

In terms of circularity, the situation radically changed within a few decades. The recycling efficiency of dry collection and treatment (N-DEC) had been steadily but slowly rising since the beginning of the nineteenth century. Thanks to improvements

Fig. 6 Sewage farm (Credit: SIAAP)

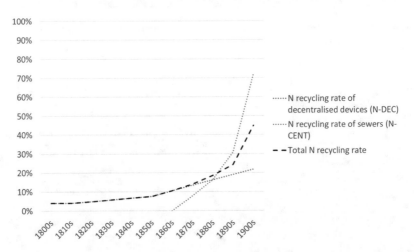

Fig. 7 Recycling rates of nitrogen in Paris: N-DEC (decentralised devices), N-CENT (sewer) and resulting total nitrogen recycling rate (1800s to 1900s) (see text for data sources)

in night soil collection and treatment techniques, N-DEC was multiplied by four, but it was still lower than 20% at the beginning of the twentieth century. In contrast, sewage farms were rapidly developed: 10 years after the *tout-à-l'égout* law, more than 70% of Paris's sewage water was actually spread on agricultural land, resulting in close to 50% total N recycling (Fig. 7).

Whereas the 1800–1868 period was marked by an intention towards circularity that failed to succeed in practice, the 1868–1905 period was characterised by a rapid shift in circularity mostly due to growing connection rates of toilets to sewers and adapted surfaces of sewage farms to be fertilised by human excreta. In 1888, Henri

Baudrillart stated that "the entire population owes its well-being to this thorough metamorphosis" [32].

3 Waste-Oriented Management Leading to Linearity (1905 to Today)

3.1 The Sacrifice of the Seine (1905–1968)

The transformation of Paris human excreta management from unsuccessful to successful circularity took place in a few decades between 1890 and 1910. However, several aspects explain that the circularity reached in the 1900s would be the maximum value for the nineteenth and twentieth centuries.

Fossil deposits of N or P were discovered in the nineteenth century, and the chemical fertiliser industry grew. In 1913, the Haber-Bosch process made it possible to produce chemical N fertiliser from the N in the air (N_2). Urban fertilisers became less and less attractive to farmers [33]. Sewage spreading required large surfaces, and the benefits of agricultural production did not cover the costs of spreading: in the 1900s, operational costs were more than ten times higher than the revenue of sewage farms [28].

In the 1900s, the population of Paris started to stabilise, in contrast to the population of the conurbation. The suburb sewage had to be managed, and sewage farms needed to be extended. When World War I began, most projects were interrupted: thereafter, sewage farms were not extended and their surface peaked in 1906.[4] In the meantime, flush toilets had been adopted by the population. The campaign of house owners against sewers, which took place in the second half of the nineteenth century in Paris, was outstripped by the demand for flush toilets and sewers by households. After Paris, flush toilets began to spread to the suburbs (Fig. 8).

In 1905, authorisation was given by the *Conseil Général de la Seine* to test the intensive bacterial treatment of sewage. Human excreta and sewage were less and less considered as a resource and acquired the new status of waste. The *Schéma général d'assainissement de la Seine*, adopted in 1929, endorsed this point of view and promoted WWTPs as the way to achieve sanitation. Sanitary aspects dominated the relationship towards human excreta and Pierre Koch from the French Higher Council for Public Health[5] stated in 1933 that "wastewater had to be destroyed" [4]. Wastewater irrigation decreased, but wastewater intensive treatment did not relay it. Circularity began to decline and pollution to rise. Whereas most pollution was widespread in the soil, underground water, rivers and air in the nineteenth

---

[4]New sewage farm extension projects were still studied after World War I, but they were not implemented.

[5]*Conseil Supérieur de l'Hygiène Publique de France.*

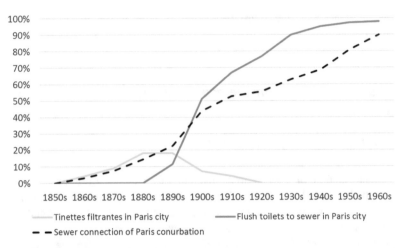

Fig. 8 Connection of toilets to the sewer system between the 1850s and the 1960s in the city of Paris and the Paris conurbation (see text for data sources)

Fig. 9 Achères wastewater treatment plant in the 1960s (Credit: Barles collection)

century due to deficient dry collection, collection of human excreta in the sewers led to a pollution concentrated in the surface water. The first large WWTP, Achères, only started operation in 1940, and its size was largely insufficient for a population that exceeded five million inhabitants in the 1950s (Fig. 9). Less than 20% of human excreta was treated in this WWTP until the 1960s. Between the 1940s and the 1960s,

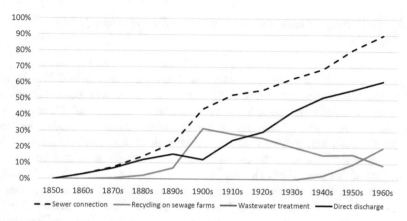

Fig. 10 Sewer connection of the Paris conurbation population and proportion of total human excreta recycled in sewage farms, directly discharged or treated in a wastewater treatment plant (1850s to 1960s) (see text for data sources)

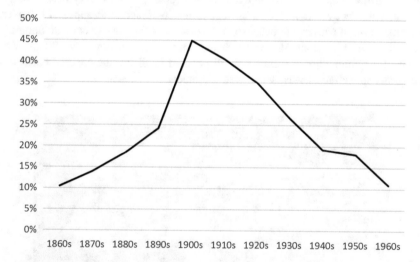

Fig. 11 Nitrogen recycling rate of the Paris conurbation (1860s to 1960s) (see text for data sources)

more than 50% of all human excreta of the Paris conurbation was directly discharged into the rivers via the sewers (Fig. 10).

This situation led to the "biological death" of the Seine, with less than 3 mg O_2 L^{-1}, from Paris to the estuary, in summer at the end of the 1960s. Circularity had peaked above 40% in the 1900s. In the 1960s, it was back to 10%, the same level as in the 1860s (Fig. 11).

3.2   Pollution Treatment by Resource Destruction (1968 to Today)

Concerns about the catastrophic quality of rivers grew after World War II, at a time when the Paris population was growing and the Seine was biologically dying. The 1960s were a turning point in the river's pollution: the national Water Law was enacted in 1964, urging recovery of surface water quality. In 1968, the Paris conurbation adopted a general wastewater management plan which aimed at treating all wastewater produced by the conurbation. It took several decades to achieve the main goal of this plan (Fig. 12).

At that time, wastewater treatment concentrated on reducing carbon (C) pollution. N and P removal in the first WWTP was quite low (less than 10% removal for N). Between the 1970s and the 1990s, oxygen levels rose, but so did N and P pollution due to lack of sufficient treatment and an ever-growing population [13]. Even if sewage sludge was mostly spread on agricultural land, wastewater treatment reinforced the linearity of human excreta management because very little N and P were actually recovered in sludge. Sewage farms continued to disappear. Moreover, contamination of wastewater because of the mixing of human excreta with domestic water, industrial water and rainwater became a matter of concern. In the 2000s, high levels of contamination in metals were discovered in all sewage-covered farmlands and direct spreading of wastewater was forbidden.

Fig. 12 The last large fish mortality event in the Paris conurbation due to wastewater discharge in the Seine. July 1994, Bougival (Credit: DRIEE-IF/Service Police de l'Eau)

N and P pollution started to be addressed with the 1978 Paris Convention (becoming the OSPAR convention in 1992) and with the European Urban Waste Water Treatment Directive (UWWTD) in 1991. The transfer of N and P from Paris's excreta via the Seine and the English Channel to neighbouring countries led these countries in 1981 to request that the quantity of N and P released by the Seine River basin be divided by two between 1985 and 1995 (PARCOM recommendation 88/2). Similarly, the UWWTD required elimination of 70% of N and 80% of P from wastewater by 1998. In 2013, 25 years after the PARCOM recommendation and 15 years after the UWWTD deadline, just as France was about to be fined by the European Union for noncompliance, compatible N and P treatment were finally implemented in the Paris conurbation.

However, in this new human excreta management system, circularity is not the prevailing concern. Like C, N is mainly lost by transfer to the atmosphere. Denitrification consists in the conversion of reactive nitrogen into the non-reactive form N_2, the opposite of the Haber-Bosch process. On one hand, chemical N fertilisers are produced from N_2 using fossil resources. On the other hand, human N contained in excreta is converted to N_2 using fossil resources. The N imprint of Paris and the industrial world in general is thus considered unsustainable [14]. Today, one-third of the N of the Paris conurbation excreta is still discharged to the rivers, mainly in the form of nitrates. Two new forms of pollution also appear: nitrous oxide (N_2O) gaseous emissions in the treatment process and nitrite (NO_2^-) discharged into the river. The global recycling rate of human N in Paris today is 5%, the same value as with the inefficient dry management methods of the beginning of the nineteenth century.

The situation for P is ambiguous. Since it is possible to fix P in sludge via precipitation, P pollution treatment can simultaneously lead to increased P recycling. However, in Paris, incineration of sludge has been gradually implemented in the new WWTPs. P contained in ash is not recovered, and, in the 2010s, a remarkable 84% treatment rate of P in WWTPs only leads to a 40% recycling rate, mostly because of sludge incineration [14] (Fig. 13).

4 Future Human Excreta Management Scenarios

4.1 Lock-In and Opportunities

Progressive adoption of the "toilet–sewer–wastewater treatment plant" combination since the end of the nineteenth century has created a sociotechnical lock-in in human excreta management. Modern comfort is associated with the disconnection of the population with their excreta and consumption of large volumes of water. With respect to human excreta management, the debate gradually shifted from the best possible management of human excreta itself to the management of mixed sewage water, then to the management of sludge and finally to the management of ash from incinerated sludge [7]. Nutrient pollution of the rivers has only been very recently

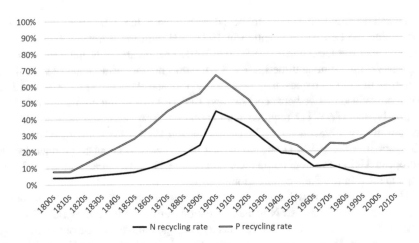

Fig. 13 Global N and P recycling rates of human excreta in the Paris conurbation (1800s to 2010s) (see text for data sources)

addressed, and the Seine's good status has still not been reached downstream of the Paris WWTPs. Circularity was only very recently conceptualised as a possible goal of human excreta management in the twenty-first century [7].

Considering the sociotechnical lock-in around "toilet–sewage–wastewater treatment plant", the less disruptive progression would be the transformation of WWTPs into recycling facilities. Since the concept of a circular economy has stepped up, the circular WWTP is very often put forward. However, it is subjected to many limitations. No technology is available today to implement large-scale N recovery from wastewater. The "circular wastewater treatment plant" is usually only circular on one element, P, but not on N, C,[6] K, etc. [7]. An unfavourable C/N ratio in mixed sewage water (too much urinary N as compared to total wastewater C) makes it very difficult to valorise both C and N in large-scale plants [34]. Today, the best technologies available do not allow for more than ~30% N recycling. The 2018 IWA-NRR conference showed a new dynamic in N recovery research in WWTPs but without clear perspectives of the technologies that would be available in the future.

Since the 1990s, consciousness of the fertilising potential of human excreta, most particularly urine, has grown [35]. Many source separation initiatives have been implemented in buildings or ecovillages in Europe, starting in Sweden at the beginning of the 1990s and gradually spreading in Scandinavia and the German-speaking European countries [7]. The main concept is to (re-)divert urine, or urine and faecal matter, from the domestic sewage water and convert it into fertiliser. A review of these alternative human excreta management systems implemented in Europe shows that, until now, only urine source separation initiatives have managed to achieve circularity in human excreta management [7]. Source separation requires

---

[6]On a denitrifying WWTP, part of the carbon can be converted to valuable methane, but most of it is lost by oxidation as carbon dioxide.

Fig. 14 Two new products related to circularity in human excreta management: urine-derived fertiliser Aurin (left) and urine-separating toilet Save! (right) (Credit: F. Esculier)

a profound change in the sociotechnical system associated with human excreta management. It still represents a niche innovation, but it is currently gaining attention in the Seine River basin. In October 2018, the Seine-Normandy Basin Committee[7] voted up to an 80% subsidy to source-separating projects.

A few models of urine-diverting flush toilets were developed in the 1990s and 2000s, and a new one has been marketed since March 2019 (Save! toilet by Laufen) (Fig. 14). Simultaneously, the first urine-derived fertiliser was officially licensed in 2018 by the Swiss Federal Office for Agriculture for use on all plants (Aurin by Vuna) (Fig. 14). Treatment and concentration of urine inside the toilet are also being developed [36]. This technology could allow changing an existing conventional toilet to a urine-diverting fertiliser-producing toilet without having to install new pipes in a building. Its deployment can thus potentially be faster than urine-diverting systems with pipe collection of urine. A socioecological transition of human excreta management towards circularity with deployment of decentralised and efficient dry management of excreta thus appears to be a plausible option.

4.2 Recovering Circularity Through Source Separation

Three scenarios have been developed for the future management of human excreta in the Paris conurbation for the 2020–2100 period.

---

[7]In each large French water basin, the Basin Committee votes for fees and subsidies, associated with water uses and implemented by the Water Agencies.

In scenario 1, called "Linearity", one can imagine that, due to difficulties in developing sewage sludge use because of micropollutant contamination and the complexity of its organisation, sewage sludge incineration increases from 50 to 100%. N and P circularity decreases to 0%, and human excreta management becomes totally linear. In this scenario, N and P pollution depends on the ability of the Paris conurbation to maintain its sewer network as well as intensive and efficient N and P treatment at the WWTP.

In scenario 2, called "End-of-pipe", one can imagine that circularity becomes a goal of human excreta management, but only end-of-pipe solutions are implemented. On one hand, with sludge P recycling, P soars to more than 80% circularity, exceeding the 1900s recycling peak. On the other hand, N recycling remains limited to 30% circularity, below the 1900s recycling peak.

In scenario 3, called "Circularity", a combination of source separation and end-of-pipe solutions is implemented. Urine diversion is compulsory in all new constructions (faecal matter can potentially be source-separated also, as long as it does not impair N and P circularity). In existing buildings, urine diversion is also implemented using toilet integrated treatment. In this scenario, we assume that the deployment speed of urine diversion in the Paris conurbation follows the same curve as flush toilet deployment in the city of Paris after the 1880s (Fig. 8). Urine diversion thus reaches 67% coverage in the 2040s and 98% coverage in the 2090s. Starting at 70% efficiency, the urine collection rate gradually increases up to 85% efficiency in the 2050s. End-of-pipe P recycling is implemented as in "End-of-pipe" scenario. Concerning N, increasing urine diversion deeply modifies the C/N ratio of sewage water. When complete separation of urine is achieved, 85% recovery of N at the sewage water treatment plants is assumed through N recovery from sludge with primary decantation and short retention time activated sludge [37]. We assume a linear progression of N recovery at the treatment plants.

With these hypotheses, N recycling already reaches 60% in the 2030s, with 25% recycling at the sewage treatment plant and 35% in decentralised facilities. In the 2060s, N and P recycling exceeds 90%, equally distributed between treatment plants and decentralised systems for P, mainly through decentralised systems for N. Consequently, N and P pollution is necessarily very low in all compartments of the environment. Figure 15 represents the N and P circularity of the Paris conurbation from 1800 to 2010 (historical data) with the addition of the 2020–2100 period with the circularity scenario.

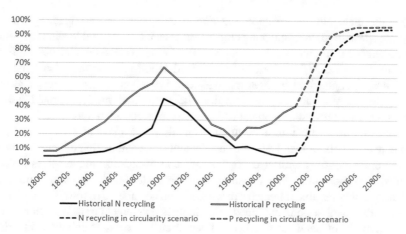

Fig. 15 Past N and P circularity of human excreta in the Paris conurbation and future N and P circularity in the "Circularity" scenario (see text for data sources)

5 Conclusion

Over the last two centuries, human excreta management has undergone great changes in Paris, some of them counterintuitive. Indeed, the driver of these transformations can be found in the goals that were assigned to this urban service: during the nineteenth century, it was as much dedicated to the improvement of public health as to the production of fertiliser, in order to prevent "hunger arising from the furrow, and disease from the stream"[8] [38]. Urban sanitation went hand in hand with rural food production; these two targets oriented technical choices towards the maximisation of nutrient recovery. Night soil was turned into *poudrette*, ammonium sulphate, urate and so on, moderately increasing recovery rates. The development of the domestic water supply during the second half of the nineteenth century raised a huge debate on the appropriateness of the combined sewer system associated with sewage farms. These were implemented during the last third of the century; sewer water spreading resulted in a huge increase in N and P recovery that rose to 50% at the beginning of the twentieth century. This moment can also be considered as the end of the circularisation phase.

The discovery of new sources of nutrients (fossil phosphates, potash mines as well as the Haber-Bosch process regarding N) disqualified urban fertilisers, and the recycling aim was abandoned. At the same time, the Paris urban area grew both in area and population, leading to an increasing production of human excreta and sewage. No longer of use, it was discharged into the river, which led to a severe pollution of the Seine River. The development of WWTPs since the 1940s was not sufficient to restore water quality, despite substantial amelioration since the 1980s (chapter "Ecological Functioning of the Seine River: From Long-Term Modelling Approaches to Highfrequency Data Analysis"). Simultaneously, the agricultural use of sewage sludge did not prevent decreasing N and P recycling rates, especially for N

---

[8]*La faim sortant du sillon et la maladie sortant du fleuve.*

(5%): the metabolic linearisation was almost complete. On one hand, agriculture became specialised, disconnecting crop and livestock farming and leading to opened biogeochemical nutrient cycles (chapter "The Seine Watershed Water-Agro-Food System: Long-Term Trajectories of C, N, P Metabolism"); on the other hand, the final stage of agro-food nutrient management, i.e. urban human excreta management, reveals yet another disconnection, between food production and human excretion, with linear management of human excreta considered as waste.

This questions the future of urban sanitation systems and their aptitude to promote circularisation and achieve sustainability. A combination of source separation of urine (possibly with faecal matter) and end-of-pipe techniques gives the best results, with more than 90% recycling rates for both N and P around 2060. This scenario is only feasible if human excreta are again considered as resources and if the aim of urban sanitation is redefined. This implies a complete transformation of the political basis of sanitation, a better integration of social metabolism in urban policies (and others), a decompartmentalisation of urban and rural policies. This also implies a change in sanitation techniques, the conception of buildings, the training of those who are in charge of these services and more generally of the material culture of urbanites. In short, a socioecological transition.

Acknowledgments The authors would like to thank all partners of the OCAPI programme (www. leesu.fr/ocapi) and the PIREN-Seine programme for their technical and financial support to this chapter. The PIREN-Seine research programme (www.piren-seine.fr) belongs to the Zone Atelier Seine part of the international Long-Term Socio-Ecological Research (LTSER) network.

References

1. Steffen W, Richardson K, Rockström J et al (2015) Planetary boundaries: guiding human development on a changing planet. Science 347(6223):1259855. https://doi.org/10.1126/science.1259855
2. Skambraks AK, Kjerstadius H, Meier M (2017) Source separation sewage systems as a trend in urban wastewater management: drivers for the implementation of pilot areas in Northern Europe. Sustain Cities Soc 28:287–296. https://doi.org/10.1016/j.scs.2016.09.013
3. Esculier F, Tabuchi JP, Créno B (2015) Nutrient and energy flows related to wastewater management in the Greater Paris: the potential of urine source separation under global change constraints. International conference on water, megacities and global change, Paris
4. Barles S (2005) L'invention des déchets urbains: France 1790–1970. Champ Vallon, Seyssel
5. Fischer-Kowalski M, Haberl H (2007) Socioecological transitions and global change: trajectories of social metabolism and land use. Edward Elgar, Cheltenham, Northampton
6. Barles S (2007) Feeding the city: food consumption and flow of nitrogen, Paris, 1801–1914. Sci Total Environ 375:48–58. https://doi.org/10.1016/j.scitotenv.2006.12.003
7. Esculier F (2018) Le système alimentation/excrétion des territoires urbains: régimes et transitions socio-écologiques. Thèse de doctorat en sciences et techniques de l'environnement. Université Paris-Est, Paris. https://hal.archives-ouvertes.fr/tel-01787854/document
8. Schmid Neset TS (2005) Environmental imprint of human food consumption: Linköping, Sweden 1870–2000. PhD thesis, Linköping studies in arts and science. Department of Water and Environmental Studies, Univ. Linköping

9. Barles S (2002) L'invention des eaux usées: l'assainissement de Paris de la fin de l'ancien régime à la seconde guerre mondiale. In: Bernhardt C, Massard-Guilbaud G (eds) Le Démon moderne. La pollution dans les sociétés urbaines et industrielles d'Europe/The Modern Demon. Pollution in Urban and Industrial European Societies. Coll. Histoires croisées, Presses de l'Université Blaise Pascal, Clermont-Ferrand, pp 129–156
10. Préfecture de la Seine (1881–1959) Annuaires statistiques de la ville de Paris. Imprimerie Nationale, Paris
11. Jacquemet G (1979) Urbanisme parisien: la bataille du tout-à-l'égout à la fin du XIXe siècle. Rev Hist Mod Contemp 26(oct-déc):505–548
12. Bellanger E (2010) Assainir l'agglomération parisienne. Histoire d'une politique interdépartementale de l'assainissement (XIXe-XXe siècles). Avec la collaboration d'Éléonore Pineau. Éditions de L'Atelier, Paris
13. Rocher V, Azimi S (2017) Evolution de la qualité de la Seine en lien avec les progrès de l'assainissement de 1970 à 2015. Johanet, Paris
14. Esculier F, Le Noë J, Barles S et al (2018) The biogeochemical imprint of human metabolism in Paris megacity: a regionalized analysis of a water-agro-food system. J Hydrol 573:1028–1045. https://doi.org/10.1016/j.jhydrol.2018.02.043
15. Carré C, Mouchel JM, Servais P (2017) Assainir la banlieue de Paris: des fosses septiques au tout-à-l'égout, quels effets sur la qualité de l'eau de la Seine. In: Lestel L, Carré C (eds) Les rivières urbaines et leur pollution. Quae, Paris
16. Vincey P (1910) L'assainissement de la Seine et les champs d'épandage de la ville de Paris. Extrait des Mémoires de la Société nationale d'agriculture de France, t. CXLII, Paris
17. Bollon J, Filali A, Fayolle Y et al (2016) N_2O emissions from full-scale nitrifying biofilters. Water Res 102:41–51. https://doi.org/10.1016/j.watres.2016.05.091
18. Bollon J, Filali A, Fayolle Y et al (2016) Full-scale post denitrifying biofilters: sinks of dissolved N_2O? Sci Total Environ 563-564:320–328. https://doi.org/10.1016/j.scitotenv.2016.03.237
19. Le Noë J, Billen G, Esculier F et al (2018) Long term socio-ecological trajectories of agro-food systems revealed by N and P flows: the case of French regions from 1852 to 2014. Agric Ecosyst Environ 265:132–143. https://doi.org/10.1016/j.agee.2018.06.006
20. Boudriot PD (1986) Essai sur l'ordure en milieu urbain à l'époque pré-industrielle. Boues, immondices et gadoue à Paris au XVIIIe siècle. Histoire, économie société 5(4):515–528
21. Pluvinage C (1912) Industrie et commerce des engrais et des anti-cryptogamiques et insecticides. J.-B. Baillière et fils, Paris. http://catalogue.bnf.fr/ark:/12148/cb31125773f
22. Pluvinage C (1927) Industrie et commerce des engrais - Engrais azotés et organiques, produits chimiques agricoles (tome 1). J.-B. Baillière et fils, Paris
23. Paulet M (1853) L'Engrais humain. Histoire des applications de ce produit à l'agriculture, aux arts industriels, avec description des plus anciens procédés de vidanges et des nouvelles réformes dans l'intérêt de l'hygiène. Comptoir des imprimeurs-unis, Veuve Comon, Paris
24. Boussingault JB (1858) Emploi des eaux de puits de Paris dans la panification. Gazette hebdomadaire de médecine et de chirurgie 5(29):507
25. Girardin J (1876) Des fumiers et autres engrais animaux, Septième édition. Garnier frères, Paris. http://catalogue.bnf.fr/ark:/12148/cb305108417
26. Héricart de Thury M (1820) Rapport à la société royale et centrale d'agriculture sur un nouvel engrais proposé sous le nom d'urate par MM. Donat et compagnie. Imprimerie de Madame Huzard, Paris
27. Beaudemoulin LA (1853) Assainissement de Paris. Revue de l'architecture et des travaux publics 11:131–138
28. Barles S (2007) Urban metabolism and river systems: an historical perspective – Paris and the Seine, 1790–1970. Hydrol Earth Syst Sci 11:1757–1769
29. Mille AA, Durand-Claye A (1869) Compte-rendu des essais d'utilisation et d'épuration. Régnier et Dourdet, Paris
30. Liger F (1875) Dictionnaire historique et pratique de la voierie, de la construction, de la police municipale et de la contiguïté: fosses d'aisances, latrines, urinoirs et vidanges. Baudry, Paris. https://gallica.bnf.fr/ark:/12148/bpt6k740549?rk=21459;2

31. Arnould J (1889) Nouveaux éléments d'hygiène, 2nd edn. J. B. Baillière et fils, Paris
32. Baudrillart H (1888) Les populations agricoles de la France. Maine, Anjou, Touraine, Poitou, Flandre, Artois, Picardie, Île-de-France, passé et présent. Librairie Guillaumin, Paris, pp 455–626
33. Barles S, Lestel L (2007) The nitrogen question: urbanization, industrialization, and river quality in Paris, 1830-1939. J Urban Hist 33:794–812. https://doi.org/10.1177/0096144207301421
34. Larsen TA, Gujer W (1996) Separate management of anthropogenic nutrient solution (human urine). Wat Sci Tech 34(3–4):87–94
35. Drangert JO (1998) Urine blindness and the use of nutrients from human excreta in urban agriculture. GeoJ 45:201–208
36. Randall DG, Naidoo V (2018) Urine: the liquid gold of wastewater. J Environ Chem Eng 6:2627–2635. https://doi.org/10.1016/j.jece.2018.04.012
37. Wilsenach JA, van Loosdrecht CM (2006) Integration of processes to treat wastewater and source-separated urine. J Environ Eng 132(3):331–341. https://doi.org/10.1061/(ASCE)0733-9372(2006)132:3(331)
38. Hugo V (1862) Les Misérables, Lacroix, Verboekhoven et Cie. English translation: I. F. Hapgood, Thomas Y. Crowell & Co, New York (1887)

How Should Agricultural Practices Be Integrated to Understand and Simulate Long-Term Pesticide Contamination in the Seine River Basin?

H. Blanchoud, C. Schott, G. Tallec, W. Queyrel, N. Gallois, F. Habets, P. Viennot, P. Ansart, A. Desportes, J. Tournebize, and T. Puech

Contents

1 Introduction 142
2 Study Sites 143
   2.1 Presentation of the Orgeval and Vesle Basins 143
   2.2 Data Acquisition 144
3 Quantification of Past Pesticide Use 145
   3.1 The Orgeval Case 146
   3.2 The Vesle Case 146
4 The Orgeval Catchment Over Time 147
   4.1 Long-Term Observed Contamination 147
   4.2 Pesticide Fate Modelling with the STICS Crop Model 149
5 The Spatial Approach in the Vesle Catchment 151
6 Pesticides at the Seine River Basin Scale 153

The copyright year of the original version of this chapter was corrected from 2019 to 2020. A correction to this chapter can be found at https://doi.org/10.1007/698_2020_667

H. Blanchoud (✉) and A. Desportes
EPHE, PSL University, UMR Metis 7619 (SU, CNRS, EPHE), Paris, France
e-mail: helene.blanchoud@ephe.psl.eu

C. Schott and T. Puech
INRA, UR ASTER, Mirecourt, France

G. Tallec, P. Ansart, and J. Tournebize
Irstea, UR HYCAR, Antony, France

W. Queyrel
UMR 1347 Agroecologie, AgroSup Dijon, Dijon, France

N. Gallois and P. Viennot
MINES ParisTech, PSL University, Centre de Géosciences, Fontainebleau, France

F. Habets
Laboratoire de Géologie, Ecole normale supérieure, PSL University, CNRS UMR 8538, Paris, France

Nicolas Flipo, Pierre Labadie, and Laurence Lestel (eds.), *The Seine River Basin*,
Hdb Env Chem (2021) 90: 141–162, https://doi.org/10.1007/698_2019_385,
© The Author(s) 2020, corrected publication 2020, Published online: 3 June 2020

7 Conclusion 156
8 Perspectives 157
References 158

Abstract Modelling long-term pesticide transfer to rivers at the catchment scale is still difficult due to a lack of knowledge of agricultural practices and poorly adapted field observation. The Orgeval experimental catchment was first investigated to validate a modelling approach. In addition to pesticide practices investigated over 20 years, directly collected from farmers, monthly integrated river samples were analysed for 10 years. To explicitly integrate agricultural practices and crop rotation, the STICS crop model was adapted to simulate pesticide transfer in soil. Annual load simulations were compared to observed pesticide fluxes in rivers. To simulate the contamination of groundwater, the STICS-Pest model was coupled to the MODCOU hydrogeological model. The results are discussed at the subbasin scale in relation to available data. To upscale the approach at the Seine River basin scale, other strategies need to be developed.

Keywords eLTER, Modelling, Monitoring, Pesticide, Phytosanitary practices, PIREN-Seine, STICS-Pest, Zone Atelier Seine

1 Introduction

River contamination by pesticides is currently observed in the Seine basin. This is mainly due to long-term uses of such substances in agricultural areas. Still today, atrazine and its metabolite deethylatrazine (DEA) are one of the most frequently detected pesticides in the Seine basin, even though atrazine has been banned since 2003. In this context, we need to better understand and simulate pesticide fate in watersheds consistent with long-term uses.

Numerous pesticide fate models are available. The main differences between them are mostly related to water flow from simple water transfer such as PELMO [1] and PRZM [2] to more physically oriented models such as PEARL [3], MACRO [4] and RZWQM [5]. However, modelling pesticides requires understanding the relationship between land management practices and the dynamic of contaminants, and only a few models are able to take into account specific agricultural practices [6].

Pesticide transfer models need to know when each active ingredient (AI) is applied and the quantity applied. However, unlike the other data needed for implementing models (climate, soil characteristics or land use), there is no database available on phytosanitary practices that can directly feed the models [6–8]. In addition, there are few attempts to synthesise these practices at the catchment scale. This deficiency led to the development of new methods of acquiring and processing data [6, 7].

In the PIREN-Seine programme, studies were conducted to better define crop rotations and landscape diversity in the Seine catchment area. Agricultural practices were investigated, and the first models simulated nitrate contamination behaviour in groundwater [9] and surface water [10]. New developments have been carried out to integrate pesticide behaviour, taking into account farmers' phytosanitary practices. Pesticide contamination at the catchment scale was first investigated in two small agricultural areas: the Orgeval and the Vesle. This paper is a synthesis of the work done in the PIREN-Seine research programme concerning the modelling of pesticide transfer. We present here how pesticide uses were collected, the modelling approach and the results for atrazine and its by-product DEA on two basins: the Orgeval and the Vesle. These studies highlight the difficulty of the spatio-temporal approach to modelling the fate of pesticides in the Seine basin.

2 Study Sites

The pesticide fate model was designed to gather input databases such as soil characteristics, climatic data, pesticide properties and phytosanitary practices. Given that the database is not available for the latter, an estimation of annual quantities of each AI, the application schedule and spatial distribution of application on the catchment must be at least documented. The effort needed for inquiries on pesticide applications is directly related to the size of the watershed. Therefore, the Orgeval site was considered for a long-term study, while the Vesle site was chosen for a spatial approach.

2.1 Presentation of the Orgeval and Vesle Basins

The Orgeval watershed (104 km^2) is located 70 km east of Paris (Fig. 1) where hydrological and biogeochemical cycles have been studied since 1962. It is situated on the Brie plateau, which is one of the most productive agricultural areas in France, between 140 and 180 m above sea level. The deep and low permeability clay loams result from the decomposition of limestones and millstones of the underlying Brie. Drainage has allowed intensive agricultural development of these silts of the Brie plateau. Consequently, this basin is dominated by agricultural areas (81%) and forests (18%). Roads and urban areas account for only 1% of the land area. Cereals are the main crops. Soft wheat accounts for more than 40% of crops and maize for about 14% for the 1988–2007 period. Some specific crops such as broad bean and flax are also cultivated in the Orgeval basin (5% and 4%, respectively), whereas sugar beet, barley and rapeseed are less widespread, compared to data for the Seine-et-Marne department.

The Vesle watershed (1,460 km^2) is located in one of the major agricultural regions of the Seine basin: the Champagne area (Fig. 1). Seventy-six percent of the basin is covered by agricultural lands, while the remaining surface is divided into forest (17%) and urban areas (7%). The city of Reims is situated in the downstream part of the basin. The basin is mainly covered by field crop systems: wheat (30%), barley (18%), sugar beet (13%) and alfalfa (12%). On this catchment, maize is mainly inserted into 5-year-long rotations such as beetroot (or pea or rapeseed) – wheat – maize – wheat – barley. Champagne grapes are a specific crop located on the

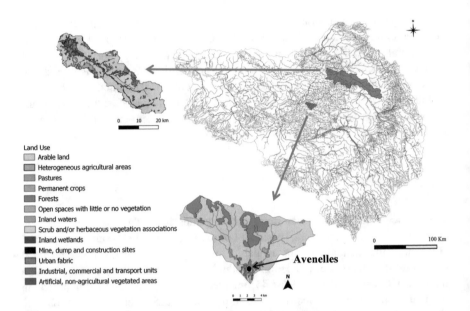

Land Use
- Arable land
- Heterogeneous agricultural areas
- Pastures
- Permanent crops
- Forests
- Open spaces with little or no vegetation
- Inland waters
- Scrub and/or herbaceous vegetation associations
- Inland wetlands
- Mine, dump and construction sites
- Urban fabric
- Industrial, commercial and transport units
- Artificial, non-agricultural vegetated areas

Fig. 1 Map of the Seine hydrographic basin, location and land use of the Orgeval and Vesle basins

slopes in the downstream part of the basin (Fig. 1). From a hydrogeological standpoint, two-thirds of the surface (upper and middle parts) is associated only with the Cenomanian limestone aquifer, while low-permeable tertiary formations have been deposited on top of it in the lower part of the basin. The Cenomanian aquifer is unconfined and drained by the Vesle River and its tributaries. The groundwater reservoir is located in a thickness layer ranging from 30 to 40 m [11] and is locally affected by fracture networks and karst systems, providing a good transmissivity. On top of the aquifer, the thickness of the unsaturated zone varies from a few metres in the valley to a few tens of metres underneath the plateau.

2.2 Data Acquisition

To study long-term contamination by pesticides, we focused on atrazine. In the Seine River and some aquifers of the basin, atrazine and its by-product DEA are currently detected in rivers and groundwater even after its ban in 2003. Atrazine is a herbicide widely used on maize as well as vineyards. It was also used on urban areas for weed control on roads and railways. In France, pesticides are monitored in rivers and aquifers for the European framework directive. However, such data is generally partial (monthly, half-yearly, annual or less), and further investigations were conducted.

In the Orgeval basin, within the PIREN-Seine programme, a flow-controlled refrigerated sampler has been installed since 2008 at the Avenelles station (see Fig. 1). Since then, samples have been collected once a week and then gathered monthly for pesticide analyses. A survey of the crop management techniques of the main farms was also carried out [12]. Pesticides were then extracted with an offline solid-phase extraction (SPE) technique with the Oasis hydrophilic lipophilic balance® (HLB, Waters) cartridge and analysed by liquid chromatography with tandem mass spectrometry (LC/MS/MS) [13, 14]. Detection limits are 1 ng L^{-1} for both atrazine and DEA.

In the Vesle basin, atrazine and DEA concentrations were extracted for 39 piezometers from the ADES database [15]. Additional specific sampling campaigns were carried out monthly from November 2003 to February 2005 on 11 piezometers by the Agence de l'Eau Seine-Normandie (Direction Territoriale des Vallées d'Oise, DTVO). This covered a larger geographic area and included more specific hydrogeological contexts.

3 Quantification of Past Pesticide Use

AIs provided at the watershed scale depend on phytosanitary practices (or treatment programmes) prescribed at the field scale by technical advisors. However, these uses vary greatly depending on pedoclimatic conditions acting on parasitism, types of crops and cultivated areas, farmers' phytosanitary strategies and regulations on pesticide use.

Most of the AI uses are registered for specific crops and often in a particular case (type of pest and/or disease). Crop rotations and their diversity (short rotations from 3 to 4 years to rotations of more than 10 years including alfalfa) ensure some diversity, except in the case of perennial crops such as vineyards. Each farmer has his/her own strategies in terms of phytosanitary practices based on the advice of technicians (cooperatives, technical institutes, etc.) and his/her personal choices (sowing date, crop variety, tillage practices, etc.) [16–18]. Then, for the same crop, pesticide use may differ from one farm to another or even from one field to another. This diversity could explain 40–60% of pesticide use [16]. Moreover, pesticide use evolves rapidly over time, depending on the regulations (banned or new AI), climate conditions (which cause diseases and pests to vary) and rotation crops [17].

The main source of information on crop rotation in the two study areas is the French Agricultural Census, which has been available at the municipal level since 1970. It is updated only every 10 years, but annual data were estimated by interpolation using annual data available at the regional scale between these dates [19–21].

With regard to pesticide use, there are few generic databases that can be used over a long time period and on large study sites. The E-Phy database [22] lists the plant protection products authorised in France, as well as their composition, homologation dates and actual registered uses and doses. However, these data cannot be directly used because farmers can choose to use one AI or another. The national sales made by distributors' database (BNV-d), which includes pesticide sales, are another source. This database could be used to spatialise pesticide use [23, 24], but data have been available only since 2008.

To better define model inputs, a "hybridization approach" using national resources and local data was developed according to two methods [6]. Since the Orgeval catchment is only about 100 km^2, all agricultural practices were monitored over a period as long as possible for the whole watershed. The Vesle catchment is too large, and therefore only herbicide use has been investigated within a 30-year period on targeted practices (weeding vineyards and corn) for which the use of these AIs was approved. Data were collected at a municipal scale.

3.1 The Orgeval Case

For the Orgeval basin, other data were available to refine the spatial accuracy of the annual rotation. First, rotation tables were provided directly by farmers, describing their crop succession on their plots (771 plots accounting for 45% of the landscape area, dating from 1970 to 2005). Moreover, data from the French Land Parcel Identification System (LPIS) geographic database [25] has been available since 2006.

Here, we chose to collect records of their past practices directly from farmers located in the study area. These documents, called "lowland logbooks" or "field logs", help to get closer to farmers' "real" practices and to reconstruct the dynamics of phytosanitary practices over 20 years for a representative sample [26]. As an example, the chart in Fig. 2 shows that atrazine was long the main AI applied on maize until its ban in 2003. However, the chart still shows some applications beyond this date, showing the authenticity and low bias of the data recorded. High-resolution data could capture spatial variability but at the cost of intensive computing [27, 28]. Therefore, we assumed a type of simplification: to keep confidential and anonymous data and also to simplify modelling input; these data were averaged by crop and by year for the whole basin in the final database.

This APOCA (Agricultural Practices of the Orgeval Catchment Area) database was used to quantify the AIs applied per crop per year since 1990. It also includes a large number of indicators to monitor changes in phytosanitary pressures [26]. Comparison of these indicators shows that the phytosanitary pressure is not decreasing overall even if applied quantities per hectare are lower.

3.2 The Vesle Case

For the Vesle study, prescriptions issued by local agricultural development agencies were used to better indicate actual farmer pesticide use. The technical magazine *Le*

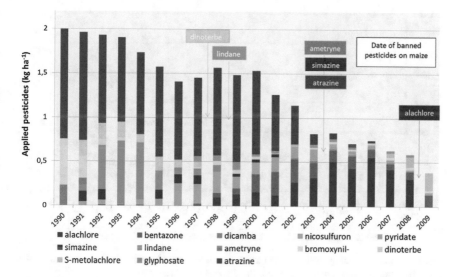

Fig. 2 Application rates of pesticides on maize from 1990 to 2009 in the Orgeval catchment

Vigneron Champenois was especially useful for the reconstitution of phytosanitary practices on Champagne vineyards since the 1970s [29]. However, because of the large number of possible treatments, it was necessary to quantify and validate it with "experts" [19–21]. Furthermore, we obtained sales for corn weed control products since the 1970s from the main cooperative in the sector (Champagne Céréales). These data, compared to those of maize surfaces collected, allowed us to estimate the quantities of active ingredients applied per hectare. These averaged values, however, remain a relatively rough estimate of current practices in this catchment area but clearly reflect the complexity of determining pesticide inputs for modelling.

We created another database called ARSEINE for the identification of the major treatment programmes by period and crop as well as the percentage of product used for each programme in relation to the whole rotation crop. In both study areas, since phytosanitary practices were not differentiated within the watershed, it is the distribution of crops that allows to determine spatial treatments [30].

4   The Orgeval Catchment Over Time

4.1   Long-Term Observed Contamination

Pesticides have been studied since 1979 in the soils, river, groundwater, air, fallout, drainage and wetland of the Orgeval basin [31–36]. Organochlorinated pesticides and then triazines, phenylureas, chloroacetanilides, neonicotinoids, sulfonylureas, etc. were successively researched over time according to new uses and marketing

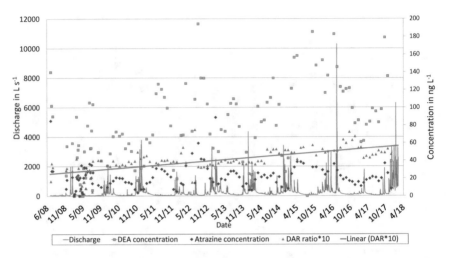

Fig. 3 Surface water flow over time (left axis), integrated monthly concentration of atrazine, DEA and DEA/atrazine ratio (DAR) with linear fitted correlation (right axis)

approval. Atrazine and its metabolite DEA were detected in all samples and were considered as the reference of past contamination for modelling.

To compare the simulations with the observed data, monitoring was adapted.

Unlike molecules still used in the watershed, atrazine found in surface river water has only slight fluctuations from 1 to 90 ng L^{-1} (Fig. 3). Since atrazine was banned in 2003, these detections correspond to the gradual elimination of stocks from soil and/or groundwater. Previous studies on groundwater contamination showed that DEA concentrations in rivers could be explained by Brie aquifer contamination where values were lower than 15 ng L^{-1} [37]. Furthermore, riverine seasonal fluctuations would also be related to the contribution of the Brie groundwater to surface water. However, the atrazine concentration is lower in groundwater than in the river, suggesting that another source could be suspected for atrazine. This could be due to desorption of atrazine from soils when the water table level rises. Although atrazine concentrations in rivers remain stable over the period studied, the DEA concentrations increase. DEA is still present in the river at concentrations up to 180 ng L^{-1}. The value of the DAR also increases and follows monotonic growth ($S = 1$, Mann-Kendall nonparametric test; [38]). Other studies also show that the maximum level of DEA contamination has not yet been reached, in particular through the gradual desorption of atrazine and its slow degradation and progression in soils [39].

To verify whether leaching of atrazine decreases over the period, annual fluxes were calculated for atrazine and DEA (Table 1). These results do not show any trend in how fluxes evolve. This is particularly due to hydrological hazards. Flows were the highest in 2016, corresponding to the exceptional flood in May 2016, which alone accounted for 35% of the annual flow. These high water flows were associated with higher concentrations, which explain the doubling of the atrazine and DEA fluxes between 2015 and 2016 (Table 1).

Table 1 Annual flow and fluxes of atrazine and DEA in g/year

Year	Flow (10^3 m^3 year^{-1})	Atrazine flux (g year^{-1})	DEA flux (g year^{-1})
2010	6,848	63.14	238.07
2011	5,698	65.89	265.62
2012	6,652	153.53	543.82
2013	9,710	286.84	528.85
2014	10,011	128.84	559.94
2015	8,164	172.05	838.84
2016	11,869	337.70	1,737.76
2017	8,012	130.19	637.58

These results demonstrate the need to develop a pesticide fate model to better forecast long-term contamination taking into account past pesticide use and soil and aquifer inertia.

4.2 Pesticide Fate Modelling with the STICS Crop Model

A pesticide transfer module was implemented in the STICS crop model [40] to better integrate cropping system specificity such as plant growth and specific agricultural practices in pesticide fate simulations. This STICS-Pest model [41] is able to simulate the main processes involved in pesticide fate, i.e. sorption, transformation and transfer through a soil profile. Liquid transfer is based on the STICS solute transfer formalism based on the mixing cell principle [42]. The model is original in that it includes a slow sorption kinetics formalism, following the Agriflux model equations according to [43]. The representation is based on the experimental works of [44, 45], who observed a time-dependent isotherm shape for sorption processes of organic compounds. This approach made it possible to better simulate long-term desorption from bound residues. A detailed description of the module is given in [41].

Five categories of parameters have to be described to implement the model: climate data, soil parameters, cultural practices (sowing and harvest dates, crop management), crop parameters (main crop rotations) and pesticide parameters. For the soil parameters, the field capacity, wilting point and bulk density are derived from the national database related to soil description and the characteristics of the French soil [46]. The organic carbon profile is related to field data and laboratory measurements [47].

Agricultural practices were extracted from the APOCA database. Atrazine inputs were calculated using phytosanitary practices provided by farmers (applied quantities on maize, application date) according to percentages of maize over the 1990–2009 period.

Pesticide characteristics are mostly extracted from international databases such as [48] or previous studies including field measurements with pedoclimatic conditions similar to the Orgeval catchment [49, 50]. Moreover, the transformation of atrazine to its metabolite DEA is estimated with the first-order kinetics and maximum occurrence fraction estimated at 0.21 [48].

Several simplifications and assumptions were made in this study. The soil characteristics were assumed to be homogeneous at the catchment scale. Given that more than 80% of the area is artificially drained, we assumed that the residence time of water is equal to or less than 1 year. Consequently, the simulation and observed data were compared on an annual scale. Moreover, since we focused on atrazine behaviour, maize was the only crop considered. Soil water content, organic matter and nitrogen content were initialised with a 1-year warm-up, while a 14-year warm-up was used for pesticide initialisation in order to take into account a stock of atrazine and DEA in the soil at the beginning of the simulation.

Since water is the main vector of pesticide transfer, the ability of STICS-Pest to simulate water transfer was checked. Cumulated water transfers simulated with STICS-Pest were therefore compared to discharge observations at the outlet of the Orgeval catchment and showed a determination coefficient of 0.99, illustrating the ability of the model to represent water transfer and its interannual variations (Fig. 4).

Concentrations simulated by the model were compared with continuous monitoring of pesticides at the outlet of the catchment (Fig. 5). Observations represented with triangles and squares (Fig. 5, left) show higher concentrations during the period of atrazine application at the beginning of the simulation than after the banishment of the active substance. The simulated atrazine concentrations fluctuated during the first 14 years linked to water outflow fluctuation followed by a strong decrease in 2004 and finally tended to stabilise at the end of the simulation. From 2008 to 2016, simulations and observations were relatively similar and followed a similar trend, as

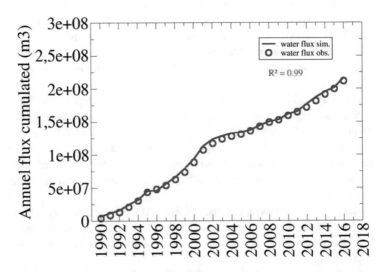

Fig. 4 Comparison of the cumulated water discharge simulated with STICS-Pest (continuous line) versus observations (circles) at the Orgeval catchment outlet

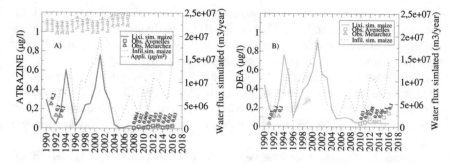

Fig. 5 Riverine observed (triangles and squares) and simulated (continuous line) annual concentration of atrazine (left) and its metabolite DEA (right) from 1990 to 2016 related to water transfer with annual outflow simulated (dotted blue line) and mean application dose of atrazine (solid blue line)

demonstrated by the 0.7 correlation coefficient, showing the ability of the model to simulate the background of atrazine 13 years after the end of the herbicide application. The observed evolution of DEA was rather similar to that of atrazine, with a lower discrepancy between the periods prior to and after its ban (Fig. 5, right). The simulated DEA concentrations were strongly related to the water discharge during the period of atrazine use. After its ban, the model tended to underestimate the DEA transfer for the last 2 years of the period. This might be due to the fact that in the model, the concentration was computed based only on the soil source of DEA, while the observation can include the source of DEA from the groundwater. In addition to water circulation in soils, integrating groundwater transfer in larger simulated basins is a step further, developed below.

5   The Spatial Approach in the Vesle Catchment

A spatial distribution of the STICS-Pest crop model was developed to take into account the inherent spatial heterogeneity associated with parameters dictating changes in water and pollutant inputs in an agro-pedo-hydrosystem, such as climate, crop rotations and management, pesticide inputs and soil properties.

This type of approach requires (1) an explicit structural description of all compartments of the hydrosystem (surface domain, vadose and saturated zones) as well as a reproduction of their respective hydrodynamic behaviours and (2) a fine spatial discretisation of transport processes throughout the entire system (spatial distribution on fine grid of water recharge, root zone concentrations in pesticides, etc.). Therefore, a coupling procedure with the spatially distributed MODCOU-NEWSAM hydrogeological model [51, 52] was developed.

The whole modelling platform was designed to simulate both pesticide fluxes coming from agricultural activity and their transport in soils as well as in underground and surface water compartments over the Vesle basin. Again, we only focussed on atrazine and DEA. From a modelling standpoint, this watershed offers advantages such as relatively simple geological and hydrogeological configurations and spatial homogeneity in soil properties and agricultural practices.

To interface STICS and the MODCOU-NEWSAM hydrogeological model, the code was integrated into a computing structure [30, 51–53], allowing it to be used in a broad spatial manner according to a bottom-up approach [54]. Consequently, this procedure integrates the variability of inputs in space and time (soil types, climate, agro-system management).

It therefore requires three kinds of input: (1) climate data from the SAFRAN atmospheric analysis system [55, 56], (2) soil data at the 1:1000000 scale [57] for which parameters are estimated through the use of pedo-transfer functions [58–61] and (3) crop rotation and management [29, 62, 63] for which pesticide use was computed with crop rotation in the ARSEINE database [21] over a 45-year period (1970–2015) in Fig. 6.

MODCOU [64, 65] is a regional spatially distributed model which describes both surface and underground water flows at a daily time step. Mainly conditioned by land use, climate and parent soil material, surface water balance calculations are based on a conceptual reservoir-based approach [66]. Infiltration leaks through a vadose zone were implemented using a set of reservoirs introducing a delay between the infiltration and the aquifer recharge [67, 68]. The finite difference resolution scheme of the 2D diffusivity equation uses the recharge as well as water pumping to compute the dynamism of piezometric heads of the saturated zone. Computed discharges in each river cell of the hydrographic net result from both stream-aquifer exchange calculations, taking into account the coupling with piezometric heads and sub-surface run-off, which is routed across the drainage network [69].

Pesticide transfer within both the vadose and saturated zones is computed by the NEWSAM module in which the substance is assumed to be fully conservative since most of the pesticide-related transformation processes take place within the first few metres of the subsoil interface [70]. A resolution scheme of the convection-only

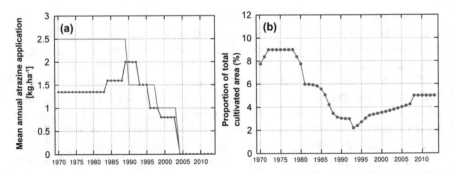

Fig. 6 Annual changes in atrazine application (in kg ha^{-1}) on maize (1971–2013 period) for one modelling unit (**a**), in red the approved dose and in blue the mean dose on the Vesle watershed and the ratio of maize (as a percentage of cultivated area) on the same modelling unit (**b**)

transport equation is run, with a 10-day time step, to compute the evolution of pesticide concentrations in each aquifer.

In this case, a regional application of the MODCOU-NEWSAM model [71] was developed, describing most of the Champagne area and covering the Vesle basin in its entirety (Fig. 1). The surface domain of the basin is discretised by over 5,000 cells within a progressive multi-scale grid of embedded square meshes. The groundwater domain is modelled by a multilayer structure of 5,900 cells. Both surface and aquifer grid cells vary from 250 m to 2 km in size. Two aquifer layers describe the main aquifers of the basin (Cenomanian Chalk aquifer and tertiary complex multilayered ensemble).

Prior to any pesticide-related simulations, hydro-dynamism was calibrated using measurements of seven hydrometric stations located on the river and four piezometers distributed along the watershed; nitrate concentration measurements from 14 boreholes were also used to calibrate solute transport.

For each cell of STICS-Pest, the spatialised model provides both water drainage and leached fluxes of atrazine and associated metabolites. Under agricultural lands only, leaching fluxes were diluted by the water drainage computed by STICS-Pest. Then computed concentrations were transmitted to the aquifer system through the unsaturated zone. Under urban and forest areas, the MODCOU water balance was used with associated pesticide concentrations set to 0. This value is justified by there being no treatment with atrazine in forests and by urban uses on impervious surfaces that directly transfer to rivers.

Figure 7 shows the location of each monitoring point used for the model calibration on the basin as well as a few sample results regarding piezometric heads (Fig. 7a), river flow (Fig. 7b) and nitrate concentration (Fig. 7c, d) at different stations and boreholes.

Good agreement was found for piezometric heads (Fig. 7a), river flow (Fig. 7b) and nitrate concentration (Fig. 7c, d) at different stations and boreholes.

The changes in atrazine at a borehole are accurately reproduced by the simulations during this period of available data (Fig. 7e). Maps of simulated concentrations for the Cenomanian aquifer layer at the end of December 2001 (deadline of authorised sale of atrazine) and in 2013 (final time step of the simulation), shown in Fig. 7, are also in agreement with observed concentrations.

A mass balance, at the scale of the entire area and over the simulation period (1971–2013), shows a 0.39% ratio between total simulated leached and applied masses of atrazine, which is close to the value previously determined, also using STICS-Pest, by [35] on the Orgeval catchment. More generally, this order of magnitude is similar in many studies reported in the literature [72–76].

6   Pesticides at the Seine River Basin Scale

At this stage of the modelling procedure, the results obtained on the Vesle basin show that it is possible to simulate the transfer of atrazine in a soil-groundwater-river system. However, considering the Seine River basin, various residence times in

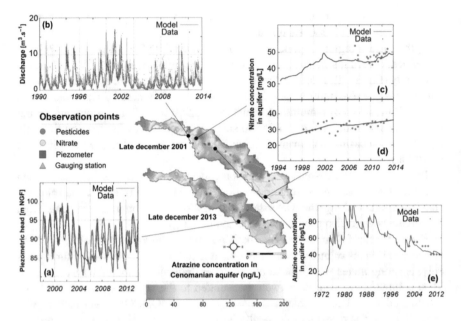

Fig. 7 Modelling results compared to measurements of groundwater levels in piezometers (**a**), river flow (**b**), nitrate (**c**, **d**) and atrazine concentrations (**e**). Maps of simulated atrazine concentrations in the Cenomanian aquifer layer (2001 and 2013). Observed data were extracted from the HYDRO and ADES databases [15, 77]

the vadose zone and aquifers make modelling more complex, and the mosaic of cropping systems must be associated with the diversity of soil types and the thickness of the unsaturated zone [21, 78].

In the first step, a prospective approach was implemented to understand pesticide contamination of groundwater in the Seine basin, again using the ADES database [15], in which pesticide contamination has been documented since 1997. As DEA is the most widely detected pesticide residue in water, it was possible to use this national monitoring database to understand the temporal evolution of the contamination. The mean concentrations of two different periods were compared: the first one from 1997 to 2003 corresponding to the use of atrazine (Fig. 8, left) and the second one from 2006 to 2014 (Fig. 8, right). Data between 2003 and 2005 were considered as a transition period when atrazine was still used by farmers (finishing stocks).

DEA was analysed on 1,436 piezometers, but many wells presented values below the limit of quantification. A value reported as below the limit of quantification was assigned a value of one-half of the limit of quantification. As this limit decreases over time from 0.1 to 0.005 $\mu g\,L^{-1}$, then the mean values can decrease over time due to the limit of quantification even if DEA was never detected. To better represent the evolution of DEA contamination, only mean values over 0.11 $\mu g\,L^{-1}$ were considered (Fig. 8). This approach allowed us to focus on contaminated wells. Mean

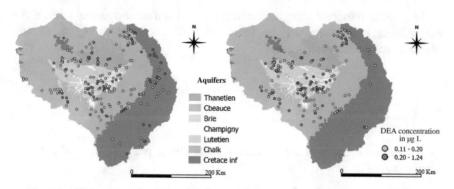

Fig. 8 Mean concentration of DEA above 0.11 µg L^{-1} in boreholes of the Seine aquifers (from ADES database), in the 1997–2003 period (left) and 2006–2014 (right)

concentrations over 0.2 µg L^{-1} are mostly observed in the Brie, Beauce and Champigny aquifers (see locations in Fig. 8) where maximum values were measured before 2003 (0.91, 1.8 and 1.42 µg L^{-1}, respectively). Some wells also show increasing concentrations of DEA over time in the Chalk and Champigny aquifers (i.e. from 0.14 to 0.18 and from 0.20 to 0.26 µg L^{-1}, respectively), whereas the Cretace aquifer shows more decreasing concentrations (from 0.21 to 0.12 µg L^{-1}). However, this cannot be directly extrapolated for the entire aquifer, because the wells did not show the same trend. For the 2006–2014 period, the mean DEA concentrations were 0.123 µg L^{-1}, 0.107 µg L^{-1} and 0.067 µg L^{-1} for the Champigny, Brie and Chalk aquifers, respectively.

A statistical approach was tested to determine the global trend of DEA contamination in each aquifer and the mean concentration in 2030 [79]. The mean values of 270 piezometers that compiled at least ten measurements throughout the period were identified. The linear tendency for each of them was calculated, and uncertainties were estimated successively using a Mann-Kendall test and a bootstrap method [38, 80]. In 2030, the estimated mean concentrations of DEA would be about 0.10 µg L^{-1} and 0.09 µg L^{-1} for the Brie and Champigny aquifers, respectively (Orgeval catchment), and about 0.01 µg L^{-1} for the Chalk aquifer (Vesle catchment). This slow decreasing concentration in the Brie and Champigny aquifers is due to a small number of boreholes (18 and 83, respectively) and occurrence of increasing contamination. In the Chalk aquifer (562 piezometers), the mean concentration would be below the limit of quantification in 2030, but this average does not reflect what could be observed locally.

This prospective method is advantageous because it presents a statistical trend of contamination. However, it is not relevant for other pesticides because data are limited in time and too scarce, even at the Seine basin scale. Therefore, the limit of quantification would have a considerable impact on mean concentrations, and this statistical approach could not be validated.

The next step for evaluating the contamination of surface and groundwaters at the Seine River basin scale will be based on mechanistic modelling, as already implemented on the Vesle basin.

7   Conclusion

Depending on the size of the watershed under study, pesticide use and database monitoring differ.

- At the plot scale, crop rotations and phytosanitary use and application conditions are known and can be combined for precise and adapted monitoring.
- In a catchment area such as Orgeval (100 km^2), the agricultural landscape is more complex. The observed contamination corresponds to major pesticide use. However, homogeneity is still observed due to a similar pedoclimatic context. The response time of the soil-groundwater-river system is fast, and contamination appears quickly and abruptly at the outlet because of drains. Major pesticide inputs can be identified using farmers' logbooks.
- In the Vesle basin ($1,000 \text{ km}^2$), the water transfer time from groundwater to the river is longer. The landscape is contrasted, resulting in heterogeneous crops in the basin (alfalfa, annual crops and vineyards). Spatialisation of uses will be of particular importance in this context, and we need to combine agricultural practices (e.g. pesticide treatment) and soil characteristics. In this case, the random distribution of crops considered in our modelling approach of basins may have a considerable impact on the model's outputs. Vineyards receive specific treatments and are well identified in the Corine land cover database; its pesticide applications were therefore correctly located. Modelling also required estimating inputs over at least 30 years, to include the transfer time to the outlet. This retrospective raised the question of the initialisation of the model, which takes a considerable part in the uncertainty of the simulation results. Nevertheless, focusing on specific weeding practices in vineyards and field crops allowed us to estimate the uses of atrazine and to model its transfer. However, working at the sub-annual level would not be realistic at this time.
- At the scale of the Seine basin ($75,000 \text{ km}^2$), the hydrogeological context is complex, with several superimposed aquifers, different response times and a patchwork of cropping systems. Monitoring databases are those assembled by the national authority for European framework directive monitoring [15, 81], and they allow only scarce and discontinuous monitoring of pesticide contamination. However, considering only the residues of atrazine, which has been widely used throughout the basin and frequently detected, it is possible to show that some aquifers still have an increasing contamination of atrazine residues despite its prohibition in 2003. Particularly vulnerable groundwaters such as the Champigny aquifer still show an increase in DEA levels, and other studies have shown

that concentrations of the percolating flow to the Chalk aquifer can reach $10 \, \mu g \, L^{-1}$ [39].

8 Perspectives

The reduction of the water resource contamination by pesticides remains a major issue. The strategy of banning molecules has only shifted the problem to new and unknown substances for which we have not yet identified the environmental effects. Improving water quality and reducing impacts are new challenges involving limiting and banning the use of these new substances. This requires a deep change in agricultural practices and systems. Organic agriculture is a typical agriculture system which bans chemical molecules of pesticides as well as mineral fertilisation. However, this agriculture accounts for only 2% of the agricultural land use in the Paris basin [82].

Even if this solution would be preferable for the environment and human health, buffer zones and grass strips can be used over the short term to reduce pesticide transfer [83–85]. In the case of the Seine catchment, about 20% of arable soils are drained, depending on soil waterlogging characteristics. Sub-surface drained areas are mainly covered with redoxic or reductic degraded luvisol. In this context, the management of sub-surface drained water is important to reducing the impact of agriculture's activities on water quality. An experiment conducted since 2012 near the Orgeval basin led to co-constructing, with local stakeholders, a set of four artificial wetlands implemented between the outlet of sub-surface buried pipes and the receiving waterbody. The ratio between the area of what the farmers accepted to convert into artificial wetland and the total cultivated areas (355 ha) was established at 0.20%. A 3-year evaluation of the removal rate of pesticides from surface water showed a reduction of about 30% of pesticides from agricultural plots and surface waterbodies by artificial wetlands, with, however, a high variability depending on (1) the pesticides' properties, (2) temperature dependency and (3) the hydraulic residence time. By extension of this in situ monitoring, the recommended ratio to adequately design artificial wetlands was evaluated at 1% for an objective of 70% pesticide removal from drained water. However, even if artificial wetlands are constructed at the recommended ratio on the Seine basin, it would not be able to solve pesticide contamination, because a large proportion of pesticides are directly transferred from fields to groundwater.

A mechanic modelling approach would help assess the risk of pesticide transfer in a watershed and explore scenarios of changing agriculture. However, several difficulties have been noted: (1) correctly representing past uses, (2) documenting the long-term behaviour of an active ingredient in a large watershed and (3) identifying the effect of changes in agricultural practices on pesticide transfer.

In that case, other parameters must be taken into account such as trajectories of pesticide uses and landscape changes regarding agricultural activities. The new national BNV-D database indicating sales of pesticides, available since 2008 (identification of substances, annual quantities at the distributor scale), could also help characterise and spatialise pesticide inputs and these changes at the regional scale [21]. This new database provides valuable perspectives for the characterisation of future uses. On the other hand, the characterisation of past practices, in particular those before the 1990s, remains difficult because of the absence of national databases and the disappearance of archival documents (re-structuring, destruction of old archives) and technical knowledge (retirement of advisors) on farms and advisory bodies.

Besides enlarging the modelling approach to the Seine River basin scale, another challenge is coupling pesticide transfer from soils and the groundwater model with the river transport model. Similar developments have been carried out for nitrate transfer in the PIREN-Seine programme (see [86]). Because specific processes for pesticide consideration were already integrated into the STICS agronomic model (especially sorption and degradation), it would also be possible to integrate them into the biogeochemical module in surface water. In any case, pesticide monitoring in rivers and groundwater is essential to compare simulated and observed data.

Acknowledgements This work was initiated by the CNRS project EC2CO Phyt'Oracle in 2008 and has since then been extended as part of the PIREN-Seine research programme, a component of the Zone Atelier Seine within the international Long-Term Socio-Ecological Research (LTSER) network (www.piren-seine.fr).

References

1. Klein M (1995) PELMO: pesticide leaching model version 2.01. Fraunhofer Institut für Umweltchemie und Okotoxi-kolgie, Schmallenberg
2. Carsel RF, Imhoff JC, Hummel PR, Cheplick JM, Donigian Jr AS (2003) PRZM-3, a model for predicting pesticide and nitrogen fate in the crop root and unsaturated soil zones: users manual for release 3.12. Center for Exposure Assessment Modeling (CEAM). U.S. Environmental Protection Agency (USEPA), Athens
3. Leistra M, van der Linden AMA, Boesten JJTI, van der Berg F (2001) PEARL model for pesticide behavior and emissions in soil_plant systems; description of the processes in FOCUS PEARL v 1.1.1. RIVM report, Alterra report 013 711,401 009. National Institute of Public Health and the Environment. Wageningen Alterra, Green World Research, Bilthoven
4. Larsbo M, Jarvis N (2003) MACRO5.0. A model of water flow and solute transport in macroporous soil. Technical description. Emergo 2003:6. Studies in the biogeophysical environment. SLU, Deptartment of Soil Science, Uppsala, p 47
5. Malone R, Ahuja WR, Ma L, Wauchope RD, Ma Q, Rojas KW (2004) Application of the root zonewater qualitymodel (RZWQM) to pesticide fate and transport: an overview. Pest Manag Sci 60(3):205–221. https://doi.org/10.1002/ps.789
6. Murgue C, Therond O, Leenhardt D (2016) Hybridizing local and generic information to model cropping system spatial distribution in an agricultural landscape. Land Use Policy 54:339–354
7. Therond O, Hengsdijk H, Casellas E, Wallach D, Adam M, Belhouchette H et al (2011) Using a cropping system model at regional scale: low-data approaches for crop management information and model calibration. Agric Ecosyst Environ 142(1–2):85–94
8. Leenhardt D, Angevin F, Biarnès A, Colbach N, Mignolet C (2010) Describing and locating cropping systems on a regional scale. A review. Agron Sustain Dev 30(1):131–138

9. Gomez E (2002) Modélisation intégrée du transfert de nitrate à l'échelle régionale dans un système hydrologique. Application au bassin de la Seine. PhD thesis, MINES ParisTech, p 291

10. Billen G, Garnier J (1999) Nitrogen transfers through the Seine drainage network: a budget based on the application of the 'Riverstrahler' model. Hydrobiologia 410:139–150

11. Mégnien C (1979) Hydrogéologie du centre du bassin de Paris – Contribution à l'étude de quelques aquifères principaux. Mémoire du BRGM n° 98. Fricotel, Epinal, p 532

12. Nicola L, Schott C, Mignolet C (2012) Dynamique de changement des pratiques agricoles dans le bassin versant de l'Orgeval et création de la base de données APOCA (Agricultural Practices of the Orgeval Catchment Area). RA 2011 PIREN-Seine, p 49

13. Masiá A, Ibáñez M, Blasco C, Sancho JV, Picó Y, Hernández F (2013) Combined use of liquid chromatography triple quadrupole mass spectrometry and liquid chromatography quadrupole time-of-flight mass spectrometry in systematic screening of pesticides and other contaminants in water samples. Anal Chim Acta 761:117–127

14. Osorio V, Schriksa M, Vughs D, de Voogtad P, Kolkmana A (2018) A novel sample preparation procedure for effect-directed analysis of micro-contaminants of emerging concern in surface waters. Talanta 186(15):527–537

15. ADES database http://www.ades.eaufrance.fr/

16. Andert S, Bürger J, Gerowitt B (2015) On-farm pesticide use in four Northern German regions as influenced by farm and production conditions. Crop Prot 75:1–10

17. Bürger J, de Mol F, Gerowitt B (2012) Influence of cropping system factors on pesticide use intensity–a multivariate analysis of on-farm data in North East Germany. Eur J Agron 40:54–63

18. Nave S, Jacquet F, Jeuffroy MH (2013) Why wheat farmers could reduce chemical inputs: evidence from social, economic, and agronomic analysis. Agron Sustain Dev 33(4):795–807

19. Mignolet C, Schott C, Benoît M (2004) Spatial dynamics of agricultural practices on a basin territory: a retrospective study to implement models simulating nitrate flow. The case of the Seine basin. Agronomie 24:219–236

20. Mignolet C, Schott C, Benoît M (2007) Spatial dynamics of farming practices in the Seine basin: methods for agronomic approaches on a regional scale. Sci Total Environ 375 (1–3):13–32

21. Puech T, Schott C, Mignolet C (2018) Evolution des bases de données pour caractériser les dynamiques des systèmes de culture sur le bassin Seine-Normandie. Technical report, INRA

22. E-phy database https://ephy.anses.fr/

23. Carles M, Cahuzac E, Guichard L, Martin P (2015) Mieux suivre spatialement l'usage des pesticides, en particulier sur les bassins versants, en s'appuyant sur un observatoire des ventes détaillé au code postal de l'utilisateur final de produit. Technical report, p 46

24. Groshens E (2014) Spatialisation des données de ventes de pesticides. Rapport sur les possibilités et limites d'une extrapolation de la démarche à l'échelle nationale. Technical report, p 37

25. Inan HI, Sagris V, Devos W, Milenov P, van Oosterom P, Zevenbergen J (2010) Data model for the collaboration between land administration systems and agricultural land parcel identification systems. J Environ Manag 91(12):2440–2454

26. Schott C, Barataud F, Mignolet C (2015) Les "carnets de plaine" des agriculteurs: une source d'information sur l'usage des pesticides à l'échelle de bassins versants? Agron Environ Soc 4(2):179–197

27. Ewert F, van Ittersum MK, Heckelei T, Therond O, Bezlepkina I, Andersen E (2011) Scale changes and model linking methods for integrated assessment of agri-environmental systems. Agric Ecosyst Environ 142(1–2):6–17

28. Zhao G, Hoffmann H, van Bussel LG, Enders A, Specka X, Sosa C et al (2015) Effect of weather data aggregation on regional crop simulation for different crops, production conditions, and response variables. Clim Res 65:141–157

29. Schott C, Mignolet C, Rat A, Ledoux E, Benoît M (2007) Modélisation des pratiques phytosanitaires sur le bassin versant de la Vesle. Pesticides: impacts environnementaux, gestion et traitements. Presses de l'école nationale des Ponts et Chaussées, pp 207–223

30. Gallois N, Puech T, Viennot P (2018) Modélisation des transferts de produits phytosanitaires vers les eaux souterraines: Cas de l'atrazine et de ses métabolites sur le bassin amont de la Vesle (Marne), Technical report, INRA-ARMINES/MINES ParisTech

31. Chevreuil M, Chesterikoff A (1979) Movement of pesticides in a watershed of the Brie region: methods and balance sheet studies. Comptes Rendus Séances Acad Agric France 65:835–845
32. Chevreuil M, Garmouma M, Fauchon N (1999) Variability of herbicides (triazines, phenylureas) and tentative mass balance as a function of stream order, in the river Marne basin (France) – Triazine and phenylurea flux and stream order. Hydrobiologia 410:349–355
33. Blanchoud H, Garban B, Ollivon D, Chevreuil M (2002) Herbicides and nitrogen in precipitation: progression from west to east and contribution to the Marne River (France). Chemosphere 47(9):1025–1031
34. Blanchoud H, Barriuso E, Nicola L, Schott C, Roose-Amsaleg C, Tournebize J (2013) La contamination de l'Orgeval par les pesticides. Dans "L'observation long terme en environnement, exemple du bassin versant de l'Orgeval". Versailles, éd QUAE, pp 159–174
35. Queyrel W (2014) Modélisation du devenir des pesticides dans les sols à partir d'un modèle agronomique: évaluation sur le long terme. PhD thesis, Université Pierre et Marie Curie. ED GRN, p 284
36. Tournebize J, Passeport E, Chaumont C, Fesneau C, Guenne A, Vincent B (2013) Pesticide de-contamination of surface waters as a wetland ecosystem service in agricultural landscapes. Ecol Eng 56:51–59
37. Blanchoud H, Bergheaud V, Nicola L, Vilain G, Bardet S, Tallec G, Botta F, Barriusio E, Schott C, Laverman A, Habets F, Ansart P, Desportes A, Chevreuil M (2010) Transfert de pesticides dans le système sol-nappe-rivière: Etude du comportement de l'atrazine et de l'isoproturon dans le bassin versant de l'Orgeval. RA 2009 PIREN-Seine, p 18
38. Thorsten P (2018) Non-parametric trend tests and change-point detection. R package
39. Chen NX, Valdes D, Marlin C, Blanchoud H, Guerin R, Rouelle M, Ribstein P (2019) Water, nitrate and atrazine transfer through the unsaturated zone of the Chalk aquifer in Northern France. Sci Total Environ 652:927–938
40. Brisson N, Mary B, Ripoche D et al (1998) STICS: a generic model for the simulation of crops and their water and nitrogen balances: theory and parametrization applied to wheat and corn. Agronomie 18(5–6):311–346
41. Queyrel W, Habets F, Blanchoud H, Ripoche D, Launay M (2016) Pesticide fate modeling in soils with the crop model STICS: feasibility for assessment of agricultural practices. Sci Total Environ 542:787–802. https://doi.org/10.1016/j.scitotenv.2015.10.066
42. Der Ploeg V, Rienk R, Ringe H, Machulla G (1995) Late fall site-specific soil nitrate upper limits for groundwater protection purposes. J Environ Qual 24(4):725–733
43. Larocque M, Banton O, Lafrance P (1998) Simulation par le modèle AgriFlux du devenir de l'atrazine et du dééthylatrazine dans un sol du Québec sous maïs sucré. Rev Sci L'eau 11 (2):191–208. http://www.erudit.org/revue/rseau/1998/v11n2/705303ar.html
44. Xing B, Pignatello JJ, Gigliotti B (1996) Competitive sorption between atrazine and other organic compounds in soils and model sorbents. Environ Sci Technol 30(8):2432–2440
45. Xing B, Pignatello JJ (1997) Dual-mode sorption of low-polarity compounds in glassy poly (vinyl chloride) and soil organic matter. Environ Sci Technol 31(3):792–799
46. Laroche B (2012) Base de Donnees Géographique Des Sols de France. INRA, Orléans. www.gissol.fr/programme/bdgsf/bdgsf.php
47. Billy C, Billen G, Sebilo M, Birgand F, Tournebize J (2010) Nitrogen isotopic composition of leached nitrate and soil organic matter as an indicator of denitrification in a sloping drained agricultural plot and adjacent uncultivated riparian buffer strips. Soil Biol Biochem 42 (1):108–117. https://doi.org/10.1016/j.soilbio.2009.09.026
48. PPDB (2013) The Pesticide Propertiezs DataBase (PPDB) developed by the Agriculture & Environment Research Unit (AERU). University of Hertfordshire, Hertfordshire. http://www.eu-footprint.org/ppdb.html
49. Baer U, Calvet R (1997) Simulation and prediction of dissipation kinetics of two herbicides in different pedo-climatic situations. Int J Environ Anal Chem 68(2):213–237. https://doi.org/10.1080/03067319708030492
50. Rat A, Ledoux E, Viennot P (2006) Transferts de Pesticides Vers Les Eaux Souterraines, Modélisation À L'échelle D'un Bassin Versant (Cas D'étude Du Bassin Amont de La Vesle). PIREN-Seine. ENSMP. http://www.sisyphe.upmc.fr/piren/webfm_send/26

51. Gallois N, Viennot P (2015) Modélisation de la pollution nitrique d'origine agricole des grands aquifères du bassin de Seine-Normandie à l'échelle des masses d'eau: Modélisations couplées hydrogéologie-agriculture. Technical report, ARMINES/MINES ParisTech
52. Beaudoin N, Gallois N, Viennot P et al (2016) Evaluation of a spatialized agronomic model in predicting yield and N leaching at the scale of the Seine-Normandie Basin. Environ Sci Pollut Res 25(24):23529–23558. https://doi.org/10.1007/s11356-016-7478-3
53. Gallois N, Viennot P (2018) Modélisation de la pollution diffuse d'origine agricole des grands aquifères du bassin Seine-Normandie: Actualisation des modélisations couplées STICS-MODCOU – Modélisation de scénarios agricoles sous changement climatique, Technical report, ARMINES/MINES ParisTech
54. Wagenet R-J, Hutson J-L (1996) Scale-dependency of solute transport modeling/GIS applications. J Environ Qual 25:495–510
55. Durand Y, Brun E, Mérindol L et al (1993) A meteorological estimation of relevant parameters for snow models. Ann Glaciol 18:65–71
56. Quintana-Segui P, Le Moigne P, Durand Y et al (2008) Analysis of near-surface atmospheric variables: validation of the SAFRAN analysis over France. J Appl Meteorol Climatol 47 (1):92–107
57. INRA (1998) Base de données géographique des sols de France à l'échelle du 1/1 000 000. INRA Infosol, Orléans
58. King D, Daroussin J, Hollis J-M et al (1994) A geographical knowledge database on soil properties for environmental studies. Final report of EC contract N°3392004. INRA, Orleans, 50 pp
59. Le Bas C, King D, Daroussin J (1997) A tool for estimating soil water available for plants using the 1:1000000 scale Soil Geographical Data Base of Europe. In: Beek KJ, de Bie KA, Driessen PM (eds) Geo-information for sustainable land management. International Journal of Aerospace Survey and Earth Sciences. ITC, Enschede, p 10
60. Donet I, Le Bas C, Ruget F et al (2001) Informations et suivi objectif des prairies. Guide d'utilisation. Agreste Chiffres et Données, 134. MAF, Paris, p 55
61. Thomasson A-J, Carter A-D (1989) Current and future uses of the UK soil water retention dataset. Proceedings of the international workshop on indirect methods for estimating the hydraulic properties of unsaturated soils, Riverside, California, pp 355–358
62. Puech T, Schott C, Mignolet C (2015) Evolution des systèmes de culture sur le bassin Seine-Normandie depuis les années 2000: construction d'une base de données spatialisée sur les pratiques agricoles. Technical report, INRA
63. Schott C, Mignolet C, Benoit M (2004) Modélisation des pratiques phytosanitaires sur le bassin de la Vesle: le cas du désherbage chimique de la vigne et du maïs de 1970 à nos jours. Technical report, PIREN-Seine/INRA
64. Ledoux E (1980) Modélisation intégrée des écoulements de surface et des écoulements souterrains sur un bassin hydrologique. PhD thesis, Ecole Nationale Supérieure des Mines de Paris
65. Ledoux E, Girard G, de Marsily G et al (1989) Spatially distributed modeling: conceptual approach, coupling surface water and groundwater. In: Morel-Seytoux HJ (ed) Unsaturated flow in hydrologic modeling – theory and practice. NATO ASI Series C. Kluwer Academic, Norwell, pp 435–454
66. Girard G, Morin G, Charbonneau R (1972) Modèle précipitations débits à discrétisation spatiale. Cahiers Orstom Ser Hydrol IX(4):35–52
67. Nash J-E, Sutcliffe J-V (1970) River flow forecasting through conceptual models, a discussion of principles. J Hydrol 10:282–290
68. Besbes M (1978) L'estimation des apports aux nappes souterraines. Un modèle régional d'infiltration efficace. PhD thesis, Université Pierre et Marie Curie-Paris VI
69. Golaz-Cavazzi C (1999) Modélisation hydrogéologique à l'échelle régionale appliquée au bassin du Rhône. PhD thesis, Ecole Nationale Supérieure des Mines de Paris
70. Arias-Estévez M, López-Periago E, Martínez-Carballo E, Simal-Gándara J, Mejuto J-C, García-Río L (2008) The mobility and degradation of pesticides in soils and the pollution of groundwater resources. Agric Ecosyst Environ 123(4):247–260

71. Viennot P, Abasq L (2013) Modélisation de la pollution nitrique des grands aquifères du bassin de Seine-Normandie à l'échelle des masses d'eau, Technical report, ARMINES/MINES ParisTech

72. Baran N, Lepiller M, Mouvet C (2008) Agricultural diffuse pollution in a chalk aquifer: influence of pesticide properties and hydrodynamic constraints. J Hydrol 358(1–2):56–69

73. Clement M, Cann C, Seux R et al (1999) Pollutions diffuses: du bassin versant au littoral. Facteurs de transfert vers les eaux de surface de quelques phytosanitaires dans le contexte agricole breton, pp 141–156

74. Hall J-K, Mumma R-O, Watts D-W (1991) Leaching and runoff losses of herbicides in a tilled and untilled field. Agric Ecosyst Environ 37(4):303–314

75. Louchart X (1999) Transfert de pesticides dans les eaux de surface aux échelles de la parcelle et d'un bassin versant viticole. Etude expérimentale et éléments de modélisation. PhD thesis, Ecole Nationale Supérieure Agronomique de Montpellier, p 270

76. Morvan X, Mouvet C, Baran N et al (2006) Pesticides in the groundwater of a spring draining a sandy aquifer: temporal variability of concentrations and uxes. J Contam Hydrol 87 (3–4):176–190

77. HYDRO database http://www.hydro.eaufrance.fr/

78. Viennot P, Gallois N (2017) Modélisation de la pollution diffuse d'origine agricole des grands aquifères du bassin de Seine-Normandie: Scénarios d'évolution climatique – Impacts et incertitudes. Technical report, ARMINES/MINES ParisTech

79. Mattei A (2017) Eléments de prospective de la contamination des cours d'eau d'Ile-de-France par les Pesticides. Confidential technical report, p 50

80. R Core Team (2014) R: a language and environment for statistical computing. R Foundation for Statistical Computing, Vienna. http://www.R-project.org/

81. SOeS database https://www.statistiques.developpement-durable.gouv.fr/

82. Anglade J, Billen G, Garnier J, Makridis T, Puech T, Tittel C (2015) Nitrogen soil surface balance of organic vs conventional cash crop farming in the Seine watershed. Agric Syst 139:82–92

83. Lacas JG, Voltz M, Gouy V, Carluer N, Gril JJ (2005) Using grassed strips to limit pesticide transfer to surface water: a review. Agron Sust Dev 25(2):253–266

84. Reichenberger S, Bach M, Skitschak A, Frede H-G (2007) Mitigation strategies to reduce pesticide inputs into ground- and surface water and their effectiveness: a review. Sci Total Environ 384(1–3):1–35

85. Tournebize J, Chaumont C, Mander U (2017) Implications for constructed wetlands to mitigate nitrate and pesticide pollution in agricultural drained watersheds. Ecol Eng 103:415–425

86. Billen G, Garnier J, Le Noë J et al (2020) The Seine watershed water-agro-food system: long-term trajectories of C, N, P metabolism. In: Flipo N, Labadie P, Lestel L (eds) The Seine River basin, Handbook of environmental chemistry. Springer, Cham. https://doi.org/10.1007/698_2019_393

Mass Balance of PAHs at the Scale of the Seine River Basin

D. Gateuille, J. Gasperi, C. Briand, E. Guigon, F. Alliot, M. Blanchard, M.-J. Teil, M. Chevreuil, V. Rocher, S. Azimi, D. Thevenot, R. Moilleron, J.-M. Brignon, M. Meybeck, and J.-M. Mouchel

Contents

1 Introduction 164
2 Material and Methods 165
   2.1 Areas Studied 165
   2.2 Dual-Scale Mass Balance Approach 167
   2.3 Data Selection and Exploration 168
   2.4 Flux Estimation Quality 169
3 Results 169
   3.1 Urban Fluxes 169
   3.2 Rural Fluxes 174
4 Discussion 180
5 Conclusions and Perspectives 182
References 183

D. Gateuille (✉)
Université Paris-Est, Laboratoire LEESU, Champs-sur-Marne, France

Université Savoie Mont Blanc, Laboratoire LCME, Chambéry, France
e-mail: david.gateuille@univ-smb.fr

J. Gasperi, C. Briand, D. Thevenot, and R. Moilleron
Université Paris-Est, Laboratoire LEESU, Champs-sur-Marne, France

E. Guigon, F. Alliot, M. Blanchard, M.-J. Teil, M. Chevreuil, M. Meybeck, and J.-M. Mouchel
Sorbonne Université, CNRS, EPHE, PSL University, UMR Metis, Paris, France

V. Rocher and S. Azimi
SIAAP – Direction Innovation et Environnement, Colombes, France

J.-M. Brignon
Institut national de l'environnement industriel et des risques, Verneuil-en-Halatte, France

Nicolas Flipo, Pierre Labadie, and Laurence Lestel (eds.), *The Seine River Basin*,
Hdb Env Chem (2021) 90: 163–188, https://doi.org/10.1007/698_2019_382,
© The Author(s) 2020, corrected publication 2020, Published online: 3 June 2020

Abstract The Seine River basin (France) is representative of the large urbanised catchments (78,650 km^2) located in Northwestern Europe. As such, it is highly impacted by anthropogenic activities and their associated emissions of pollutants such as polycyclic aromatic hydrocarbons (PAHs). These compounds, originating from household heating and road traffic, are responsible for serious environmental issues across the basin. This study aims at establishing and using mass balance analyses of PAHs at the Seine River basin scale as an efficient tool for understanding PAH pathways in the environment. A dual-scale approach (urban vs. rural areas) was used successfully, and mass balances provided useful knowledge on the environmental fate of PAHs. In urban areas, runoff and domestic and industrial discharges contributed similarly to the PAH supply to the sewer system. During the wastewater treatment process, PAHs were mainly eliminated through sludge removal. At the basin scale, substantial amounts of PAHs were quantified in soils, and the limited annual inputs and outputs through atmospheric deposition and soil erosion, respectively, suggest that these compounds have long residence times within the basin. While wastewater and runoff discharges from urban areas account for a substantial part of PAH urban fluxes to the Seine River, soil erosion seems to be the predominant contributor at the basin scale. Overall, the PAH flux at the basin outlet was greater than supplies, suggesting that the Seine River system may currently be undergoing a decontamination phase.

Keywords eLTER, Environmental fluxes, Mass balance, PIREN-Seine, Pollutant fate, Polycyclic aromatic hydrocarbons, Seine River basin, Urban fluxes, Zone Atelier Seine

1 Introduction

The Seine River basin (78,650 km^2), located in Northwestern France, has been studied since 1990 within the PIREN-Seine programme and can be considered as representative of river basins exposed to the impacts of intense human activity [1, 2]. This basin accommodates a combination of strong human pressures (17 million people, with approximately 10 million aggregated within the Paris conurbation; 30% of French industrial and agricultural production) with very limited dilution by the Seine River, due to its low flow (median flow, 300 m^3 s^{-1}); the basin therefore is structurally vulnerable, and its river course downstream of the Paris conurbation shows heavy contamination [1]. Similarly to other river basins in Europe, the Seine-Normandie Water Agency reported that more than 50% of waterbodies were in poor chemical status, mostly due to polycyclic aromatic hydrocarbons (PAHs).

PAHs are a group of widespread organic compounds. Due to their high toxicity and their known carcinogenic properties for animals and humans [3], they constitute an environmental threat. Besides natural sources, PAHs are mainly emitted by anthropogenic activities such as combustion of fuels, coal and biomass or the use of bitumen- and petroleum-containing products [4] or by domestic sources such as tobacco smoke and cooking [5]. In Europe, PAHs (nine congeners) were included in the initial list of 33 priority pollutants of the Water Framework Directive (WFD) (2000/60/EC), establishing the water policy at the river basin scale. In the European Union, water policy is based on a number of specific directives defining community-wide emission limit values and quality objectives in surface and coastal waters. The implementation of such directives requires member states to reach "good chemical" status. Based on the last report (2012, European waters – assessment of status and pressures, EEA Report), PAHs appear as a widespread cause of poor status in rivers and are identified as problematic by 11 member states.

In this context, accurate knowledge of PAH sources and fate in aquatic systems at the catchment scale was proven necessary to (1) better assess the spatial and temporal dynamics of PAH input and output fluxes over an entire catchment and (2) identify possible pollutant reduction measures. Although numerous studies were carried out on PAHs in different environmental compartments, i.e. total atmospheric deposition [6], soils [7], surface water [8] or sediment [9], there have been few attempts in the literature to draw up a total PAH budget at the catchment scale [10–14]. The published studies did not integrate all environment compartments, nor were they carried out considering the spatial urbanisation variability across the catchment at the scale of a major river basin.

2 Material and Methods

This study was based on the integrative approach previously developed to investigate the heavy metal contamination within the Seine River basin [1, 2]. To study the fate of PAHs in this basin, contamination data in various environmental compartments were gathered from previous studies, and the mass balance was established at two nested scales. The calculations of individual fluxes are detailed below. To evaluate the robustness of the mass balance calculation, each flux was associated with both an uncertainty and a grade assessing the quality of the database used to estimate such fluxes.

2.1 Areas Studied

The PAH fluxes were assessed at two complementary scales: the Paris conurbation and the rural areas (Fig. 1).

All parameters concerning the Paris conurbation sewer system were provided by the SIAAP (Paris conurbation wastewater treatment authority). Hydrological data and information about the dredging operations were provided by the Seine-Normandie Water Agency (AESN), the French department of waterways (VNF) or the Île-de-France Regional Department for Equipment and Planning (DRIEAIF). The meteorological parameters came from either the studies cited herein or Météo-France. Land use was determined using ArcMap (ver10.5) with the Corine Land Cover 2012 database. Information about the road network was obtained from the Route500® database published by the French National Geographic Institute (IGN).

The Paris conurbation covered more than 1,830 km^2 including the most populated parts of the Seine River basin. The sewer system annually collects 900 Mm^3 of water

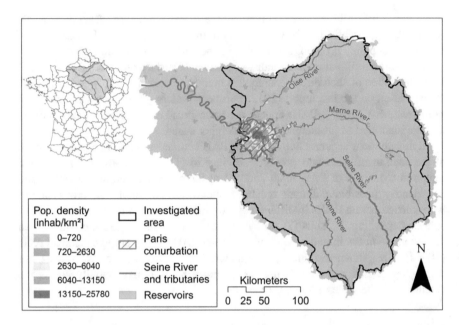

Fig. 1 Map delimiting the two areas considered in the dual approach. The black line delimits the investigated area including the uppermost parts of the basin downstream to the city of Triel-sur-Seine. The red hatched zone indicates the Paris conurbation

including industrial and domestic wastewater (about ten million inhabitants), runoff, parasite water, groundwater infiltration and street cleaning water. Depending on the location, the Paris conurbation was drained by combined sewers (30% of the surface area), separate sewers (56%) or a combination of separate and combined sewers (14%). However, due to the high population density in the historical centre, the daily volume of wastewater mainly originated from combined sewers (75%). In 2014, over 112,000 tons dry weight of sludge was produced during wastewater treatment. It was used for energy purposes (41%), spread over agricultural lands (29%), composted further before spreading (22%) or stored as hazardous material (8%) [15]. Across the conurbation, an average of 24% of the surface was impervious, while the remaining had a limited infiltration capacity leading to a mean runoff coefficient of 0.7 [16]. In 2014, the annual rainfall over the Paris conurbation was 713 mm.

The entire investigated area had a surface of 61,300 km^2 (78% of the whole Seine River basin area), mainly covered by agricultural land (66%), forested land (27%) and urban areas (7%). It accommodated 82% of the entire population of the Seine River basin. The outlet considered (Triel-sur-Seine) of the investigated area was chosen far enough downstream of the WWTP discharges to include the Paris conurbation emissions into the Seine River flux. However, it was also set upstream of the Seine River mouth in order to avoid tidal effects. The mass balances were established using the meteorological and hydrological parameters (e.g. Seine River discharge, suspended sediment concentrations, etc.) measured in 2014. This year was selected because of its representativeness in terms of meteorology and hydrology. The annual average rainfall was 743 mm, and the mean discharge of the Seine River at Paris was 325 m^3 s^{-1} (321 m^3 s^{-1} for the 2006–2018 period). A high-flow period was observed from December to late March, and the rest of the year was characterised by a low flow rate. The annual water and suspended sediment discharges at the outlet of the investigated area were estimated at 1.4×10^{10} m^3 year^{-1} and 3.1×10^5 tons year^{-1}, respectively, based on measurements on both the Seine River at Paris and the Oise River. Previous studies on the sediment yield in the Seine River basin [17, 18] showed that the average erosion rates over agricultural lands, forests and urban areas were assumed equal to 18.4, 2.0 and 0.9 tons km^{-2} year^{-1}, respectively. Thus, the estimated sediment yield to the rivers at the investigated area scale was 7.6×10^5 tons year^{-1}, in good agreement with the sediment flux estimated at the outlet considering that a fraction of eroded sediment was deposited on the river bed [18]. Approximately 1.0×10^5 tons year^{-1} of sediment are trapped in the reservoirs or on the floodplains upstream of Paris [2]. In addition, 1.2×10^5 tons year^{-1} of river bed sediment were removed during dredging operations. Either the collected sediment was used for agriculture or bank reinforcement or it was sent to landfill when the contaminant concentrations exceed the legal standards. A mass balance analysis of the sediment flux considering the erosion rate, the deposition rate on the floodplain and in the reservoirs and the sediment flux at the outlet of the basin provided an estimation of the amount of sediment stored within the river bed: 2.3×10^5 tons year^{-1}.

2.2 Dual-Scale Mass Balance Approach

To quantify the fluxes at the investigated area scale, a two-step method was applied. First, all urban fluxes were estimated for the Paris conurbation that constitutes a well-defined and densely urbanised area located in the downstream sector of the investigated catchment. Thus, the urban fluxes at the basin scale were assessed either (1) using the same calculation method as for the Paris conurbation but with an adapted database including all the data available for the investigated area or (2) multiplying the flux estimated for the Paris conurbation by 1.4 considering the change in the population from 10 (Paris conurbation) to 14 million inhabitants (the entire investigated area). This dual approach was chosen as it reduced the number of complex retroactions, thereby facilitating the consideration of a high level of detail in the flux charts. In the following, the fluxes are named according to their estimation method, namely, from direct measurement (F), economic data (E) or a combination of other fluxes (D).

Urban fluxes included the emissions to the atmosphere (E5a, E5b, E6), the atmospheric deposition (F12) and the runoff fluxes (F12e, F12f). The domestic and industrial releases to the sewer system (F22a, E22h) or to the Seine River (E22g) were also considered. The discharges of the sewer system to the river system during dry weather (F22d) or wet weather (F22c) and removal during sewer deposit cleaning processes (F22i) were taken into account. Special attention was paid to the WWTP-related fluxes including the inflow from the sewer system (F22e), the outflow to the Seine River (F22f) and urban sludge (F13a, F13b, F13c, E13d).

Rural fluxes included atmospheric fallout over agricultural land (F10) and the spreading of urban sludge (F13) or sediment (F25b) from river dredging. In addition, the inputs to forested soils through atmospheric deposition (F11a) and the forest filter effect (F11b) [19] were estimated. Road runoff outside of urban areas (E14) was also considered. The stocks in agricultural and forested soils were quantified (S10, S11). The erosion fluxes from agricultural, forested, urban and industrial land were estimated (F15, F16, F17 and F18) and combined to quantify the inputs into the Seine River (F19). In addition, storage in the reservoirs (F23a), onto the river bed (F23b) and on the floodplains (F24) was estimated as was the removal related to sediment dredging operations (F25a). Finally, the flux at the outlet of the investigated area was estimated (F21).

2.3 Data Selection and Exploration

The PAH content databases used to compute the flux estimations were gathered from about 40 previous studies investigating PAH contamination across the Seine River basin. Whenever possible, only data collected within the investigated area (Fig. 1) were considered. Otherwise, databases from other studies in France or Western Europe were used, therefore weakening the estimation of the quality grade (see below). A specific effort was made to avoid data duplication by cross-checking the sampling sites and dates in the various studies. When only summarised data were available, data were reconstructed assuming a log-normal distribution.

Fluxes were estimated for the sum of 15 compounds (Σ15) including acenaphthylene [ACY], acenaphthene [ACE], fluorene [FLU], phenanthrene [PHE], anthracene [ANT], fluoranthene [FLH], pyrene [PYR], chrysene [CHR], benzo(a)anthracene [BaA], benzo(b)fluoranthene [BbF], benzo(k)fluoranthene [BkF], benzo(a)pyrene [BaP], indeno(c,d) pyrene [IcdP], dibenzo(a,h)anthracene [DahA] and benzo(g,h,i)perylene [BghiP]. When data for some compounds were lacking, fluxes were quantified for the sum of the remaining compounds, and the missing compounds are specified in the text.

Data were systematically tested for spatial or temporal trends and for their relationship with external factors such as meteorological parameters. Whenever a parameter had a significant influence on the PAH contents, the flux calculation was improved by integrating this parameter into the calculation. Statistical investigations were made using R (ver3.4.4; R Core Team, 2016) software with the RStudio interface (v1.1.453). Except for the fluxes estimated from economic data, the calculations were carried out by multiplying the flux of the matrix considered (e.g. runoff) by randomly drawn PAH content values in the appropriate database. The estimation was made such that the number of draws was equal to the number of samples in the database and draws were made with replacements. Thus, each sample had the same probability of occurrence. To ensure a representative value of the flux and to quantify the uncertainties, the calculation was repeated 10,000 times. Then the flux value and its associated mathematical uncertainty were estimated as the mean value and the standard deviation over the 10,000 computations after having ensured that the calculations converged.

2.4 Flux Estimation Quality

The mathematical uncertainties estimated as previously described failed to reflect the quality of the database used for the flux calculation. To express this quality, grades were attributed according to five parameters including the database size, the number of studies, the sampling date in comparison with 2014, the spatial and temporal representativeness of the samples and the quality of the parameters that were used in the estimations (e.g. runoff). Each parameter was rated from 0 to 4 (low to high quality). The global quality grade was estimated as the lowest rate for individual parameters. The fluxes based on a numeric model or economic data were awarded a quality grade of 0 because no actual measurement could back up the results.

3 Results

3.1 Urban Fluxes

All urban fluxes are summarised in Fig. 2.

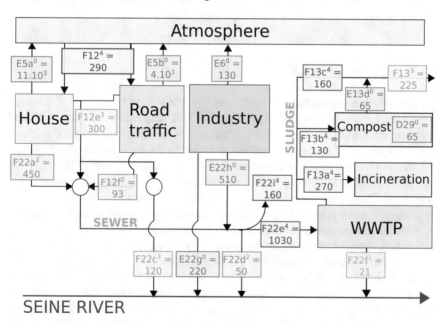

Fig. 2 PAH urban fluxes (kg year^{-1}) in the Paris conurbation. F flux calculation based on actual PAH content measurement in the area studied, E based on economic data or pollutant quantification in a similar environment, D calculated from the linear combination of other fluxes. Quality grade in superscript from 0 (worst, in red) to 4 (best, deep blue). *E5a* house heating, *E5b* road traffic, *E6* industrial emissions, *F12* atmospheric deposition, *F12e* runoff, *F12f* street cleaning, *F13* sludge spreading, *F13a* sludge incineration, *F13b* composted sludge, *E13d* compost remains, *F22a* domestic wastewater, *F22c* untreated runoff, *F22d* untreated wastewater, *F22e* WWTP inflow, *F22f* WWTP effluent, *E22g* industrial releases to the river, *E22h* industrial discharge to the sewer system, *F22i* sewer deposit removal, *D29* compost degradation

3.1.1 Emissions to the Atmosphere

Urban and industrial PAH emissions to the atmosphere were estimated based on the INERIS (French National Institute for Environmental Technology and Hazards) database. A recurrent spatial inventory of atmospheric pollutant emissions including PAHs has been made since 2003 by the INERIS. It includes emissions from all reported sources, whether anthropogenic or natural. Since source types were specified, the PAH emissions related to household heating, road traffic and industrial activities were estimated independently. Thus, household heating and road traffic, respectively, amounted to E5a = 11 tons year^{-1} ($\Sigma16$, naphthalene included) and E5b = 4 tons year^{-1} ($\Sigma16$) of PAHs, leading to total urban emissions of E5 = 15 tons year^{-1} ($\Sigma16$). At the same time, industries emitted E6 = 130 kg year^{-1} ($\Sigma16$). As for all fluxes that were not supported by environmental data, the quality grade was set at 0. In particular, although the source inventory was quite exhaustive, the emission coefficients used to quantify the emissions could hardly be confirmed. Therefore, the value of the estimated flux is provided as a rough estimate. However, the order of magnitude was in agreement with a study reported by the French Center for Atmospheric Pollution Study (CITEPA) that estimated the emissions of eight PAHs (FLH, BaA, BaP, BbF, BkF, BghiP, IcdP and DahA) at 72 tons year^{-1} for all of metropolitan France (a sevenfold higher population).

3.1.2 Atmospheric Deposition

To quantify the atmospheric fallout over the Paris conurbation, PAH contents were taken from four studies gathering 95 bulk deposition samples collected between 2002 and 2014 [6, 20–22]. The seasonality in the atmospheric fallout was instigated, and the concentrations of PAHs in the bulk deposition displayed significant monthly changes (Kruskal-Wallis test, p-value < 0.05). The rise was related to household heating during winter leading to significantly higher PAH emissions. The annual deposition was consequently estimated as the sum over the cold and the warm periods of the rainwater volumes over 4-day periods multiplied by random concentration values drawn in the corresponding database. The 4-day period was chosen to ensure a drawn probability for each sample of approximately one out of the number of samples in the database ($n = 95$). Following this method, the atmospheric deposition totalled F12 = 290 ± 70 kg year^{-1} ($\Sigma15$). This fallout flux (160 g km^{-2} year^{-1}) was lower than previously reported values within the range 208–234 g km^{-2} year^{-1} [12, 23] measured over urban areas across the Seine River basin from 1999 to 2002. These results suggest a decrease in the atmospheric deposition over the past few decades, in agreement with the overall reduction of about 44% in PAH emissions reported by the CITEPA between 2000 and 2014. Because the database gathered a large number of samples representative in time and space of the PAH atmospheric deposition in the Seine River basin, the flux estimation was awarded the quality grade of 4.

3.1.3 Runoff

The PAH flux related to runoff on roads and roofs (F12e) was quantified considering the volume of runoff collected in the sewer system. To estimate this flux, the PAH concentrations were gathered from four studies depicting 31 samples collected in separate sewer systems of urban areas [10, 24–26]. The samples covered an extended area across the Paris conurbation, providing a good spatial representativeness of the database. Moreover, they were in good agreement with previously reported values for similar urban or suburban areas [27, 28]. However, the temporal representativeness was not as good. Indeed, samples were collected at various times over the year, but the relationship between PAH concentrations and the characteristics of the rain event considered (rainfall intensity or duration of the previous dry periods) could not be investigated. Consequently, this flux was estimated as the sum over the year of 12-day periods of runoff volumes multiplied by PAH concentrations randomly drawn within the runoff database. The flux totalled F12e $= 300 \pm 40$ kg year^{-1} (Σ15). This result was in good agreement with fluxes reported for other French urban areas [29]. However, the quality grade of this flux was reduced to 3 because of the lack of information on the temporal changes in the runoff contamination.

In addition, the PAH flux during the street cleaning process (F12f) was quantified. According to the SIAAP, the amount of water used for street cleaning reached 59 Mm^3 year^{-1}. A previous study showed that the PAH concentration in water samples ranged from 0.1 to 2.2 μg L^{-1} (Σ13) [30]. This study used data from 21 samples collected at various places within the Paris city limits. Although the spatial database was spatially representative, all samples were collected from May to June 2002, and neither temporal changes nor the influence of meteorological parameters could be investigated. Consequently, and although the order of magnitude of the contamination was supported by the concentrations reported in runoff samples, the quality rate was downgraded to 2. The flux estimated in this way was F12f $= 93 \pm 3$ kg year^{-1} (Σ16, naphthalene included), and it amounted to 32% of the atmospheric deposition and 29% of the PAH runoff from roads and roofs. However, the influence of meteorological parameters could not be tested, and a redundancy between F12e (street runoff) and F12f (street cleaning) cannot be excluded because both remobilise pollutants from roads.

3.1.4 Domestic and Industrial Wastewater

The flux related to domestic wastewater (F22a) was quantified using 35 measurements either in domestic effluents or in combined sewers during dry weather periods [30, 31]. Specific attention was paid to using only samples collected close to domestic sources to avoid dilution by infiltration waters. Using data from combined sewers provided a better spatial representativeness, especially within the city of Paris itself, where there were exclusively combined sewers. However, this might have led to a flux overestimation because part of the pollution within the combined sewer is due to the remobilisation of runoff contaminants deposited during wet weather. Consequently, the quality rate was downgraded to 2. The flux was estimated by summing the daily volume of domestic wastewater over the year multiplied by a concentration value randomly drawn within the database; this led to an averaged value of F22a = 450 ± 50 kg year^{-1} (Σ15). However, the measurement of the PAH concentrations in domestic wastewater alone collected in house disposals would be necessary to better assess this flux.

According to the SIAAP, industries discharged a volume of 25 Mm^3 year^{-1}. Given that no data on PAH concentration in industrial wastewater in the Seine River basin was available, the flux was estimated using data from the literature [32] because it was reported to be similar to that measured in the Paris area [33]. The PAH flux released by industries into the sewer system was E22h = 510 ± 220 kg year^{-1} (Σ15). However, this result could not be backed up by actual measurements in the Paris conurbation, and the quality rate was downgraded to 0. In addition, 30% of industrial wastewater was released directly into the Seine River without any preliminary treatment [2], leading to a direct discharge of PAHs into the Seine River of E22g = 220 ± 90 kg year^{-1} (Σ15).

3.1.5 Discharges from the Sewer System

During the past few years, the chemical quality of storm water in the Paris area has been extensively investigated, especially within the OPUR research programme [10, 26]. The resulting database was described in "Runoff" above, leading to the same quality grade (3). The volume of runoff water collected in the separate sewer system was 95 Mm^3 for 2014. Based on these data, the flux of PAHs directly discharged into the Seine River by storm water was F22c = 120 ± 12 kg year^{-1} (Σ15).

PAH input into the river system also occurred from combined sewers during storms or during sewer system dysfunction. In 2014, 20.5 Mm^3 of combined sewer overflows (wet weather), and 18.2 Mm^3 of untreated wastewater (dry weather) were discharged. To quantify the flux of PAHs related to these overflow events, concentrations of PAHs in samples collected in combined sewers were matched to the corresponding water volumes depending on the weather conditions. The databases were constituted of 22 samples from three different studies [24, 34, 35] for dry weather conditions and 99 samples from five studies [16, 24, 30, 36, 37] for wet weather conditions. A total flux of F22d = 50 ± 4 kg year^{-1} (Σ15) was estimated, nearly 80% being released during wet weather conditions. The database used to quantify this flux summarised data from various sites and was therefore spatially representative. However, the lack of information on intra-event variability in PAH concentrations resulted in a quality grade of 2.

The last PAH flux related to the sewer system was the removal of highly contaminated sewer deposits during pipe cleaning. In 2014, 7,170 tons of sediment was removed from the sewer system. The contamination of sewer sediment has been documented in four studies [31, 38–40] and by the Paris Sanitation Department between 2000 and 2014. The corresponding database gathered PAH measurements from 406 samples, thereby ensuring good spatial and temporal representativeness. Consequently, the flux estimation was awarded a quality grade of 4. The PAH flux related to sediment cleaning in the sewer system was $F22i = 160 \pm 10$ kg year^{-1} ($\Sigma15$). Since sewer sediment is not suitable for amending agricultural land, it was stored as hazardous material.

3.1.6 Wastewater Treatment Plants

To quantify the total flux of PAHs entering WWTPs (F22e), the period investigated was divided between dry weather periods (208 days in 2014, average inflow, 2.2 Mm^3 day^{-1}) and wet weather periods (157 days, inflow, 2.8 Mm^3 day^{-1}). The PAH input was estimated as the daily collected water volumes from each sewer type multiplied by corresponding PAH concentrations measured in separate or combined sewers. For this estimation, care was taken to use data from samples collected at downstream parts of the sewer system. It was assumed that these samples were representative of the mix of waters originating from different sources including infiltration water. Consequently, entirely distinct databases were used to compute the F22a, E22h, F12e and F22e fluxes. The data of 121 samples from nine studies, including wastewater from the four main WWTPs, were used. The annual load of PAHs was $F22e = 1{,}030 \pm 50$ kg year^{-1} ($\Sigma15$). The order of magnitude was also consistent with a previous study that reported an annual flux of about 476 kg year^{-1} for six compounds (FLH, BaP, BbF, BkF, BghiP and IcdP) [36]. Consequently, this flux estimation was awarded a quality grade of 4.

Once they have entered WWTPs, PAHs may be degraded during the treatment process, removed from the effluent through sorption onto sludge or discharged into the Seine River at the outlet of the WWTP. To quantify the amount of PAHs removed by the sludge removal process, the daily mass of sludge was multiplied by the random PAH contents measured in this matrix. The database comprised data from a previous study [36] or was provided by the SIAAP. It gathered data for 16 samples collected at three WWTPs treating 74% of the Paris conurbation population. Thus, the annual PAH amount removed from the WWTP was 560 ± 40 kg year^{-1} ($\Sigma13$, ACE and ACY excluded). The largest part was reduced by the sludge thermal process ($F13a = 270 \pm 20$ kg year^{-1}, $\Sigma13$), the rest being either composted ($F13b = 130 \pm 10$ kg year^{-1}, $\Sigma13$) or spread over agricultural land without further treatment ($F13c = 160 \pm 10$ kg year^{-1}, $\Sigma13$). Considering that PAH content in sludge has been steady over the past few years [15], good spatial and temporal representativeness was expected, and the estimations of sludge-related fluxes were awarded a quality grade of 4. No data for PAHs in composted sludge was available (E13d). Therefore, the PAH degradation rates during the composting process reported elsewhere for French sludge [41] were used to quantify the flux of the remaining PAHs (E13d = 65 kg year^{-1}, Σ13) after the composting process. Given that the composted sludge was also spread, the total flux of PAHs towards agricultural lands totalled F13 = F13c + E13d = 225 kg year^{-1} (Σ13).

Finally, the annual load of PAHs released to the Seine River by the WWTPs was estimated based on PAH concentrations measured in the effluent and considering the annual volume of treated water discharged into the river. Although PAH concentrations in WWTP effluent have been monitored for many years, only six samples collected at two WWTP outlets [34] were available. When compared to previous studies elsewhere in Europe [42–44], significant differences were observed (Wilcoxon-Mann-Whitney test, $p < 0.05$) for similar wastewater treatment processes. This result may be due to the high efficiency in the suspended matter removal at one of the WWTPs (Seine-Centre). Based on data from this study, the PAH flux related to WWTP effluent discharge reached F22f = 21 ± 4.1 kg year^{-1} (Σ15). Because of the lack of data and the differences that were observed between these data and the literature, the quality rate of this estimation was downgraded to 1.

3.2 Rural Fluxes

All the rural fluxes are summarised in Fig. 3.

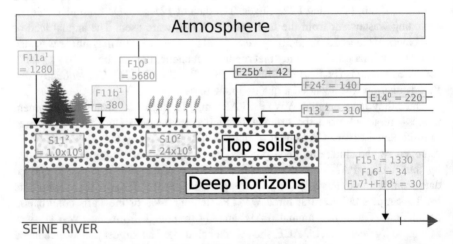

Fig. 3 PAH rural fluxes (kg year^{-1}) in the Seine River basin. *F* flux calculation based on actual PAH content measurements in the Seine River basin, *E* based on economic data or pollutant quantification in similar environment, $F_\#$ *or* $E_\#$ flux based on the estimation at the Paris conurbation scale. Quality grade in superscript from 0 (worst, in red) to 4 (best, deep blue). *F10 and F11a* atmospheric deposit over agricultural and forests, *F11b* forest filter effect, *F13 and F25b* spreading of urban sludge and sediment from river dredging, *E14* road runoff, *F15, F16, F17 and F18* erosion from agricultural, forested, urban and industrial lands, respectively, *F24* flood deposits, *S10 and S11* stocks (kg) in agricultural and forested areas

3.2.1 Atmospheric Deposition

To quantify the PAH atmospheric deposition on rural areas, the investigated area was divided by discriminating between agricultural and forested land. Indeed, specific deposition processes over vegetated areas have been previously reported due to the forest filter effect [19, 45, 46], thereby justifying different approaches. To quantify the fallout flux over agricultural areas, a large database gathering 92 samples from two studies was available [11, 47]. A significant difference was measured between the warm and cold periods (Wilcoxon-Mann-Whitney test, $p < 0.0001$). This temporal variation was included in the flux estimation, as for F12. Thus, the atmospheric deposit over agricultural land reached F10 = 5,680 $\pm$ 860 kg year^{-1} (Σ15). Because of the representative database, this estimation was awarded a grade of 3. To estimate the atmospheric deposition over forests, only the data of five samples from one study were available [47]. The same calculation method as for other fallout fluxes (F12 and F10) was used, although not enough data were available to distinguish between the warm and cold period fallouts. Thus, the atmospheric deposition over forested areas reached 1,280 $\pm$ 240 kg year^{-1} (Σ15). Due to the number of samples, the spatial and temporal representativeness was limited, however. Consequently, the quality rate of this flux estimation was downgraded to 1. The forest filter effect was quantified based on the following values reported in the literature: average litter production of 390 $\pm$ 50 g m^{-2} year^{-1} [45, 48] and average leaf contaminations of 30 $\pm$ 20 ng g^{-1} in remote forests [49, 50]. In addition, data from a previous study gathering 77 samples collected within the Seine River basin in forests directly exposed to PAH anthropogenic sources, including roads, were used [51]. Thus, the annual flux related to defoliation was F11b = 380 $\pm$ 130 kg year^{-1} (Σ15). Since the remote forest contamination could not be supported by actual measurement, the flux quantification was awarded a grade of 1.

3.2.2 Road Runoff Outside of Urban Areas

An additional runoff flux specific to the road network outside of urban areas was estimated. The surface concerned was estimated from the IGN database based on the road length and number of lanes provided. The road surface totalled 236 km^2 in the investigated area, and a runoff coefficient of 0.9 was assumed. Since no data were available for rural road runoff, this volume was matched to the database of PAH concentrations in urban runoff water ($n = 31$), and the flux totalled E14 = 220 $\pm$ 30 kg year^{-1}. This flux reached E14$_\#$ = 310 $\pm$ 50 kg year^{-1} when considering runoff flux in urban areas ending up in permeable soils. However, this database was representative of a rural environment, and this quantification was awarded a grade of 0.

3.2.3 Stocks in Soils

Due to their lipophilic properties, PAHs are mainly stored at the surface layer of soils [52]; consequently, the contamination depth depends on land use. For undisturbed soils such as forests, previous studies have shown that PAHs were mainly concentrated within the 8–10 cm of the topmost layer of soil [7, 53]. In agricultural areas, regular ploughing usually resulted in homogenisation of the physico-chemical properties of the surface layer [54], and the PAH content was assumed to be constant through the ploughed layer. In the Seine River basin, a 25-cm-deep contaminated layer was assumed. Therefore, the masses of contaminated soils were estimated as the product of the surface considered by the contamination depth by an average dry density of 1,350 kg m^{-3} [7]. Concerning the stocks in the urban and industrial areas, only the permeable fraction of the surface was expected to accumulate PAHs and was therefore taken into account. Since the thickness of the contaminated layer varied greatly depending on the site's history, an average value of 10 cm was used, as for undisturbed areas. Consequently, the estimation only constituted a lower limit of the PAH stocks in urban areas. Contaminant stocks were estimated by multiplying the masses of soils by random values of PAH contents drawn among 130 samples from two studies [7, 55]. A significant difference was measured between the soil PAH contents depending on their land use (Wilcoxon-Mann-Whitney test, $p < 0.0001$), especially between forest and agricultural samples and urban and industrial samples. The samples collected in forested and agricultural areas underwent an additional procedure to discriminate between the samples depending on their distance to the road network with a 150-m threshold. Indeed, previous studies reported specific traffic-related contamination within this range [7, 56]. The stocks on both sides were quantified independently. Thus, 22% of agricultural land and 13% of forest area were situated within the critical distance to the road network. The PAH stocks in agricultural lands reached S10 = 24 ± 17 × 10^6 kg ($n = 61$, Σ15), equally disseminated between the road vicinity and the more remote areas. High relative uncertainty due to the heterogeneity of the PAH content database for agricultural lands was observed. In forested areas, the PAH stock amounted to S11 = 1.0 ± 0.3 × 10^6 kg ($n = 26$) with a smaller portion (12%) located close to roads. For urban and industrial areas, the PAH amount reached S12 = 1.8 ± 0.6 × 10^6 kg ($n = 45$). Because further investigation on the relationship between the road traffic and the additional PAH stocks in the vicinity would be necessary to correctly quantify the pollutant amounts in the areas concerned, the stock estimation was awarded a grade of 2. In addition, the flux of PAHs spread with urban sludge was estimated based on the population, leading to a total amount of F13$_{\#}$ = 310 kg year^{-1} (Σ13).

3.2.4 Soil Erosion

The flux of PAH transferred from topsoils to the river was estimated by matching the land use map, the average PAH contents and erosion rates estimated across the area [18]. Thus, the PAH erosion fluxes reached F15 = 1,330 ± 1,010 kg year^{-1} (Σ15) for agricultural lands, F16 = 34 ± 10 kg year^{-1} (Σ15) for forests and F17 + F18 = 30 ± 9 kg year^{-1} (Σ15) for urban and industrial areas. Consequently, the overall erosion flux from soils was about 1,400 ± 1,030 kg year^{-1} (Σ15). As for the soil stock estimation, a high mathematical uncertainty due to the heterogeneity of the PAH content database for agricultural lands was observed. Moreover, this erosion flux was estimated from PAH content in soils instead of eroded particles because no other data were available. This method did not consider a potential enrichment process during erosion due to particle sorting. Yet, previous laboratory experiments reported PAH-enriched eroded particles by a factor up to 3 [57, 58]. However, this enrichment process has never been investigated at the catchment scale, and it could not be taken into account in this estimation. Consequently, this flux estimation constitutes a rough estimation, and it was awarded a quality grade of 1. In addition, the PAH transfer to the deep soil horizons and to groundwater was not quantified. However, because of the PAH retention in topsoil horizon [52, 53], vertical flux of PAHs within the soils was ignored.

3.2.5 Sediment Dredging

To estimate the fluxes related to sediment dredging in the main stem of the Seine River basin, the mass of collected material was multiplied by values of the PAH contamination of sediment. The data of 95 samples were provided by the French department of waterways, which is responsible for supervising dredging in the Seine River basin. The flux of PAHs removed from the stems with the dredged material amounted to F25a = 660 ± 70 kg year^{-1} with 6% (F25b = 40 ± 4 kg year^{-1}) of it being spread over agricultural areas. Since the volume of sediment, the percentage of dry matter, the PAH content in the sediment and the purpose of the dredge material were known for each operation, the flux estimation was awarded a grade of 4.

3.2.6 Storage in Reservoirs and Floodplains

To quantify the amount of PAHs (Fig. 4) annually stored in reservoirs (F23a), on river beds (F23b) and over floodplains (F24), the sedimentation rates were multiplied by the PAH content in suspended sediment (SS) randomly drawn in the database. As the floodplain and the reservoirs were located upstream of the Paris conurbation, the database was limited to SS samples collected in the uppermost parts of the Seine River basin. Finally, 104 samples from four studies were available [9, 11, 59, 60]. The annual flux of PAHs from the rivers towards the reservoirs and

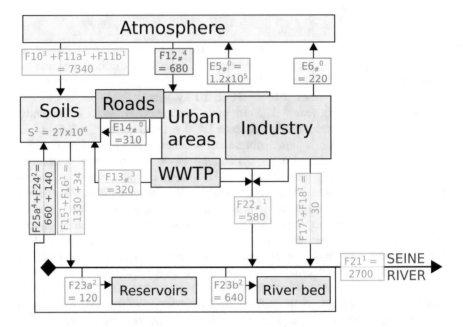

Fig. 4 PAH fluxes at the Seine River basin scale (kg year^{-1}). F flux calculation based on actual PAH content measurement in Paris conurbation, E based on economic data or pollutant quantification in a similar environment, $F_{\#}$ or $E_{\#}$ flux based on the estimation at the Paris conurbation scale. Quality grade in superscript from 0 (worst, in red) to 4 (best, deep blue). *E5 and E6* domestic and industrial emissions, *F10, F11a and F12* atmospheric deposit over agricultural, forested and urban lands, *E11b* forest filter effect, *S* stocks (kg) in soils, *F13* spreading of urban sludge, *E14* road runoff outside of urban areas, *F15, F16, F17 and F18* erosion from agricultural, forested, urban and industrial lands, *F22* releases from the sewer system and the WWTP to the Seine River, *F23a and F23b* storage in reservoirs and on the river bed, *F24* flood deposits, *F25* river dredging

the floodplains totalled F23a $= 120 \pm 10$ kg year^{-1} and F24 $= 140 \pm 11$ kg year^{-1} ($\Sigma15$), respectively. The entire database of PAH contents in sediment ($n = 183$) was used to quantify the storage on river beds, although a greater weight was given to samples collected upstream of the Paris conurbation because 92% of the waterbody area was located upstream of the urban areas. The estimated amount of PAHs stored on the river bed was F23b $= 640 \pm 50$ kg year^{-1} ($\Sigma15$). The flux estimations were marred by uncertainty due to the difficulty properly quantifying the amount of sediment being deposited every year. However, the deposition rates were consistent with the amount of material delivered to the Seine River through erosion and the SS flux at the study area outlet. Moreover, the database was complete enough to ensure the good spatial and temporal representativeness of the PAH contents in upstream SS. Therefore, the flux estimations were awarded a grade of 2.

3.2.7 Flux Transported by the Seine River

To investigate the PAH flux within the Seine River, calculations were based on concentrations measured independently in the particulate and the aqueous phases. The databases for the particulate and the dissolved fluxes of PAHs contained 15 samples from two studies [9, 60] and 11 samples from two studies [8, 60]. The annual PAH flux was estimated by attributing random values of the PAHs drawn from the databases to the monthly aqueous and solid fluxes. Finally, the total PAH flux reached F21 = 2,700 ± 550 kg year^{-1} (Σ15), the largest part (77%) of it being in the particulate phase. This result was consistent with a previous publication that reported an estimation of the PAH inputs to the Seine River mouth ranging from 2000 to 2,535 kg year^{-1} over the 2009–2013 period [61]. However, the limited number of samples did not allow for a correct investigation of the relationships between the PAH contents and the hydrological parameters, especially SS concentrations. Due to sampling difficulties, few data were available to quantify PAHs in the dissolved and particulate phases for the flood period downstream of the Paris conurbation. Contamination data measured during the low-flow period were therefore used for the flood period. Yet, previous studies have reported an increase in PAH contents in river water during floods [62] by a factor 2 for upstream sampling sites. This flux might therefore have been underestimated by about 30%. Considering the low number of samples, this flux estimation was awarded a grade of 1.

3.2.8 Estimation of Urban Fluxes at the Basin Scale

PAH atmospheric emissions (Fig. 4) were quantified based on source spatial inventory established by INERIS, as previously described. The emissions from domestic and industrial sources were estimated at approximately E5$_{\#}$ = 122 tons year^{-1} and E6$_{\#}$ = 220 kg year^{-1}, respectively. The atmospheric deposition over urban areas was estimated considering the whole urban surface (4,020 km^2) and by integrating complementary data collected at other urban areas [11] into the bulk deposition database ($n = 115$). No significant change was measured in the concentrations between the four sampling sites (Kruskal-Wallis test, $p = 0.27$). The total fallout over urban areas totalled F12$_{\#}$ = 680 ± 153 kg year^{-1} (Σ15).

Scarce data were available to quantify other urban fluxes at the basin scale. In particular, only a few studies investigated the PAH concentrations in the sewer systems outside of the Paris conurbation [12, 63]. However, the concentrations measured in the Paris conurbation were consistent with data reported elsewhere in France and Europe [32]. Since these fluxes were estimated based on the average wastewater produced per capita, it was assumed that the calculations made at the Paris conurbation scale could be used to estimate the urban fluxes at the entire basin scale by taking into account the population considered at the different spatial scales. This approach was reasonable for most urban fluxes, except for the runoff contribution because changes in the population density led to a lower runoff coefficient but also to greater vehicle use. Accordingly, it was assumed that the urban runoff flux at the entire basin scale could be estimated proportionally to the total population but that the runoff specifically related to the road runoff outside of urban areas should be quantified independently (detailed below). Finally, at the basin scale investigated, the urban releases were $F22c_{\#} = 170 \pm 20$ kg year^{-1} ($\Sigma 15$) and $F22d_{\#} = 70 \pm 10$ kg year^{-1} ($\Sigma 15$) for separate and combined sewer discharge, respectively. The direct discharges of PAHs from industries to rivers were estimated at $E22g_{\#} = 310 \pm 90$ kg year^{-1} ($\Sigma 15$). Finally, the PAH flux within the WWTP effluents reached $F22f_{\#} = 30 \pm 6$ kg year^{-1} ($\Sigma 15$). Thus, the amount of PAHs released from the urban areas to the river system totalled $F22_{\#} = 580 \pm 140$ kg year^{-1} ($\Sigma 15$).

4   Discussion

The estimated fluxes at both area scales investigated are summarised in Fig. 4.

The PAH flux to the atmosphere was dominated by household heating and road traffic (E5). The atmospheric deposition estimated at the Paris conurbation scale (F12) reached 2% of the emission quantification based on economic data (E5 + E6). At the entire basin scale, the atmospheric fallout (F10 + F11a + F11b + F12#) reached 8 tons year^{-1} and accounted for only 6.5% of the emissions. This disparity between emissions and atmospheric deposition was unexpected, and it may stem from the difference in the calculation methods, where the estimation of emissions from economic data required the use of emission factors that may be poorly defined, and the measurement of atmospheric deposition may be biased due to underestimation of the gaseous exchanges [64]. In addition, environmental processes such as long-range transportation outside of the investigated area or photo-oxidation resulting in PAH degradation in the atmosphere [65, 66] could also partly explain that the deposition flux is far lower than the emissions flux.

The flux related to urban runoff across the investigated area (F12e) was similar to atmospheric deposition. When considering only the impervious area, the fallout flux only accounted for 22% of the estimated runoff flux, thereby suggesting that remobilisation of pollutants deposited on roofs and roads was the main process of runoff contamination. This result was consistent with previously published work [10] depicting runoff flux as four times higher than deposition flux for a small residential catchment.

The mass balance of PAHs at the Paris conurbation sewer system scale was very consistent. Indeed, the sum of incoming PAH fluxes (F22a + F12e + F12f + E22h) to the sewer system was $1,350 \pm 310$ kg year^{-1}, while the sewer outflows (F22c + F22d + F22i + F22e) amounted to $1,360 \pm 20$ kg year^{-1} (Fig. 2). This approach was made possible by the SIAAP's thorough knowledge of the collected and treated wastewater volumes across the Paris conurbation. The result showed that the three main PAH sources, i.e. the domestic and industrial wastewater and the runoff on impervious surfaces, contributed equally to the PAH supplies to the sewer system. In the mass balance analysis for urban areas, the fluxes of PAHs released by industries constituted the main uncertainty. To refine these results, further investigations on the PAH concentrations in industrial effluents are required.

A specific investigation on the WWTP incoming flux (F22e) showed that separate sewers only accounted for 9% of annual PAH load to the WWTPs, while supplies from the combined sewer system reached 15% and 76% during dry and wet weather flows, respectively. It was assumed that the wet weather supply of PAHs was directly related to sewer sediment remobilisation rather than runoff [67].

When estimated at the WWTP scale, the fluxes related to sludge and effluents, respectively, amounted to 55% (Σ13) and 2% (Σ15) of the incoming flux, suggesting that approximately 43% of PAHs were degraded or volatilised during the treatment process. This proportion of PAHs in sludge was in agreement with previously published results [68], but the amount of PAHs released through effluent discharge was about 15 times lower. This result could be explained because the database was in part collected at the Seine-Centre WWTP characterised by high suspended matter removal efficiency, which may not be fully representative of the other WWTPs.

The overall discharge of PAHs into the Seine River system, including both WWTP effluents and overflows from the sewer system (F22c + F22d + F22f + E22g) within the Paris conurbation, reached 410 ± 110 kg year^{-1}. The main uncertainties on this flux estimation lay in data related to the characterisation of industrial and WWTP effluents. At the entire basin scale, the quality of this estimation was degraded because of the lack of information on the wastewater volumes and on the efficiency of the small-capacity WWTPs. This observation can be generalised for all the environmental fluxes since data were scarce for the rural part of the basin. A substantial study would be necessary to gather all the required information and properly estimate the flux at this scale.

Agricultural lands, forests and urban areas accumulated 72%, 19% and 9%, respectively, of the total atmospheric deposition. The forest filter effect (F11b) totalled 20% of the direct deposition flux over forest areas. In comparison, the PAH stocks in soils were distributed between agricultural lands (89%), urban areas (7%) and forests (4%). The low values of the inputs to the stock ratio in soils suggest that the PAH accumulation occurred over a long period of time and that biodegradation processes were not significant. Consequently, PAHs constitute a very persistent pollution within the River Seine basin.

The erosion-related flux was estimated at 1400 kg year^{-1}, but the relationship between the PAH contents in soil and in eroded particles could not be specifically investigated. A potential enrichment process occurring during erosion could not be quantified. Yet, a previous study has shown that the PAH content is much higher in suspended sediment than in the surrounding soils even in the most remote rural catchment [11]. The overall flux of PAHs stored within the river system (F23a + F23b) was estimated at 760 ± 70 kg year^{-1}, and it reached about 28% of the PAH flux carried downstream of the study area outlet (F21).

Overall, the soil erosion was the main source of PAHs in the Seine River, while urban release constituted a significant but smaller source. The PAH inputs to the Seine River (F15 + F16 + F17 + F18 + F22$_{\#}$) were about twofold lower than the PAH flux carried downstream of the study area outlet (F21). Several hypotheses can be considered to explain this gap. Firstly, PAH fluxes from the urban areas could have been underestimated. More specifically, the change in spatial scale constitutes a major source of uncertainty because most of the calculation parameters were poorly known at the entire basin scale. Secondly, and as previously described, an enrichment process that would occur during erosion could drastically increase the PAH flux from soil to rivers. Thus, an average enrichment by a factor 3 [57] would balance the PAH inputs and outputs. Finally, the mass balance could be right and the Seine River basin may currently be undergoing a decontamination phase involving the remobilisation of ancient and more contaminated sediment. This process can explain the short-term difference between pollutant inputs and outputs, as was already suggested for metals [69].

5 Conclusions and Perspectives

For the first time in the literature, a mass balance at a large spatial scale, i.e. the upper basin of the Seine River, was attempted for PAHs. This approach was enlightening in terms of both the environmental fate of PAHs and the priority research areas to complete these investigations.

Except for PAH emissions, all the estimated fluxes had consistent orders of magnitude. Although mistakes in the flux estimations could not be excluded, the difference between the amounts of PAHs emitted to the atmosphere and deposited from it suggests that physico-chemical processes still need to be investigated. In particular, the fate of PAHs in the atmosphere is poorly known. A number of studies have focused on the process responsible for chemical oxidation and photo-oxidation of atmospheric PAHs, but the phenomenon has never been quantified at a large spatial scale. Such an investigation would be challenging but would provide significant knowledge on a process that may drastically limit environmental PAH contamination. From a public perspective, it would be advantageous to quantify the deposition of the compounds arising from these degradation processes because they have been shown to constitute a health issue.

At the Paris conurbation scale, the sewer system inputs and outputs closely matched given the well-established wastewater and sludge mass balances. The result showed the importance of runoff and wastewater management to deal with PAH fluxes. In particular, the treatment processes implemented at the WWTPs have proven their efficiency in PAH removal, thereby drastically limiting the flux being released through WWTP effluents. However, two major sources of uncertainty remained. Firstly, the PAH fluxes related to industrial discharge to the sewers and rivers were poorly known. Since neither the wastewater volume nor the PAH concentrations were quantified, considerable work would be required to overcome this issue despite the current deindustrialisation of the basin. Because various industries are reported be responsible for different effluent contaminations, a specific inventory would be necessary. Secondly, more data on the PAH concentrations in the WWTP effluents would be necessary to validate or improve the quantification of the flux discharged to the Seine River. This investigation would even be crucial at the Seine River basin scale because small-capacity WWTPs vary greatly in efficiency. A similar consideration can be made for most of the urban fluxes because various inconsistencies appeared in the mass balance at the Seine River basin scale due to the lack of representative databases for such an extended area. Consequently, research is still required to precisely quantify the PAH flux in the urban parts of the basin other than the Paris conurbation. In particular, fluxes related to the sewer system and the WWTP are poorly known in less densely populated areas. This difficulty was highlighted by the road runoff estimation. To properly quantify this flux, one should investigate the relationship between the PAH concentrations in the runoff water, the runoff volume and the daily vehicle traffic.

In addition, certain fluxes could not be estimated based on the compiled databases, especially the environmental degradation fluxes. However, according to our PAH budgets, only degradation within the atmosphere would be significant. Indeed, the long-term contaminations of the soils which constituted the main PAH reservoir suggest that no significant degradation occurs in this compartment. This result could be reinforced by long-term monitoring of soils. Similarly, PAH volatilisation from soils was not quantified. More generally, the gaseous exchanges at the ground-plant-atmosphere interface required further investigation.

Finally, there is a serious need to investigate the enrichment process to refine the PAH flux related to soil erosion at the Seine River basin scale. Indeed, this parameter plays a crucial role in determining if the Seine River basin is currently balanced in terms of PAH input and outputs or if it is currently undergoing a decontamination phase. Contaminated sediment stored in the river bed and in reservoirs is likely to be remobilised during massive flood events such as the one that occurred in June 2016 [69]. A long-term survey of the Seine River would be required to precisely quantify the average annual fluxes in the Seine system. In particular, closer monitoring of flood events would drastically improve the accuracy of the estimation of the PAH flux being carried by the Seine River downstream of Paris (F21).

Acknowledgements This work was conducted in the framework of the PIREN-Seine research programme (www.piren-seine.fr), a component of the Zone Atelier Seine within the international Long Term Socio-Ecological Research (LTSER) network.

References

1. Meybeck M, Lestel L, Bonté P et al (2007) Historical perspective of heavy metals contamination (Cd, Cr, Cu, Hg, Pb, Zn) in the Seine River basin (France) following a DPSIR approach (1950–2005). Sci Total Environ 375:204–231
2. Thévenot DR, Moilleron R, Lestel L et al (2007) Critical budget of metal sources and pathways in the Seine River basin (1994–2003) for Cd, Cr, Cu, Hg, Ni, Pb and Zn. Sci Total Environ 375:180–203

3. Mastrangelo G, Fadda E, Marzia V (1996) Polycyclic aromatic hydrocarbons and cancer in man. Environ Health Perspect 3:1166–1170
4. Bouloubassi I, Saliot A (1991) Composition and sources of dissolved and particulate Pah in surface waters from the Rhone Delta (nw Mediterranean). Mar Pollut Bull 22:588–594
5. Ravindra K, Sokhi R, Vangrieken R (2008) Atmospheric polycyclic aromatic hydrocarbons: source attribution, emission factors and regulation. Atmos Environ 42:2895–2921
6. Moreau-Guigon E, Alliot F, Gaspéri J et al (2016) Seasonal fate and gas/particle partitioning of semi-volatile organic compounds in indoor and outdoor air. Atmos Environ 147:423–433
7. Gateuille D, Evrard O, Lefevre I et al (2014) Combining measurements and modelling to quantify the contribution of atmospheric fallout, local industry and road traffic to PAH stocks in contrasting catchments. Environ Pollut 189:152–160
8. Tusseau-Vuillemin M-H, Gourlay C, Lorgeoux C et al (2007) Dissolved and bioavailable contaminants in the Seine River basin. Sci Total Environ 375:244–256
9. Ollivon D, Garban B, Chesterikoff A (1995) Analysis of the distribution of some polycyclic aromatic hydrocarbons in sediments and suspended matter in the river Seine (France). Water Air Soil Pollut 81:135–152
10. Bressy A, Gromaire M-C, Lorgeoux C et al (2012) Towards the determination of an optimal scale for stormwater quality management: micropollutants in a small residential catchment. Water Res 46:6799–6810
11. Gateuille D, Evrard O, Lefevre I et al (2014) Mass balance and decontamination times of Polycyclic Aromatic Hydrocarbons in rural nested catchments of an early industrialized region (Seine River basin, France). Sci Total Environ 470–471:608–617
12. Motelay-Massei A, Ollivon D, Garban B et al (2007) Fluxes of polycyclic aromatic hydrocarbons in the Seine estuary, France: mass balance and role of atmospheric deposition. Hydrobiologia 588:145–157
13. Rodenburg LA, Valle SN, Panero MA et al (2010) Mass balances on selected polycyclic aromatic hydrocarbons in the New York–New Jersey Harbor. J Environ Qual 39:642–653
14. Froger C, Quantin C, Gasperi J et al (2019) Impact of urban pressure on the spatial and temporal dynamics of PAH fluxes in an urban tributary of the Seine River (France). Chemosphere 219:1002–1013
15. Mailler R (2015) Devenir des micropolluants prioritaires et émergents dans les filieres conventionnelles de traitement des eaux résiduaires urbaines des grosses collectivités (filières eau et boues), et au cours du traitement tertiaire au charbon actif. Université Paris-Est, Champs-sur-Marne. https://hal-enpc.archives-ouvertes.fr/tel-01226483/
16. Kafi M, Gasperi J, Moilleron R et al (2008) Spatial variability of the characteristics of combined wet weather pollutant loads in Paris. Water Res 42:539–549
17. Delmas M, Cerdan O, Mouchel J-M, Garcin M (2009) A method for developing a large-scale sediment yield index for European river basins. J Soils Sediments 9:613–626
18. Delmas M, Cerdan O, Cheviron B et al (2012) Sediment export from French rivers to the sea. Earth Surf Process Landf 37:754–762
19. Terzaghi E, Wild E, Zacchello G et al (2013) Forest filter effect: role of leaves in capturing/releasing air particulate matter and its associated PAHs. Atmos Environ 74:378–384
20. Ollivon D, Garban B, Blanchard M et al (2002) Vertical distribution and fate of trace metals and persistent organic pollutants in sediments of the Seine and Marne rivers (France). Water Air Soil Pollut 134:57–79
21. Blanchard M, Teil M-J, Guigon E et al (2007) Persistent toxic substance inputs to the river Seine basin (France) via atmospheric deposition and urban sludge application. Sci Total Environ 375:232–243
22. Teil M-J, Moreau-Guigon E, Blanchard M et al (2016) Endocrine disrupting compounds in gaseous and particulate outdoor air phases according to environmental factors. Chemosphere 146:94–104
23. Garban B, Blanchoud H, Motelay-Massei A et al (2002) Atmospheric bulk deposition of PAHs onto France: trends from urban to remote sites. Atmos Environ 36:5395–5403

24. Gasperi J, Sebastian C, Ruban V et al (2014) Micropollutants in urban stormwater: occurrence, concentrations, and atmospheric contributions for a wide range of contaminants in three French catchments. Environ Sci Pollut Res 21:5267–5281
25. Zgheib S, Moilleron R, Chebbo G (2011) Influence of the land use pattern on the concentrations and fluxes of priority pollutants in urban stormwater. Water Sci Technol 64:1450–1458
26. Zgheib S, Moilleron R, Chebbo G (2012) Priority pollutants in urban stormwater: part 1 – case of separate storm sewers. Water Res 46:6683–6692
27. Ngabe B, Bidleman TF, Scott GI (2000) Polycyclic aromatic hydrocarbons in storm runoff from urban and coastal South Carolina. Sci Total Environ 255:1–9
28. Menzie CA, Hoeppner SS, Cura JJ et al (2002) Urban and suburban storm water runoff as a source of polycyclic aromatic hydrocarbons (PAHs) to Massachusetts estuarine and coastal environments. Estuaries 25:165–176
29. Hannouche A, Chebbo G, Joannis C et al (2017) Stochastic evaluation of annual micropollutant loads and their uncertainties in separate storm sewers. Environ Sci Pollut Res 24:28205–28219
30. Gasperi J, Rocher V, Moilleron R, Chebbo G (2005) Hydrocarbon loads from street cleaning practices: comparison with dry and wet weather flows in a parisian combined sewer system. Polycycl Aromat Compd 25:169–181
31. Gasperi J (2006) Introduction et transferts des hydrocarbures à différentes échelles spatiales dans le réseau d'assainissement parisien. Ecole Nationale des Ponts et Chaussées, Champs-sur-Marne. https://pastel.archives-ouvertes.fr/pastel-00002103
32. Sánchez-Avila J, Bonet J, Velasco G, Lacorte S (2009) Determination and occurrence of phthalates, alkylphenols, bisphenol A, PBDEs, PCBs and PAHs in an industrial sewage grid discharging to a Municipal Wastewater Treatment Plant. Sci Total Environ 407:4157–4167
33. Blanchard M, Teil MJ, Ollivon D et al (2004) Polycyclic aromatic hydrocarbons and polychlorobiphenyls in wastewaters and sewage sludges from the Paris area (France). Environ Res 95:184–197
34. Mailler R, Gasperi J, Rocher V et al (2013) Biofiltration vs conventional activated sludge plants: what about priority and emerging pollutants removal? Environ Sci Pollut Res 21:5379–5390
35. Teil M-J, Alliot F, Blanchard M, et al (2008) Contamination de l'Orge et de la Seine par des micropolluants organiques: PBDE, phtalates, alkylphénols et HAP sous différentes conditions hydrologiques. https://hal.archives-ouvertes.fr/hal-00766528
36. Blanchard M, Teil M-J, Ollivon D et al (2001) Origin and distribution of polyaromatic hydrocarbons and polychlorobiphenyls in urban effluents to wastewater treatment plants of the Paris area (FRANCE). Water Res 35:3679–3687
37. Gasperi J, Zgheib S, Cladière M et al (2012) Priority pollutants in urban stormwater: part 2 – case of combined sewers. Water Res 46:6693–6703
38. Rocher V, Azimi S, Moilleron R, Chebbo G (2004) Hydrocarbons and heavy metals in the different sewer deposits in the 'Le Marais' catchment (Paris, France): stocks, distributions and origins. Sci Total Environ 323:107–122
39. Rocher V, Garnaud S, Moilleron R, Chebbo G (2004) Hydrocarbon pollution fixed to combined sewer sediment: a case study in Paris. Chemosphere 54:795–804
40. Gasperi J, Moilleron R, Chebbo G (2006) Spatial variability of polycyclic aromatic hydrocarbon load of urban wet weather pollution in combined sewers. Water Sci Technol 54:185–193
41. Barret M, Carrère H, Delgadillo L, Patureau D (2010) PAH fate during the anaerobic digestion of contaminated sludge: do bioavailability and/or cometabolism limit their biodegradation? Water Res 44:3797–3806
42. Busetti F, Heitz A, Cuomo M et al (2006) Determination of sixteen polycyclic aromatic hydrocarbons in aqueous and solid samples from an Italian wastewater treatment plant. J Chromatogr A 1102:104–115
43. Vogelsang C, Grung M, Jantsch TG et al (2006) Occurrence and removal of selected organic micropollutants at mechanical, chemical and advanced wastewater treatment plants in Norway. Water Res 40:3559–3570

44. Fatone F, Di Fabio S, Bolzonella D, Cecchi F (2011) Fate of aromatic hydrocarbons in Italian municipal wastewater systems: an overview of wastewater treatment using conventional activated-sludge processes (CASP) and membrane bioreactors (MBRs). Water Res 45:93–104

45. Horstmann M, McLachlan MS (1998) Atmospheric deposition of semivolatile organic compounds to two forest canopies. Atmos Environ 32:1799–1809

46. Simonich SL, Hites RA (1994) Vegetation-atmosphere partitioning of polycyclic aromatic hydrocarbons. Environ Sci Technol 28:939–943

47. Moreau-Guigon E, Alliot F, Gasperi J, et al (2015) Contamination de l'atmosphère par les composés perturbateurs endocriniens en Ile-de-France. https://hal-enpc.archives-ouvertes.fr/hal-01162348

48. Augusto L, Ranger J, Binkley D, Rothe A (2002) Impact of several common tree species of European temperate forests on soil fertility. Ann For Sci 59:233–253

49. Nadal M, Schuhmacher M, Domingo JL (2004) Levels of PAHs in soil and vegetation samples from Tarragona County, Spain. Environ Pollut 132:1–11

50. Kipopoulou AM, Manoli E, Samara C (1999) Bioconcentration of polycyclic aromatic hydrocarbons in vegetables grown in an industrial area. Environ Pollut 106:369–380

51. Moreau-Guigon E, Gaspéri J, Alliot F, et al (2012) Contamination de l'air par les contaminants organiques, bioindication et conséquences sur la contamination des sols. http://piren16.metis.upmc.fr/?q=webfm_send/1164

52. Enell A, Reichenberg F, Warfvinge P, Ewald G (2004) A column method for determination of leaching of polycyclic aromatic hydrocarbons from aged contaminated soil. Chemosphere 54:707–715

53. Krauss M, Wilcke W, Zech W (2000) Polycyclic aromatic hydrocarbons and polychlorinated biphenyls in forest soils: depth distribution as indicator of different fate. Environ Pollut 110:79–88

54. Doick KJ, Klingelmann E, Burauel P et al (2005) Long-term fate of polychlorinated biphenyls and polycyclic aromatic hydrocarbons in an agricultural soil. Environ Sci Technol 39:3663–3670

55. Gaspéri J, Ayrault S, Moreau-Guigon E et al (2018) Contamination of soils by metals and organic micropollutants: case study of the Parisian conurbation. Environ Sci Pollut Res 25:23559–23573

56. Crépineau C, Rychen G, Feidt C et al (2003) Contamination of pastures by polycyclic aromatic hydrocarbons (PAHs) in the vicinity of a highway. J Agric Food Chem 51:4841–4845

57. Zheng Y, Luo X, Zhang W et al (2012) Enrichment behavior and transport mechanism of soil-bound PAHs during rainfall-runoff events. Environ Pollut 171:85–92

58. Luo X, Zheng Y, Wu B et al (2013) Impact of carbonaceous materials in soil on the transport of soil-bound PAHs during rainfall-runoff events. Environ Pollut 182:233–241

59. Gasperi J, Garnaud S, Rocher V, Moilleron R (2009) Priority pollutants in surface waters and settleable particles within a densely urbanised area: case study of Paris (France). Sci Total Environ 407:2900–2908

60. Gaspéri J, Moreau-Guigon E, Labadie P, et al (2010) Contamination de la Seine par les micropolluants organiques: évolution selon les conditions hydriques et l'urbanisation

61. Seine-Aval G, Fisson C (2015) Flux en contaminants à l'estuaire de la Seine. 34. https://www.seine-aval.fr/wp-content/uploads/2017/01/Fisson-2015-Flux.pdf

62. Agence de l'eau Seine-Normandie, Chevreuil M, Programme interdisciplinaire de recherche sur l'environnement de la Seine (2009) La micropollution organique dans le bassin de la Seine maîtriser l'impact des molécules créées par l'homme. Agence de l'eau Seine-Normandie, Nanterre. http://piren16.metis.upmc.fr/?q=webfm_send/824

63. Motelay-Massei A, Garban B, Tiphagne-larcher K et al (2006) Mass balance for polycyclic aromatic hydrocarbons in the urban watershed of Le Havre (France): transport and fate of PAHs from the atmosphere to the outlet. Water Res 40:1995–2006

64. Demircioglu E, Sofuoglu A, Odabasi M (2011) Particle-phase dry deposition and air–soil gas exchange of polycyclic aromatic hydrocarbons (PAHs) in Izmir, Turkey. J Hazard Mater 186:328–335

65. Ringuet J, Albinet A, Leoz-Garziandia E et al (2012) Reactivity of polycyclic aromatic compounds (PAHs, NPAHs and OPAHs) adsorbed on natural aerosol particles exposed to atmospheric oxidants. Atmos Environ 61:15–22

66. Ringuet J, Albinet A, Leoz-Garziandia E et al (2012) Diurnal/nocturnal concentrations and sources of particulate-bound PAHs, OPAHs and NPAHs at traffic and suburban sites in the region of Paris (France). Sci Total Environ 437:297–305

67. Gasperi J, Gromaire MC, Kafi M et al (2010) Contributions of wastewater, runoff and sewer deposit erosion to wet weather pollutant loads in combined sewer systems. Water Res 44:5875–5886

68. Qiao M, Qi W, Liu H, Qu J (2014) Occurrence, behavior and removal of typical substituted and parent polycyclic aromatic hydrocarbons in a biological wastewater treatment plant. Water Res 52:11–19

69. Le Gall M, Ayrault S, Evrard O et al (2018) Investigating the metal contamination of sediment transported by the 2016 Seine River flood (Paris, France). Environ Pollut 240:125–139

Ecological Functioning of the Seine River: From Long-Term Modelling Approaches to High-Frequency Data Analysis

J. Garnier, A. Marescaux, S. Guillon, L. Vilmin, V. Rocher, G. Billen, V. Thieu, M. Silvestre, P. Passy, M. Raimonet, A. Groleau, S. Théry, G. Tallec, and N. Flipo

Contents

1   Introduction 190
2   A Modelling Approach for a Comprehensive Understanding of the Ecological
    Functioning of the Seine River 192
3   Long-Term Trends in Water Quality 194
    3.1   Changes in Organic Pollution from Urban Point Sources 194
    3.2   Long-Term Nutrient Contamination and Algal Growth 199
    3.3   Modelling Nutrients and Phytoplankton Biomass 201
4   River Metabolisms: Autotrophy, Heterotrophy and CO_2 Saturation 201
    4.1   High-Frequency Analysis of Oxygen Data Along the Main Seine Stem 203
    4.2 CO_2 Supersaturation in the Seine River: Carbon Sources, Fate and Budget 204

The copyright year of the original version of this chapter was corrected from 2019 to 2020. A correction to this chapter can be found at https://doi.org/10.1007/698_2020_667

J. Garnier (✉), A. Marescaux, G. Billen, V. Thieu, and M. Raimonet
SU CNRS EPHE UMR 7619 Metis, Paris, France
e-mail: josette.garnier@sorbonne-universite.fr

S. Guillon and N. Flipo
Geosciences Department, MINES ParisTech, PSL University, Fontainebleau, France

L. Vilmin
Department of Earth Sciences, Utrecht University, Utrecht, The Netherlands

V. Rocher
SIAAP, Direction Innovation, Colombes, France

M. Silvestre, P. Passy, and S. Théry
CNRS SU FR 3020 FIRE, Paris, France

A. Groleau
Université de Paris, IPGP, CNRS, UMR 7154, Paris, France

G. Tallec
Irstea, UR HYCAR, Antony, France

Nicolas Flipo, Pierre Labadie, and Laurence Lestel (eds.), *The Seine River Basin*, Hdb Env Chem (2021) 90: 189–216, https://doi.org/10.1007/698_2019_379,
© The Author(s) 2020, corrected publication 2020, Published online: 3 June 2020

5   Further Improvement of Water Quality 207
     5.1   Progress Made in Wastewater Treatments 207
     5.2   Acting on Diffuse Agricultural Sources to Improve Water Quality 207
     5.3   Impact of the Seine River Nutrient Fluxes at the Coastal Zone 208
     5.4   The Context of Climate Change 209
6   Conclusions and Perspectives 210
References 211

Abstract At the start of the PIREN-Seine program, organic pollution by the effluent of the Parisian conurbation was responsible for episodic anoxia in the lower Seine River, while nutrients from both point and diffuse sources are used to cause eutrophication, a nuisance for drinking water production from surface water and biodiversity. The implementation of the EU Water Framework Directive led to a drastic decrease of organic carbon, phosphorus and ammonium concentrations in surface waters starting in the early 2000s and to a reduction of the frequency and the amplitude of phytoplankton blooms. However, nitrate contamination from fertiliser-intensive agriculture continued to increase or at best levelled off, threatening groundwater resources and causing unbalanced nutrient ratios at the coastal zone where eutrophication still results in harmful algal blooms. High-frequency O_2 data combined with models, which have been developed for 30 years, can help discriminate the contribution of auto- vs. heterotrophic metabolism in the CO_2 supersaturation observed in the Seine River. Despite the impressive improvement in water quality of the Seine River, episodic crises such as summer low-flow conditions still threaten the good ecological status of both river and coastal waters. Modelling scenarios, including further wastewater treatments and structural changes in agriculture and future changes in hydrology under climate changes, provide the basis for a future vision of the ecological functioning of the Seine River network.

Keywords Aquatic continuum, Ecological functioning, Long-term socioecological study, Riverstrahler model, Scenarios

1   Introduction

The concept of ecosystem functioning lies at the root of ecological sciences [1, 2]. It was at the origin of the International Biological Program, which coordinated large-scale ecological and environmental studies during the 1964–1974 period. Regarding aquatic ecosystems, the first research efforts were devoted to stagnant systems (lakes, sandpit lakes, reservoirs) where discharged point sources (domestic and industrial) were recognised as a major cause of water quality alteration through eutrophication and organic pollution. Many comprehensive studies started at that time, showing, based on long-term whole-ecosystem experiments, the strict control of primary production by phosphorus [3, 4], while silica (Si) depletion occurring with excess phosphorus (P) and/or nitrogen (N) limits the growth of siliceous algae such as diatoms, possibly leading to severe and undesirable ecosystem shifts [5].

Among the four-dimensional nature of lotic ecosystems [6], the longitudinal upstream–downstream dimension was the leading factor, inspired by the River Continuum Concept (RCC) describing the longitudinal auto- vs. heterotrophic metabolism pattern [7]. A river system can be considered as a network of connected river stretches of different stream orders [8], each characterised by a common scheme of ecological and biogeochemical processes, whose intensities depend on local hydro-morphological features and on inputs received from the upstream network and the watershed. The metabolism in each of the network's ecosystems is therefore dependent on its position in the network (its stream order for river stretches), and all ecosystems are collectively influenced by the water runoff flowing through the system as well as by the point and diffuse sources of nutrients originating from the watershed. At the outlet of the system, water and nutrients are exported to marine coastal zones, the functioning of which can be strongly influenced by these inputs.

Taking into account the role of connected stagnant ecosystems [9–11], as well as denitrification in riparian zones [12, 13], allows the investigation of the lateral dimension in the sense of water interactions with drained terrestrial parts of a watershed [6]. Whereas [14–17] investigated the vertical dimension, i.e. water and nutrient exchanges between aquifers and surface water [18, 19] in the context of a multi-scale view of stream–aquifer interfaces [20], the fourth dimension, which provides the time scale, has long been a major concern of the PIREN-Seine program, both in terms of past and future scenario analysis [21, 22].

The Seine River system is a textbook example illustrating this overall vision. It drains one of the most intensive agricultural areas in the world, as well as one of the largest European metropolises. The population is mostly concentrated along the main downstream stretch (seventh order), 150 km upstream from the beginning of the estuarine zone. Water quality problems related to urban organic pollution and eutrophication were very acute at the beginning of the PIREN-Seine program and have been largely repaired over the last few decades.

In this chapter we first provide a short summary of the methodologies used in our studies of the Seine River metabolism, based on a combination of field measurements and modelling approaches. We then describe the long-term changes in organic matter and nutrient contamination of the system, followed by a discussion on the metabolism of river ecosystems in terms of auto- and heterotrophy as well as gas production. Finally, a prospective view of the river functioning concludes the chapter.

2   A Modelling Approach for a Comprehensive Understanding of the Ecological Functioning of the Seine River

The Riverstrahler model, first developed in the 1990s [23, 24], is still one of the few tools available to model nutrient cycling and ecological functioning at the scale of an entire drainage network based on the network's morphological characteristics (e.g. length and depth of the rivers), the hydro-meteorological conditions and the human activities in the watershed (point and diffuse pollution sources related to domestic and agricultural water usages). It combines a generic model that mimics ecological and biogeochemical processes (RIVE) with a water flow routing scheme through the drainage network partly based on its Strahler ordination [8] (Fig. 1a).

Starting from a conceptualisation with one community of phytoplankton taking up only N and P, and one bacterial community degrading organic matter, the RIVE model (www.fire.upmc.fr/rive) has gradually evolved over the last 30 years (Fig. 1b). Among the significant developments, the introduction of diatoms and silica uptake allowed us to specifically examine river primary production and oxygen production since diatoms are the dominant species in rivers, specifically in the Seine River [24]. Besides building up an organic stock mineralised by bacteria, phytoplankton is grazed by microzooplankton, which is also represented in the model [9]. Nitrifying bacteria and nitrification were additionally introduced for a realistic simulation of ammonium dynamics, a typical element of point sources [25–27], taking into account N_2O as an intermediate of nitrification [28]. Phosphorus, coming from both point and diffuse sources, was specifically studied for a parametrisation of its adsorption–desorption dynamics [29–31]. The model also includes a calculation of nutrient exchanges across the sediment–water interface as a result of a given sedimentation flux of organic material, taking into account organic matter degradation, oxygen consumption and N, P and Si processes, mixing in the interstitial and solid phases and accretion of the sedimentary column by inorganic matter sedimentation [15, 32, 33] (Fig. 1c).

Recent developments concern the introduction of inorganic forms of carbon and greenhouse gas emissions (carbon dioxide, CO_2 [34]; methane, CH_4; nitrous oxide, N_2O). The current version of RIVE comprises 30 state variables.

The RIVE model applied to the whole upstream–downstream network with the same parameters implies the recognition of the unicity of the kinetics formalisation of ecological processes, whatever the location in the network, a strong assumption that has proved realistic, not only along a land-to-sea continuum in a gradient of watershed sizes (the Seine River [23]; the Danube River [35]) but also in stagnant systems [9, 10] and in a variety of regions (sub-tropical [36]; subarctic [37]). The RIVE model also performs fairly well in the case of transitory events such as combined sewer overflows [38, 39]. For the study of urban systems, it is implemented in the ProSe model [32, 38, 40], which was used to study the plurennial impact of transient events on the carbon and oxygen cycles [41, 42], as well as nutrients [27, 31, 43]. The implementation of RIVE in the ProSe model simulates

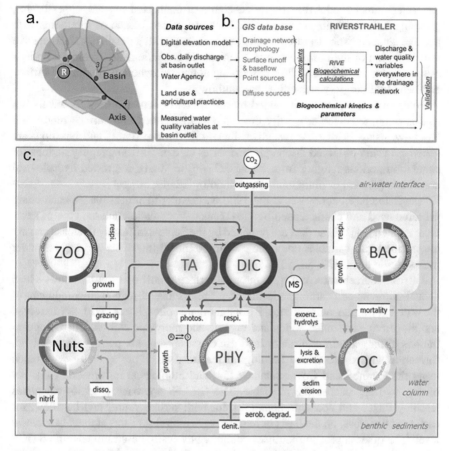

Fig. 1 Principles of the Riverstrahler model. (**a**) Representation of the river objects in Riverstrahler (basin; *A* axis, *R* reservoirs and ponds) following Strahler ordination [23]. (**b**) Conceptual framework of the Riverstrahler model implementation. (**c**) Schematic representation of RIVE processes (nitrification (nitrif), dissolution (disso), denitrification (denit), grazing, respiration (respi.), mortality, lysis and excretion, sedimentation (sedim) and erosion, aerobic degradation (aerob. degrad), outgassing and different variables (zooplankton (ZOO), phytoplankton (PHY), bacteria (BAC), nutrients (Nuts), organic carbon (OC), total alkalinity (TA) and dissolved inorganic carbon (DIC))

sediment processes properly [20, 44] as well as periphyton scouring [32], which are both key processes for a proper estimation of water quality [41, 42].

The Riverstrahler model requires a detailed description of the hydrographic network for which elementary basins (EBs), defined as portions of the watershed drained by a segment of river between two confluences or between a spring and the first confluence, are characterised by their slope, width and length and their position within the upstream–downstream scheme of river confluences, i.e. their stream order [8]. This structure of the Seine River's drainage network was obtained from the IGN database (CarTHAgE®) and from an elevation model (50 m, http://professionnels.ign.fr/bdalti). According to the required resolution of the application, the Riverstrahler model runs on an ensemble of connected objects, either basins, grouping EBs according to an idealised scheme of confluence of tributaries with average characteristics by stream order, or branches, with a detailed and exhaustive representation of the morphology at 1 km spatial resolution. Reservoirs and ponds are also taken into account with a description of their morphology (depth and surface area) and hydrology (water inflow and outflow).

Hydrological inputs are provided by the HYDRO database (http://www.hydro.eaufrance.fr), averaged at a daily time step and separated into surface runoff and baseflow, using Eckhardt's recursive filter [45]. Surface runoff and baseflow are generated using observed discharge time series at, e.g. 50–100 gauging stations for simulation periods ranging from 1 year [23, 46] to yearly contrasted hydrological conditions [47, 48] or plurennial time-windows [22, 49, 50].

The Seine basin comprises about 1900 wastewater treatment plants, each being georeferenced and characterised by the connected population and the type of treatment applied [51]. These data are provided by the Seine–Normandy database and are transformed into variables compatible with the model (nutrients and carbon under their specific forms [37, 52, 53]).

Diffuse sources enter the river through surface runoff and the baseflow. Mean annual nutrients and carbon concentrations are associated with each component of the water: N (nitrate, ammonium), P (total inorganic phosphorus, TIP), Si (dissolved and biogenic), suspended solids and organic and inorganic carbon. These concentrations differ according to major land use classes (cropland, permanent grassland, forests, urbanised areas) and lithological features, which are spatially determined using the Corine Land Cover database [54] and lithological information [51]. For nitrate, two approaches are available. The GRAFS approach [55, 56] calculates leaching fluxes and concentrations [57], for both surface runoff and baseflow, for each land use class, on the basis of regional agricultural statistics at the *département* level. The other approach mobilises complex physically based models: the STICS agronomic model and the MODCOU hydrogeological model [18, 58]. Inorganic carbon was recently added to the modelling approach [34, 59], the aquifer's lithology being a major controlling factor.

The Riverstrahler model calculates seasonal variations of water quality and ecological functioning for any tributary of the river system given point source and diffuse source forcings.

3 Long-Term Trends in Water Quality

3.1 Changes in Organic Pollution from Urban Point Sources

From about 11 million inhabitants in 1955, the population of the Seine basin had increased to 16 million in 2018 [51], with an uneven geographical distribution, with 75% concentrated in the Paris metropolis and along the main Seine branch from Paris to the estuary.

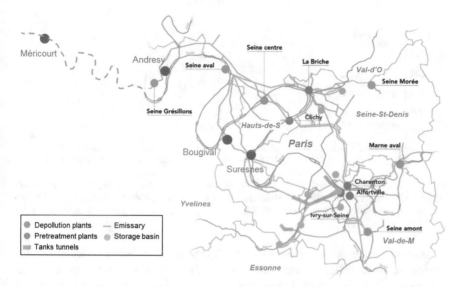

Fig. 2 Map of the Parisian conurbation indicating the WWTPs (blue circles and names in black, Seine Aval [SAV] previously Achères), the pretreatment plants (purple) and the storage tunnels/basins (orange and green), as provided by SIAAP. The four high-frequency O₂ monitoring stations are also indicated (red). Paris and conurbation *départements* (grey)

The history of domestic Paris wastewater management since the beginning of the twentieth century, marked by technical choices for sewage management (Fig. 2), is that of a race between increasing wastewater production and treatment capacity. Farmland application, the destination of nearly all the collected sewage in 1900, was rapidly insufficient to absorb the volumes produced. At the end of the 1960s, around 60% of the wastewater collected was discharged into the river without any treatment, leading to oxygen depletion and fish mortality (the species richness was reduced to only three species in 1970 [60]). The first biological wastewater treatment plant was impounded in 1940 in Achères, 70 km downstream of the Paris city centre. Sanitation centralisation started with the construction of large wastewater collectors from 1954 to 1972. In 1970, most raw waters reached the Seine River through large spillways in Clichy, 30 km downstream of Paris. Changes in sanitation strategy were implemented in three steps [60]:

– Development of the largest European WWTP in Achères (nowadays called Seine Aval, SAV) from 1940 until the 1980s
– Construction of new WWTPs to increase treatment capacities until the 1990s
– Since then, deployment of new processes of wastewater treatment for achieving complete removal not only of carbon but also of nitrogen and phosphorus

All these improvements favoured the recolonisation of the aquatic environment by fish populations: today 32 fish species are found in the Seine River stretch crossing Paris and its conurbation [61].

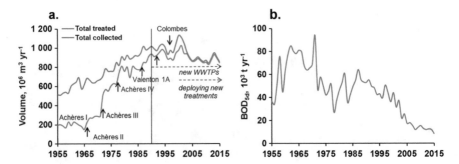

Fig. 3 Long-term trajectories 1955–2015: (**a**) of collected and treated volumes in the SIAAP stations of the Paris conurbation, (**b**) of BOD load for the volume treated (source; SIAAP). *NB* the decrease in collected wastewater (after 2005 in (**a**)) stems from a decrease in domestic water use

Data on wastewater volumes in the Paris agglomeration show the increase of the amount of water collected and treated in WWTPs before discharge into the river system (Fig. 3a). The volume treated represents the polluting wastewater volume and can be expressed in biological oxygen demand (BOD_{5d}) (Fig. 3b).

Even though almost 100% of the wastewater collected is currently treated, the combined sewer system still discharges significant amounts of raw water into the river system during large rainfall events. Such short-term events typically last 1–20 h (90% in less than 10 h) and can lead to fast, temporary decreases in oxygen concentrations in the receiving river of up to ~2 mgO_2 L^{-1} at low flow [62].

ProSe has simulated the effect of these successive changes in the sewage system since the 1970s on oxygen fluxes in the main branch of the Seine River downstream of Paris during low-flow conditions (Table 1). Metabolism (production and respiration; see Sect. 3 for further detail) is calculated in terms of oxygen budgets for the 1970s, 1980s and 2012 and for a prospective scenario that considers the full treatment and the smooth release over time of combined sewer overflows (CSOs). Emissions for the 1970s and 1980s were characterised based on average water discharges and concentrations of sediments, ammonium and BOD monitored by the SIAAP in 1969–1971 and 1979–1981, respectively. These data are available for major WWTP effluents of the Paris conurbation and in the largest spillways for untreated wastewater, located in Clichy and Achères (Fig. 2).

For low-flow conditions, outside of algae bloom periods, changes in the conurbation's sewage management have had little effect on primary production but clear impacts on ecosystem respiration (Table 1). In the 1970s, the Seine River system was highly heterotrophic upstream of the SAV WWTP, due to the release of large volumes of organic matter in Clichy. In 1980, ~95% of the wastewater was piped towards Achères, out of which almost 90% was treated before emission. This significantly helped improve oxygen conditions upstream of the WWTP, reducing net oxygen consumption by almost one order of magnitude. However, heterotrophic respiration downstream increased by 50%. The continuous improvements in treatment capacity and technology since the 1990s have significantly improved oxygen

Table 1 Oxygen budgets simulated with the ProSe model for 2012 low-flow conditions (30 consecutive driest days), with typical wastewater emissions from the 1970s and 1980s, real emissions from 2012 and a scenario with full treatment of all collected waters

		Period/scenario			
	Oxygen fluxes (tonnes O_2 day^{-1})	1970	1980	2012	2012 smoothed CSOs
Paris-SAV (77 km)	Photosynthesis	8.0	8.8	8.0	8.3
	Autotrophic respiration	−1.7	−1.7	−1.6	−1.6
	Heterotrophic respiration	−89.0	−14.0	−9.4	−8.4
	Benthic respiration	−12.0	−4.0	−3.3	−2.4
	NEP	−95.0	−11.2	−6.3	−4.1
	Reaeration	68.0	13.0	12.0	15.0
SAV-Poses (142 km)	Photosynthesis	73.0	69.0	67.0	73.0
	Autotrophic respiration	−18.0	−17.0	−16.0	−18.0
	Heterotrophic respiration	−72.0	−106.0	−34.0	−31.0
	Benthic respiration	−57.0	−59.0	−19.0	−19.0
	NEP	−74.0	−113.0	−2.0	5.0
	Reaeration	149.0	191.0	37.0	47.0

NEP = net ecosystem production = photosynthesis + total ecosystem respiration

conditions along the entire river stretch. Simulations further show that increasing treatment capacity to totally prevent combined sewer overflows during rain events would reduce the total ecosystem respiration upstream of SAV, where most of the large overflows occur, and would possibly lead to slightly autotrophic conditions downstream of SAV.

The rest of the Seine basin experienced a similar evolution of wastewater management, although delayed by several decades. Nowadays, the Seine basin comprises about 1900 WWTPs [51] with rather efficient organic matter and phosphorus treatments. The largest of them (>10,000 inhab equivalent) are equipped with N treatment units.

Oxygen and ammonium longitudinal profiles are determined based on 50-year data series available along the lower Seine River, historically impacted by the city of Paris (Fig. 4). The profiles shown in Fig. 4 stand for rather low water situations (150–280 m^3 s^{-1}) currently observed from mid-spring to mid-autumn, with low dilution capacity of point source pollution that actively metabolised by microorganisms. In the 1970s, two areas were subject to oxygen deficits, one immediately downstream of the Paris conurbation (50–100 km) and the other 200 km downstream, at the entrance of the fluvial estuary.

The Riverstrahler modelling approach is able to reproduce the major trends observed. It indicates that the first oxygen deficits were linked to the heterotrophic degradation of the organic load discharged by the SAV WWTP, while the second was caused by the autotrophic nitrification of the ammonium load discharged at the same location but oxidised after a 150 km transit, which corresponds to the time required by the nitrifying bacteria to build up a large enough population [25, 26, 28].

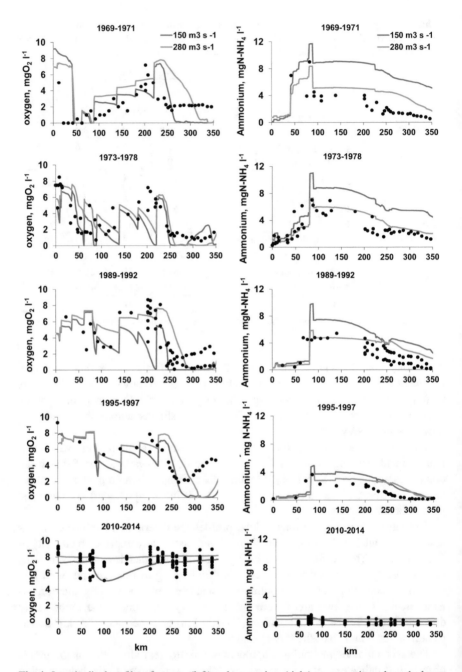

Fig. 4 Longitudinal profiles of oxygen (left) and ammonium (right) concentrations along the lower Seine River (0 km at Paris). Black dots are observations; coloured lines are simulations by the Riverstrahler model for two levels of water flow. Time periods are shown from top to bottom (from the oldest to the most recent)

At this time discharged ammonium led to concentrations up to 8–10 mg N-NH$_4$ L^{-1} downstream of SAV, decreasing to 2 mg N-NH$_4$ L^{-1} in the estuary (Fig. 4). Oxygen depletion immediately downstream of Paris occurred until the early 1990s, when the proportion of treated effluent increased (Fig. 3a). Oxygen depletion farther in the estuary was only interrupted by the implementation of a nitrification treatment in the SAV WWTP in 2007. All these efforts had a very positive effect on the oxygenation of the river, which today can boast a good ecological status with respect to oxygen concentrations [62].

Exactly as simulated in Garnier et al. [28], nitrifying treatment, implemented in WWTPs since 2007, considerably lowered the ammonium discharged in the river (a tenfold decrease) [27] and favoured full reoxygenation of the estuary (Fig. 4) [29]. In addition, molecular methods show that ammonia-oxidising bacteria (AOB), (*Nitrosomonas oligotropha* and *Nitrosomonas ureae-like bacteria*), introduced by the WWTP effluents, survived and actively participated in in-river NH$_4$ oxidation far downstream of the WWTP outflow [27, 63].

Although such changes in water treatment were mandatory [64, 65], the studies carried out as a part of the PIREN-Seine program have helped decision-makers over the past 30 years.

3.2 Long-Term Nutrient Contamination and Algal Growth

As seen above, improved water treatments reduced ammonium levels and increased oxygenation of the lower Seine River (Fig. 5). Although water treatments have also efficiently reduced phosphates since the mid-1990s, phytoplankton blooms (>100 µg Chl*a* L^{-1}) still occurred until 2005, causing nuisance for the drinking water production and producing, after bloom decline, large organic loadings, which contributed to oxygen depletion [66].

A reduction of phosphorus load by a factor 10, reached only after 2005, has been necessary to significantly decrease algal growth and avoid eutrophication. Nowadays, algal blooms as high as observed 20 years ago have disappeared. Despite these spectacular improvements, the Seine River remains fragile. For example, low summer water curtails the dilution of point source pollutions, increasing nutrient concentrations above growth limitation levels. Additionally, malfunctioning of WWTPs for maintenance purposes occasionally occurs, possibly accentuating temporary degradation of water quality [29].

While point source reductions led to a significant reduction of ammonium and phosphate pollution, nitrate from diffuse agricultural sources steadily increased from an average of winter month concentrations of about 3 mgN-NO$_3$ L^{-1} in 1971 to 6.3 mgN-NO$_3$ L^{-1} in 2011, i.e. more than 3 mgN-NO$_3$ L^{-1} over 40 years. The recent trend is towards stabilisation, and even a decrease in the recent years, although NO$_3$ concentrations peaked again at 8 mgN-NO$_3$ L^{-1} during the mild and dry 2017 winter (Fig. 5).

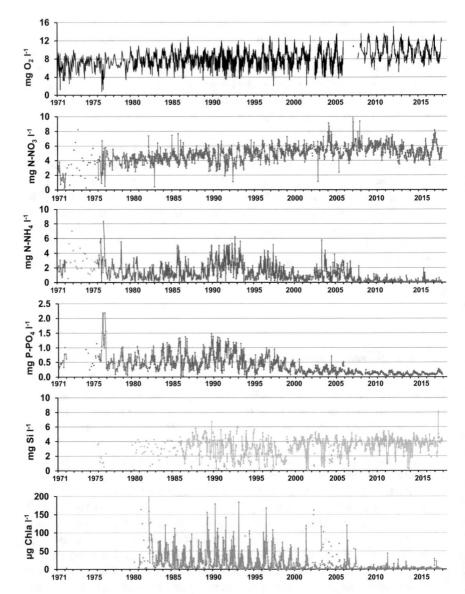

Fig. 5 Long-term variations in observed concentrations of water quality (1971–2017) at the outlet of the Seine basin (Poses): oxygen, nitrate, ammonium, phosphates, dissolved silica, phytoplankton biomass (in chlorophyll a, Chl*a*)

Regarding silica, also diffuse but originating from the natural rock weathering process, its concentrations oscillated around 3.8 mg Si L^{-1} over the last 50 years. Distinct periods of Si depletion are observed corresponding to uptake by diatoms during blooms. In agreement with the observation of a strong reduction of blooms after 2007, these silica depletions have been much less pronounced in recent years than they were 20 years ago [24] (Fig. 5).

Suspended matter dynamics (not shown) did not change significantly over the period, except for a decrease during low flow, from 20 mg L^{-1} until the mid-2000s to less than 10 mg L^{-1} in recent years. This change might be due to the decrease in phytoplankton biomass and the implementation of grass strips for stream protection from erosion and pollution, rendered compulsory in 2005. River navigation is the main contributor for maintaining particles in suspension during low flow through a release of energy into the system [44]. The evolution of fluvial traffic since the 2000s must therefore be taken into account to balance the effect of grass strip implementation.

3.3 Modelling Nutrients and Phytoplankton Biomass

Long-term simulations with Riverstrahler match the observations (Fig. 6 for 1984–1990 [50] and 2007–2012 [29]). Once their formulations have been defined and parameters determined based on experimental field and lab studies, no further adjustment is carried out, and the comparison between simulations and observations constitutes a validation test for the model.

Temporally refined simulations performed with the ProSe model across the Paris urban area over the 2007–2012 period also validate the formalisms of the RIVE model [27, 41, 42], even at the second order with a geostatistical analysis [31, 43, 44]. Discrepancies between observations and simulations raise new scientific questions, either on an incomplete representation of processes in the model or on the accuracy of the forcing data (diffuse and point sources) or even on the quality of the observed data [43].

4 River Metabolisms: Autotrophy, Heterotrophy and CO_2 Saturation

The metabolism of any ecosystem is well characterised by the intensity of their two basic functions of autotrophy and heterotrophy, i.e. photosynthesis and respiration, which directly interact with the carbon and oxygen cycles. In river systems, according to the RCC [7], a distinct longitudinal pattern is predicted with ecosystem metabolism shifting from dominant heterotrophy in small rivers receiving most of their energy from allochthonous organic matter from the watershed to autotrophy in

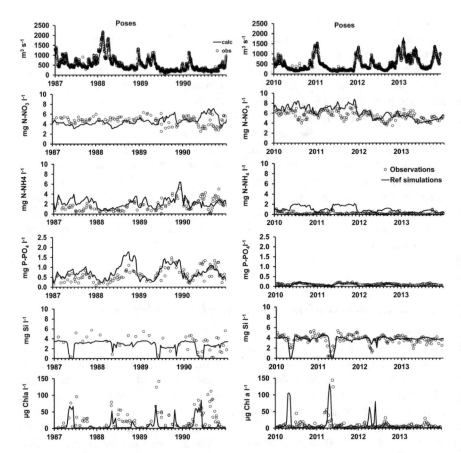

Fig. 6 Riverstrahler simulations of water quality (1987–1990 and 2010–2013) at the outlet of the Seine basin (Poses). Simulations (lines) are compared to the observations (dots). From top to bottom: discharge, nitrate, ammonium, phosphates, dissolved silica and phytoplankton (in chlorophyll a concentrations)

large rivers favourable to phytoplankton development and again to heterotrophy in estuaries, where organic matter accumulation and increased depth and turbidity reduced the possibility of photosynthesis. While this scheme is valid for pristine or slightly human-impacted rivers, it is largely affected by anthropogenic nutrient and organic matter contamination [67]. Downstream of the Paris urban area, Europe's largest WWTP's outflows into the Seine River can contribute up to one-third of the total river discharge during extreme low-flow periods.

4.1 High-Frequency Analysis of Oxygen Data Along the Main Seine Stem

Dissolved oxygen was long used as an indicator of quality for streams and rivers, mainly based on threshold values, e.g. the well-known concentration of 4 $mgO_2 L^{-1}$ under which aquatic life slows down [68]. Classically, dissolved oxygen data are also used to calibrate biogeochemical models [22, 41]. Since the 2010s, there has been considerable development of numerical routines for direct quantification of metabolism from high-frequency dissolved oxygen monitoring [69, 70], constituting a functional indicator of river ecological functioning.

High-frequency monitoring of dissolved oxygen has been carried out since the 1990s by the SIAAP (Syndicat Interdépartemental pour l'Assainissement de l'Agglomération Parisienne) given the MeSeine monitoring network (see Fig. 2). Sensors were first based on polarographic electrodes (Evita Oxy 4150) [71] and since the 2010s on optodes (LDO Hach Lange, LXV416.99.20001) [72], which allows for more stable and robust measurements. High-frequency oxygen data are available at four stations downstream of Paris and upstream and downstream of the SAV wastewater treatment plant, for two periods: 2002–2004 and 2015–2017. Data typically exhibited a marked seasonal cycle, with low oxygen concentrations (under-saturation) during summer. The seasonal dynamics is first controlled by the temperature dependence of oxygen solubility and second by the increase of microbial activity with temperature [73]. From a minimum concentration around 2 mg L^{-1} during 2002–2004, the situation has been significantly improved over the last 15 years. Oxygen concentrations are now above the 4 mg L^{-1} threshold.

Sub-hourly oxygen measurements at plurennial scales show typical seasonal patterns and circadian cycles allowing for the calculation of ecosystem metabolism [69, 70, 74, 75]. The quantification of the daily metabolism at a monitoring station is based on the analysis of oxygen circadian variations, from which values of ER (ecosystem respiration, defined as negative), GPP (gross primary production) and NEP (net ecosystem production; NEP = GPP + ER) are deduced. Each term ranged from 1 to 30 gO_2 m^{-2} day^{-1} following a seasonal cycle with the lowest values in winter and the highest in summer, within a range similar to values found in isolated eutrophic ponds [74] but slightly higher than those found in freshwaters for low and high Strahler orders [15, 41, 69, 76, 77].

For the two periods and all the stations, NEP is on average negative in this urbanised sector, indicating heterotrophic functioning, which is fully consistent with previous estimates [41, 69, 78]. NEP seasonal variability is more intense during the 2002–2004 period than during the 2015–2017 period, due to the reduction in algal development and its respiration and to a reduction of bacterial heterotrophic respiration following the decrease in organic matter emissions from WWTPs (Fig. 3).

The spatial and temporal dynamics of metabolism along the reach and between the two periods studied are characterised in terms of average daily metabolism for each entire period (Fig. 7). During 2002–2004, the absolute value of ER and GPP increased continuously from upstream to downstream, roughly doubling metabolism

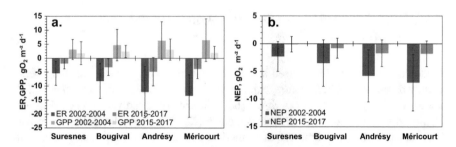

Fig. 7 Average intensity of daily metabolism during 2002–2004 and 2015–2017 for the four stations studied (with Suresnes upstream to Méricourt downstream). GPP, ER (**a**) and NEP (**b**) are in g O_2 m^{-2} day^{-1}. Metabolism is calculated for a reaeration coefficient $k_{600} = 2$ m/day and a mixing depth $Z_{mix} = 5$ m

intensity from Suresnes to Méricourt and hence a similar trend for NEP. These increases indicate substantial input of allochthonous organic carbon from this highly populated area, especially downstream of Bougival, and production of autochthonous algal organic carbon.

During the 2015–2017 period, the metabolism intensity still increased from Suresnes to Andrésy (Fig. 7), even though ER and GPP intensities are three times lower than during the 2002–2004 period (Fig. 7). However, between Andrésy and Méricourt, ER and GPP intensities decreased, because allochthonous carbon inputs had already been mineralised upstream and autochthonous algal biomass had been reduced.

On average, there is a two- to threefold decrease in ER and GPP metabolism intensity between the two periods, underlining the positive and significant impact of the evolution of treatments in the SAV WWTP, as well as in all other smaller plants. Indeed, between the two periods, the pressure exerted on the lower Seine River by the Paris conurbation was divided by three. This decrease is in agreement with the decrease in BOD load (see above) for all the SIAAP WWTPs, as well as with total phosphorus and ammonium loads from SAV, which were reduced by factors of 3.4 and 4.4, respectively, over the 15-year time span.

4.2 CO$_2$ Supersaturation in the Seine River: Carbon Sources, Fate and Budget

Several aquatic carbon sources and processes are in use. Besides in-stream autotrophy and heterotrophy, inputs of dissolved inorganic carbon (including carbon dioxide) from the basin contribute to supersaturation of CO_2 in the river with respect to atmospheric concentrations.

Long-term data from the Water Agency and from the SIAAP allow the reconstruction of partial pressure of CO_2 (pCO$_2$) and of the biodegradable organic loadings since 1970 [34]. Going back to the 1970s, pCO$_2$ is estimated given the pH, water temperature and total alkalinity [34]. Because the lower Seine is the most impacted sector of the river, pCO_2 values were compared upstream of Paris and downstream at Poses (the outlet of the river). Upstream of Paris, pCO_2 oscillated around 4,000 ppm during the 1970–2015 period. A maximum supersaturation of 10,000–12,000 ppm was reached downstream of Paris during the period of maximum organic pollution (1985–2000). BOD emitted from WWTPs indeed followed the exact same pattern as pCO_2 [34] and therefore directly contributed to the CO_2 budget through direct outflows in the river system. Improvements of wastewater treatments in WWTPs therefore had a positive effect not only on water quality in terms of oxygenation, nutrients and eutrophication (see Fig. 5) but also on pCO_2 and hence on CO_2 emissions by the river network.

To determine diffuse sources, values from the ADES database (www.ades. eaufrance.fr, last accessed on 2018/11/05) and our own pCO_2 measurements in piezometers were used. Data provided at the scale of the whole Seine basin were grouped by water body units according to the lithology and stratigraphy [34, 59].

In-river CO_2 concentrations simulated with Riverstrahler match the CO_2 estimated with the Water Agency database [34, 79] (Fig. 8). The values in the Marne River, a major tributary of the Seine River upstream of Paris, indicate an upstream "background level" of 1.2 mgC L^{-1}, which is above the CO_2 saturation value of around 0.2 mgC L^{-1} (393 ppmv) at 10°C, whereas WWTPs outflows lead to a supersaturation of >2 mgC L^{-1} in the lower Seine River from Paris to the outlet.

Based on these simulations, an annual average budget (2010–2013) of the Seine River is calculated for inorganic (IC) and organic (OC) carbon (Table 2). Inorganic and organic carbon budgets differ by one order of magnitude, with total inputs to the river system of 17,300 and 1,600 kg C km^{-2} $year^{-1}$, respectively. The contribution of groundwater to IC is the highest (57.5%), although inputs from the subroot zone are also significant (34.4%) [79]. Compared to diffuse sources, the contribution of

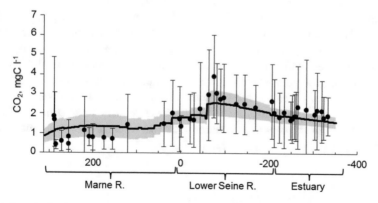

Fig. 8 Longitudinal profile of CO_2 in the Marne River and lower Seine to the estuary for the 2010–2013 period. Black dots are average pCO_2 estimations (see text). Bars are standard deviations. The black line is the averaged Riverstrahler simulation for the same period. The simulation envelope (grey area) represents standard deviations of CO_2

Table 2 Average inorganic (a) and organic (b) carbon budgets in the Seine hydrosystem ($kgC\ km^{-2}\ year^{-1}$) calculated by Riverstrahler (2010–2013)

(a) 2010–2013	Processes involved in inorg C budget	$kgC\ km^{-2}\ year^{-1}$	%
Input to river	Diffuse sources from subroot	5,963	34.4
	Diffuse sources from groundwater	9,968	57.5
	Urban point sources	1,135	6.6
	Heterotrophic respiration	266	1.5
	Denitrification	0	0
Output from river	Delivery to the outlet	12,483	68.4
	Ventilation	5,619	30.8
	Nitrification	37	0.2
	NPP	105	0.6
(b) 2010–2013	Processes involved in org C budget	$kgC\ km^{-2}\ year^{-1}$	%
Input to river	Diffuse sources from subroot	870	53.9
	Diffuse sources from groundwater	227	14.1
	Urban point sources	375	23.2
	Nitrification	37	2.3
	NPP	105	6.5
Output from river	Delivery to the outlet	1,086	65.7
	Heterotrophic respiration	110	6.7
	Net sedimentation	456	27.6

WWTPs to the IC and OC inputs is comparatively low and accounts for 6.5% and 23.2%, respectively.

Total carbon exports at the outlet in Poses are reported to account for 69% and 65% of IC and OC total inputs. CO_2 ventilation is an important physical process in this IC budget, i.e. 30.8% of overall losses (Table 2). The high inorganic diffuse sources brought to the river network are clearly responsible for the supersaturation in CO_2 and hence CO_2 evasion. It confirms the potential contribution of freshwater to the global CO_2 emissions [80] through abiotic processes. Even though Cole et al. [81] pointed out the importance of riverine CO_2 sources in the global carbon budget, biotic processes under the Seine's temperate climate were negligible in carbon emissions. Heterotrophic respiration by microorganisms accounts for only 1.5% of IC inputs. Similarly, losses through net primary production (NPP) also accounted for a small proportion in terms of IC (0.5%) and even less for nitrification (0.2%) for the OC balance (Table 2).

5 Further Improvement of Water Quality

5.1 Progress Made in Wastewater Treatments

Riverstrahler is now used to explore various future wastewater treatment scenarios. Although considerable improvements have been made in the largest WWTPs of the Seine basin, some of the smallest WWTPs of the 1900 spread over the basin still have to implement the Urban Wastewater Directives [64]. Compared to a reference situation (2002–2014), nitrogen and phosphorus fluxes at the outlet of the Seine River could be decreased by 10 and 20%, respectively, when current regulations of tertiary treatments (nitrification and denitrification and dephosphatation) are generalised. For silica, a diffuse source pollutant, no visible change would occur with further wastewater treatment (Fig. 9a). An appreciation of these efforts is provided by a backward scenario simulating the situation of wastewater treatment as it was in the 1980s. The difference with the reference scenario is very telling in terms of the distance covered over the last 30 years. These results also indicate that any relaxation in the treatment of domestic effluents, e.g. following a disengagement of the national government, could lead to a dramatic situation. Slight changes in silica fluxes (Fig. 9a) result from silica consumption with more or less nutrient (N, P) availability for diatom growth.

Despite all the efforts made in WWTPs, which strongly improved the water quality in the Seine River, nutrient fluxes delivered at the coastal zone are still responsible for eutrophication in the Seine Bight, with toxic algal blooms [82] and fishery problems. Moreover, nitrate contamination of groundwater [58] and closure of drinking water wells are a real problem for the basin's water agencies and drinking water suppliers. Pesticide contamination, associated with intensive agricultural practices, is also an important concern in the basin [83].

5.2 Acting on Diffuse Agricultural Sources to Improve Water Quality

Point source pollutions are today almost completely under control. Simultaneously, diffuse agricultural sources, particularly of nitrogen, have become the subject of particular concern because of their cascading effect on many compartments of the environment [84]. Two contrasted scenarios of the future of the agro-food system have been established and compared [85, 86]. The first one consists of the continuation of the current trends of opening and specialisation (O/S) of the conventional cropping systems following regulation but without questioning the basic structural features of the current production system. The second scenario (A/R/D) involves a radical reorganisation of agricultural production, generalising organic farming practices, reconnecting livestock and crop farming [57, 87] and changing human consumption towards a Demitarian diet, i.e. cutting consumption of animal proteins by

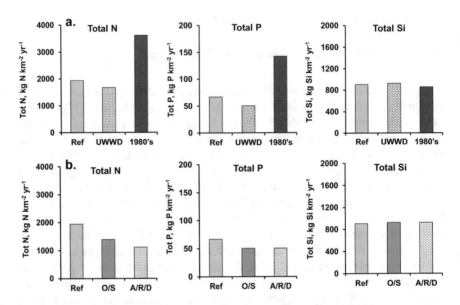

Fig. 9 Total nitrogen (N), phosphorus (P) and silica (Si) fluxes at the outlet of the Seine River calculated by Riverstrahler for the 2002–2014 reference situation (Ref.): (**a**) for a completion of the UWWD, and for a situation that would mimic the condition in the 1980s with low wastewater treatment, (**b**) for the O/S and A/R/D scenarios, respectively, agriculture opening and specialisation and autonomy, livestock reconnection and Demitarian diet. Both O/S and A/R/D scenarios also include the improvement of wastewater treatments

half. The O/S scenario shows a decrease in nutrient fluxes, partly linked to the improvement of wastewater treatment, which is also included in the scenarios (Fig. 9b).

A possible further reduction in N fluxes is demonstrated with the A/R/D scenario (Fig. 9b). Regarding P fluxes, the marginal changes observed are linked to the regulation compliance of the WWTP scenario, while silica fluxes, essentially coming from rock weathering, do not change.

5.3   Impact of the Seine River Nutrient Fluxes at the Coastal Zone

At the coastal zone, the Seine River waters flowing north in the Channel and to the Southern North Sea have been recognised as a major threat for eutrophication. France has therefore been convicted several times by the EU Court of Justice for not sufficiently reducing its N fluxes [82, 88].

An indicator for assessing the risk of coastal eutrophication potential (ICEP), based on the nutrient imbalance in riverine fluxes with respect to the algae requirement, has been calculated. The N- and P-ICEP [89] are based on the C:N:P:Si molar

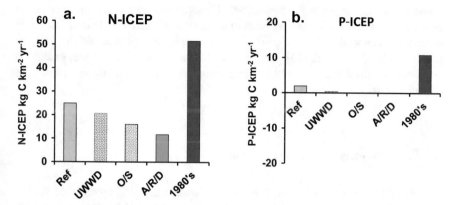

Fig. 10 Indicator of coastal eutrophication potential (ICEP) for N and P for the reference and the different scenarios. (**a**) N-ICEP, (**b**) P-ICEP

ratios of 106:16:1:20 [90]. The ICEP can be positive or negative and represents the excess or the deficit of either N or P, respectively, with respect to silica (Fig. 10b). In the 1980s conditions, both N and P were in excess compared to silica. The reference situation shows that P is now much less in excess and nearly balanced compared to Si, whereas N is systematically far above a balanced N/Si situation. Improving agricultural practices, which would further decrease N-ICEP, would reduce the risk of coastal eutrophication. The most radical agricultural scenario (A/R/D) has the advantage of also banishing pesticides.

The scenarios of modified agricultural practices and urban sewage treatments highlighted that changes, not only in agricultural practices but also in the agricultural production system, are needed to reduce exports of nitrogen to the coastal zone and associated eutrophication risks. However, these changes in human practices might take place in a context of climate change that also impacts water quality.

5.4 The Context of Climate Change

Scenarios were developed to evaluate the impact of climate change on river water quality and eutrophication indices [91]. For all scenarios, eutrophication indices remained high, indicating that climate change is not expected to decrease eutrophication risks in the future and confirming that changes in human practices are needed.

When investigating changes landward in the river system, water quality is expected to be mostly modified in summer. At the river outlet, climate change has been shown to increase nitrate and phosphate concentrations during low flow due to a decreasing river discharge. This indicates a higher sensitivity to urban release from WWTPs during low flow in the future and a higher sensitivity to WWTP malfunctioning. These increases in nitrate and phosphate concentrations are particularly high in stream orders greater than 4 (up to +19% and 42%, respectively, at the outlet for the driest RCP 8.5 projection). Changes in water temperature are even expected to superimpose with potential modifications of phytoplankton biomass and composition as well as oxygen concentrations, as shown by the sensitivity analysis performed by Wang et al. [73]. Acting on diffuse sources and, to a lesser extent, on point sources (as discussed above) is therefore even more necessary in a context of climate change.

6 Conclusions and Perspectives

The PIREN-Seine program is a rare example of an interdisciplinary scientific project successfully spanning 30 years. Owing to its longevity, it has been able to document the long-term changes experienced by the Seine River system and its metabolism.

At the time the program was launched, in the late 1980s, the major water quality issue was the oxygen deficits caused by discharge of insufficiently treated urban wastewater, particularly in the main branch of the Seine downstream from Paris. The heterotrophic degradation of the organic matter and the nitrification of the reduced nitrogen brought by these point sources of pollution were responsible for severe oxygen depletion in these sectors.

One decade later, the problem of algal blooms in the large tributaries of the Seine upstream from Paris appeared as a major threat for drinking water production from surface water (about half the supply of the Paris agglomeration) and also as a significant contributor to organic matter loading in the downstream sectors. The abundance of phosphorus, coming from point sources, and the resulting lack of any limitation of algal growth appeared as the major cause of this river's eutrophication problem. The strong contrast between autotrophy in these upstream sectors and the heterotrophy of the downstream sectors of the river network was particularly striking.

In the early 2000s, major progress in wastewater treatment, including the drastic reduction of organic matter loading, as well as the treatment of phosphorus, completely changed the situation. Good oxygen levels were recovered nearly everywhere, while algal blooms were considerably reduced as a result of the limitation of phosphorus.

While the WWTP outflows of nutrients were drastically reduced following the European regulations of the Water Framework Directive [65], diffuse pollutions have not yet produced the same success story. Nitrate (and pesticide) contamination is still a major threat for groundwater quality and the drinking water supply, for river water quality and the diversity of its biological communities and for marine coastal zones and the occurrence of toxic algal blooms.

In essence, controlling diffuse pollutions is much more difficult to achieve than containing point source pollutions. It requires not only technical measures but also a complete overhaul of the structure of the agro-food system and hence of the socioecological metabolism of the watershed. Moreover, inertia of groundwater transfer time delays improvements, which is most often difficult for farmers to accept, in view of the efforts required of them.

In addition to the PIREN-Seine program, another parallel long-term program on the Seine Estuary has made it possible to quantify the role of the Seine estuary in the transfer and transformation of the nutrients and carbon fluxes from the upstream hydrological network [92, 93]. Riverstrahler has already combined with coastal zone models in a chain that has quantified the impact of the Seine River nutrient deliveries at the coastal zone, showing that nutrient imbalance, i.e. a large excess of N over Si and P, was the cause of coastal eutrophication, with toxic algae events in the Seine Bight and fishing closure [94]. However, the often neglected buffer role of the estuary is now being included in the model chain.

Acknowledgements This work is a contribution to the PIREN Seine research program (www. piren-seine.fr), within the framework of the Zone Atelier Seine, a site of the international Long Term Socio Ecological Research (LTSER) network. This work was partly funded by several other projects (Thresholds -EU FP6-, Aware -EU FP7- Seas-Era Emosem -EU FP7 ERA-NET-, Liteau FLAM -French Ministry of Ecology and Water Agency-, ITN C-Cascades -EU Horizon 2020-).

References

1. Odum EP (1959) Fundamental of ecology. W.B. Saunders Company, Philadelphia
2. Wetzel RG (1983) Limnology2nd edn. Saunders College Publishing, Philadelphia
3. Schindler DW (1974) Eutrophication and recovery in experimental lakes: implications for lake management. Science 184:897–899. https://doi.org/10.1126/science.184.4139.897
4. Schindler DW (1998) Replication versus realism: the need for ecosystem-scale experiments. Ecosystems 1:323–334. https://doi.org/10.1007/s100219900026
5. Schelske CL, Stoermer EF (1971) Eutrophication, silica depletion, and predicted changes in algal quality in Lake Michigan. Science 173:423–424. https://doi.org/10.1126/science.173.3995.423
6. Ward JV (1989) The four-dimensional nature of lotic ecosystems. J N Am Benthol Soc 8:2–8
7. Vannote RL, Minshall GW, Cummins KW et al (1980) The river continuum concept. Can J Fish Aquat Sci 37:130–137. https://doi.org/10.1139/f80-017
8. Strahler AN (1957) Quantitative analysis of watershed geomorphology. Geophys Union Trans 38:913–920
9. Garnier J, Billen G (1994) Ecological interactions in a shallow sand-pit lake (Lake Créteil, Parisian Basin, France): a modelling approach. Hydrobiologia 275/276:97–114
10. Garnier J, Billen G, Sanchez N, Leporcq B (2000) Ecological functioning of the Marne reservoir (Upper Seine Basin, France). Regul Rivers Res Manage 16:51–71
11. Passy P, Garnier J, Billen G et al (2012) Restoration of ponds in rural landscapes: modelling the effect on nitrate contamination of surface water (the Seine River Basin, France). Sci Total Environ 430:280–290. https://doi.org/10.1016/j.scitotenv.2012.04.035
12. Billen G, Ramarson A, Thieu V et al (2018) Nitrate retention at the river–watershed interface: a new conceptual modeling approach. Biogeochemistry 139:31–51. https://doi.org/10.1007/s10533-018-0455-9
13. Flipo N, Even S, Poulin M et al (2007) Modelling nitrate fluxes at the catchment scale using the integrated tool CaWaQS. Sci Total Environ 375:69–79. https://doi.org/10.1016/j.scitotenv.2006.12.016

14. Flipo N, Jeannée N, Poulin M et al (2007) Assessment of nitrate pollution in the Grand Morin aquifers (France): combined use of geostatistics and physically-based modeling. Environ Pollut 146:241–256. https://doi.org/10.1016/j.envpol.2006.03.056

15. Flipo N, Rabouille C, Poulin M et al (2007) Primary production in headwater streams of the Seine basin: the Grand Morin case study. Sci Total Environ 375:98–109. https://doi.org/10.1016/j.scitotenv.2006.12.015

16. Newcomer ME, Hubbard SS, Fleckenstein JH et al (2016) Simulating bioclogging effects on dynamic riverbed permeability and infiltration. Water Resour Res 52:2883–2900. https://doi.org/10.1002/2015WR018351

17. Newcomer ME, Hubbard SS, Fleckenstein JH et al (2018) Influence of hydrological perturbations and riverbed sediment characteristics on hyporheic zone respiration of CO2 and N2. J Geophys Res Biogeosci 123:902–922. https://doi.org/10.1002/2017JG004090

18. Billen G, Garnier J, Le Noë J et al (2020) The Seine watershed water-agro-food system: long-term trajectories of C, N, P metabolism. In: Flipo N, Labadie P, Lestel L (eds) The Seine River basin. Handbook of environmental chemistry. Springer, Cham. https://doi.org/10.1007/698_2019_393

19. Flipo N, Gallois N, Labarthe B et al (2020) Pluri-annual water budget on the Seine basin: past, current and future trends. In: Flipo N, Labadie P, Lestel L (eds) The Seine River basin. Handbook of environmental chemistry. Springer, Cham. https://doi.org/10.1007/698_2019_392

20. Flipo N, Mouhri A, Labarthe B et al (2014) Continental hydrosystem modelling: the concept of nested stream-aquifer interfaces. Hydrol Earth Syst Sci 18:3121–3149. https://doi.org/10.5194/hess-18-3121-2014

21. Billen G, Garnier J (1997) The Phison River plume: coastal eutrophication in response to changes in land use and water management in the watershed. Aquat Microb Ecol 13:3–17. https://doi.org/10.3354/ame013003

22. Billen G, Garnier J, Ficht A, Cun C (2001) Modeling the response of water quality in the Seine River estuary to human activity in its watershed over the last 50 years. Estuaries 24:977–993

23. Billen G, Garnier J, Hanset P (1994) Modelling phytoplankton development in whole drainage networks: the RIVERSTRAHLER model applied to the Seine river system. Hydrobiologia 289:119–137

24. Garnier J, Billen G, Coste M (1995) Seasonal succession of diatoms and chlorophycae in the drainage network of the river Seine: observations and modelling. Limnol Oceanogr 40:750–765

25. Brion N, Billen G (1998) Une réevaluation de la méthode de mesure de l'activité nitrifiante autotrophe par la méthode d'incorporation de bicarbonate marqué H14CO3 - et son application pour estimer des biomasses de bactéries nitrifiantes. Rev des Sci l'Eau 11:283–302

26. Brion N, Billen G, Guézennec L, Ficht A (2000) Distribution of nitrifying activity in the Seine River (France) from Paris to the estuary. Estuar Coasts 23:669–682. https://doi.org/10.2307/1352893

27. Raimonet M, Vilmin L, Flipo N et al (2015) Modelling the fate of nitrite in an urbanized river using experimentally obtained nitrifier growth parameters. Water Res 73:373–387. https://doi.org/10.1016/j.watres.2015.01.026

28. Garnier J, Billen G, Cébron A (2007) Modelling nitrogen transformations in the lower Seine river and estuary (France): impact of wastewater release on oxygenation and N2O emission. Hydrobiologia 588:291–302. https://doi.org/10.1007/s10750-007-0670-1

29. Aissa-Grouz N, Garnier J, Billen G et al (2015) The response of river nitrification to changes in wastewater treatment (The case of the lower Seine River downstream from Paris). Ann Limnol Int J Limnol 51:351–364. https://doi.org/10.1051/limn/2015031

30. Némery J, Garnier J (2007) Origin and fate of phosphorus in the Seine watershed (France): agricultural and hydrographic P budgets. J Geophys Res 112:1–14. https://doi.org/10.1029/2006JG000331

31. Vilmin L, Aissa-Grouz N, Garnier J et al (2015) Impact of hydro-sedimentary processes on the dynamics of soluble reactive phosphorus in the Seine River. Biogeochemistry 122:229–251. https://doi.org/10.1007/s10533-014-0038-3

32. Flipo N, Even S, Poulin M et al (2004) Biogeochemical modelling at the river scale: plankton and Periphyton dynamics - Grand Morin case study, France. Ecol Model 176:333–347

33. Thouvenot-Korppoo M, Billen G, Garnier J (2009) Modelling benthic denitrification processes over a whole drainage network. J Hydrol 379:239–250. https://doi.org/10.1016/j.jhydrol.2009.10.005

34. Marescaux A, Thieu V, Borges AV, Garnier J (2018) Seasonal and spatial variability of the partial pressure of carbon dioxide in the human-impacted Seine River in France. Sci Rep 8:13961. https://doi.org/10.1038/s41598-018-32332-2

35. Garnier J, Billen G, Hannon E et al (2002) Modeling transfer and retention of nutrients in the drainage network of the Danube River. Estuar Coast Shelf 54:285–308

36. Le TPQ, Billen G, Garnier J, Chau VM (2014) Long-term biogeochemical functioning of the Red River (Vietnam): past and present situations. Reg Environ Chang 15:329–339. https://doi.org/10.1007/s10113-014-0646-4

37. Sferratore A, Garnier J, Billen G et al (2006) Diffuse and point sources of silica in the Seine River watershed. Environ Sci Technol 40:6630–6635. https://doi.org/10.1021/es060710q

38. Even S, Poulin M, Mouchel J-M et al (2004) Modelling oxygen deficits in the Seine River downstream of combined sewer overflows. Ecol Model 173:177–196

39. Even S, Mouchel J-M, Servais P et al (2007) Modeling the impacts of combined sewer overflows on the river Seine water quality. Sci Total Environ 375:140–151. https://doi.org/10.1016/j.scitotenv.2006.12.007

40. Even S, Poulin M, Garnier J et al (1998) River ecosystem modelling: application of the ProSe model to the Seine river (France). Hydrobiologia 373:27–37

41. Vilmin L, Flipo N, Escoffier N et al (2016) Carbon fate in a large temperate human-impacted river system: focus on benthic dynamics. Glob Biogeochem Cycles 30:1086–1104. https://doi.org/10.1002/2015GB005271

42. Vilmin L, Flipo N, Escoffier N, Groleau A (2018) Estimation of the water quality of a large urbanized river as defined by the European WFD: what is the optimal sampling frequency? Environ Sci Pollut Res 25:23485–23501. https://doi.org/10.1007/s11356-016-7109-z

43. Polus E, Flipo N, de Fouquet C, Poulin M (2011) Geostatistics for assessing the efficiency of distributed physically-based water quality model. Application to nitrates in the Seine River. Hydrol Process 25:217–233. https://doi.org/10.1002/hyp.7838

44. Vilmin L, Flipo N, De Fouquet C, Poulin M (2015) Pluri-annual sediment budget in a navigated river system: The Seine River (France). Sci Total Environ 502:48–59. https://doi.org/10.1016/j.scitotenv.2014.08.110

45. Eckhardt N (2008) A comparison of baseflow indices, which were calculated with seven different baseflow separation methods. J Hydrol 352:168–173

46. Billen G, Garnier J (2000) Nitrogen transfers through the Seine drainage network: a budget based on the application of the "Riverstrahler" model. Hydrobiologia 410:139–150

47. Garnier J, Némery J, Billen G, Théry S (2005) Nutrient dynamics and control of eutrophication in the Marne River system: modelling the role of exchangeable phosphorus. J Hydrol 304:397–412

48. Thieu V, Billen G, Garnier J (2009) Nutrient transfer in three contrasting NW European watersheds: the Seine, Somme, and Scheldt Rivers. A comparative application of the Seneque/Riverstrahler model. Water Res 43:1740–1754. https://doi.org/10.1016/j.watres.2009.01.014

49. Billen G, Garnier J, Némery J et al (2007) A long-term view of nutrient transfers through the Seine River continuum. Sci Total Environ 375:80–97. https://doi.org/10.1016/j.scitotenv.2006.12.005

50. Passy P, Gypens N, Billen G et al (2013) A model reconstruction of riverine nutrient fluxes and eutrophication in the Belgian Coastal Zone since 1984. J Mar Syst 128:106–122

51. Flipo N, Lestel L, Labadie P et al (2020) Trajectories of the Seine River basin. In: Flipo N, Labadie P, Lestel L (eds) The Seine River basin. Handbook of environmental chemistry. Springer, Cham. https://doi.org/10.1007/698_2019_437

52. Garnier J, Laroche L, Pinault S (2006) Determining the domestic specific loads of two wastewater plants of the Paris conurbation (France) with contrasted treatments: a step for exploring the effects of the application of the European directive. Water Res 40:3257–3266. https://doi.org/10.1016/j.watres.2006.06.023

53. Servais P, Garnier J, Demarteau N et al (1999) Supply of organic matter and bacteria to aquatic ecosystems through waste water effluents. Water Res 33:3521–3531

54. Bossard M, Feranec J, Otahel J (2000) CORINE land cover technical guide: addendum 2000. Union Eur Brussels, Brussels

55. Billen G, Lassaletta L, Garnier J (2014) A biogeochemical view of the global agro-food system: nitrogen flows associated with protein production, consumption and trade. Glob Food Sec 3:209–219. https://doi.org/10.1016/j.gfs.2014.08.003

56. Le Noë J, Billen G, Garnier J (2017) How the structure of agro-food systems shapes nitrogen, phosphorus, and carbon fluxes: the generalized representation of agro-food system applied at the regional scale in France. Sci Total Environ 586:42–55. https://doi.org/10.1016/j.scitotenv.2017.02.040

57. Anglade J, Billen G, Garnier J et al (2015) Nitrogen soil surface balance of organic vs conventional cash crop farming in the Seine watershed. Agric Syst 139:82–92. https://doi.org/10.1016/j.agsy.2015.06.006

58. Ledoux E, Gomez E, Monget JM et al (2007) Agriculture and groundwater nitrate contamination in the Seine basin. The STICS-MODCOU modelling chain. Sci Total Environ 375:33–47

59. Marescaux A, Thieu V, Garnier J (2018) Carbon dioxide, methane and nitrous oxide emissions from the human-impacted Seine watershed in France. Sci Total Environ 643:247–259. https://doi.org/10.1016/j.scitotenv.2018.06.151

60. Rocher V, Azimi S (2017) Evolution de la qualité de la Seine en lien avec les progrès de l'assainissement. Editions Johannet, Paris

61. Azimi S, Rocher V (2016) Influence of the water quality improvement on fish population in the Seine River (Paris, France) over the 1990-2013 period. Sci Total Environ 542:955–964. https://doi.org/10.1016/j.scitotenv.2015.10.094

62. Vilmin L (2014) Modélisation du fonctionnement biogéochimique de la Seine de l'agglomération parisienne à l'estuaire à différentes échelles temporelles. MINES ParisTech, Paris

63. Cébron A (2004) Nitrification, bactéries nitrifiantes et émission de N2O. Université Pierre et Marie Curie, Paris

64. Parliament Council of the European Union (1991) Council Directive 91/271/EEC concerning urban waste-water treatment

65. Parliament Council of the European Union (2000) Council Directive 2000/60/EC establishing a framework for community action in the field of water policy

66. Garnier J, Servais P, Billen G et al (2001) Lower Seine river and estuary (France) carbon and oxygen budgets during low flow. Estuaries 24:964–976. https://doi.org/10.2307/1353010

67. Billen G, Décamps H, Garnier J et al (1995) Atlantic river systems of Europe (France, Belgium, The Netherlands). In: Cushing C, Cummins K, Minshall G (eds) River and stream ecosystems. Elsevier, Amsterdam, pp 389–418

68. Warren CE, Doudoroff P, Shumway DL (1973) Development of dissolved oxygen criteria for freshwater fish. US EPA, Washington

69. Escoffier N, Bensoussan N, Vilmin L et al (2018) Estimating ecosystem metabolism from continuous multi-sensor measurements in the Seine river. Environ Sci Pollut Res 25:23451–23467. https://doi.org/10.1007/s11356-016-7096-0

70. Needoba JA, Peterson TD, Johnson KS (2012) In: Tisuia-Arashiro SM (ed) Molecular biological technologies for ocean sensing. Humana Press, New York, pp 73–101

71. Kanwisher J (1959) Polarographic oxygen electrode. Limnol Oceanogr 4:210–217. https://doi. org/10.2307/2832703
72. Klimant I, Meyer V, Kühl M (1995) Fiber-optic oxygen microsensors, a new tool in aquatic biology. Limnol Oceanogr 40:1159–1165. https://doi.org/10.4319/lo.1995.40.6.1159
73. Wang S, Flipo N, Romary T (2018) Time-dependent global sensitivity analysis of the C-RIVE biogeochemical model in contrasted hydrological and trophic contexts. Water Res 144:341–355. https://doi.org/10.1016/j.watres.2018.07.033
74. Guillon S, Thorel M, Flipo N et al (2019) Functional classification of artificial alluvial ponds driven by connectivity with the river: consequences for restoration. Ecol Eng 127:394–403. https://doi.org/10.1016/j.ecoleng.2018.12.018
75. Odum HT (1956) Primary production in flowing waters. Limnol Oceanogr 1:795–801
76. Dodds WK, Veach AM, Ruffing CM et al (2013) Abiotic controls and temporal variability of river metabolism: multiyear analyses of Mississippi and Chattahoochee River data. Freshwater Sci 32:1073–1087. https://doi.org/10.1899/13-018.1
77. Battin TJ, Kaplan LA, Findlay S et al (2008) Biophysical controls on organic carbon fluxes in fluvial networks. Nat Geosci 1:95–100
78. Garnier J, Billen G (2007) Production vs. Respiration in river systems: an indicator of an "ecological status". Sci Total Environ 375:110–124. https://doi.org/10.1016/j.scitotenv.2006. 12.006
79. Marescaux A (2018) Carbon cycling across the human-impacted Seine River basin: from the modelling of carbon dioxide outgassing to the assessment of greenhouse gas emissions. Sorbonne Université, Paris
80. Le Quéré C, Andrew RM, Friedlingstein P et al (2018) Global carbon budget 2018. Earth Syst Sci Data 10:2141–2194. https://doi.org/10.5194/essd-10-2141-2018
81. Cole JJ, Prairie YT, Caraco NF et al (2007) Plumbing the global carbon cycle: integrating inland waters into the terrestrial carbon budget. Ecosystems 10:171–184
82. Passy P, Le Gendre R, Garnier J et al (2016) Eutrophication modelling chain for improved management strategies to prevent algal blooms in the Bay of Seine. Mar Ecol Prog Ser 543:107–125. https://doi.org/10.3354/meps11533
83. Blanchoud H, Schott C, Tallec G et al (2020) How agricultural practices should be integrated to understand and simulate long-term pesticide contamination in the Seine River basin? In: Flipo N, Labadie P, Lestel L (eds) The Seine River basin. Handbook of environmental chemistry. Springer, Cham. https://doi.org/10.1007/698_2019_385
84. Sutton MA, Howarth CM, Erisman JW et al (2011) The European nitrogen assessment: sources, effects and policy perspectives. Cambridge University Press, Cambridge
85. Billen G, Thieu V, Garnier J, Silvestre M (2009) Modelling the N cascade in regional watersheds: the case study of the Seine, Somme and Scheldt rivers. Agric Ecosyst Environ 133:234–246. https://doi.org/10.1016/j.agee.2009.04.018
86. Garnier J, Billen G, Vilain G et al (2014) Curative vs. preventive management of nitrogen transfers in rural areas: lessons from the case of the Orgeval watershed (Seine River basin, France). J Environ Manag 144:125–134. https://doi.org/10.1016/j.jenvman.2014.04.030
87. Garnier J, Anglade J, Benoit M et al (2016) Reconnecting crop and cattle farming to reduce nitrogen losses to river water of an intensive agricultural catchment (Seine basin, France): past, present and future. Environ Sci Pol 63:76–90. https://doi.org/10.1016/j.envsci.2016.04.019
88. Lancelot C, Thieu V, Polard A et al (2011) Cost assessment and ecological effectiveness of nutrient reduction options for mitigating Phaeocystis colony blooms in the Southern North Sea: an integrated modeling approach. Sci Total Environ 409:2179–2191. https://doi.org/10.1016/j. scitotenv.2011.02.023
89. Billen G, Garnier J (2007) River basin nutrient delivery to the coastal sea: assessing its potential to sustain new production of non-siliceous algae. Mar Chem 106:148–160. https://doi.org/10. 1016/j.marchem.2006.12.017
90. Redfield AC, Ketchum BH, Richards FA (1963) The sea. Ideas and observations on progress in the study of the seas. The composition of the sea-water comparative and descriptive oceanography. Interscience Publishers, New York, pp 26–77

91. Raimonet M, Thieu V, Silvestre M et al (2018) Landward perspective of coastal eutrophication potential under future climate change: the Seine River case (France). Front Mar Sci 5:136. https://doi.org/10.3389/fmars.2018.00136
92. Garnier J, Beusen A, Thieu V et al (2010) N:P:Si nutrient export ratios and ecological consequences in coastal seas evaluated by the ICEP approach. Glob Biogeochem Cycles 24: GB0A05. https://doi.org/10.1029/2009GB003583
93. Romero E, Le Gendre R, Garnier J et al (2016) Long-term water quality in the lower Seine: lessons learned over 4 decades of monitoring. Environ Sci Pol 58:141–154. https://doi.org/10.1016/j.envsci.2016.01.016
94. Desmit X, Thieu V, Billen G et al (2018) Reducing marine eutrophication may require a paradigmatic change. Sci Total Environ 635:1444–1466. https://doi.org/10.1016/j.scitotenv.2018.04.181

Aquatic Organic Matter in the Seine Basin: Sources, Spatio-Temporal Variability, Impact of Urban Discharges and Influence on Micro-pollutant Speciation

G. Varrault, E. Parlanti, Z. Matar, J. Garnier, P. T. Nguyen, S. Derenne, V. Rocher, B. Muresan, Y. Louis, C. Soares-Pereira, A. Goffin, M. F. Benedetti, A. Bressy, A. Gelabert, Y. Guo, and M.-A. Cordier

Contents

1 Introduction 218
2 Sources and Variability of Organic Matter in the Seine River Watershed 220
  2.1 Context and Objectives 220
  2.2 Methods for Characterising Organic Matter in River Water 220
  2.3 Sources and Spatio-Temporal Variability of OM in the Seine River Basin 223
  2.4 Impact of the Paris Conurbation Discharges on DOM Quantity and Quality
    in the Seine River 227
3 Influence of Organic Matter on the Fate of Micro-pollutants and the Role Played by Urban
  Organic Matter 228
  3.1 Context 228
  3.2 Material and Methods 229
  3.3 Characterisation of DOM Trace Metal-Binding Ability 230

The copyright year of the original version of this chapter was corrected from 2019 to 2020. A correction to this chapter can be found at https://doi.org/10.1007/698_2020_667

G. Varrault (✉), Z. Matar, B. Muresan, Y. Louis, C. Soares-Pereira, A. Goffin, and A. Bressy
LEESU, Université Paris-Est Créteil, École des Ponts ParisTech, Université Paris-Est, Créteil, France
e-mail: varrault@u-pec.fr

E. Parlanti, P. T. Nguyen, Y. Guo, and M.-A. Cordier
Univ. Bordeaux, CNRS, UMR 5805 EPOC, Talence, France

J. Garnier and S. Derenne
SU CNRS EPHE UMR 7619 METIS, Paris, France

V. Rocher
SIAAP, Direction Innovation, Colombes, France

M. F. Benedetti and A. Gelabert
Equipe Géochimie des Eaux, Univ. Paris Diderot, Sorbonne Paris Cité, Institut de Physique du Globe - UMR 7154, Paris, France

Nicolas Flipo, Pierre Labadie, and Laurence Lestel (eds.), *The Seine River Basin*, Hdb Env Chem (2021) 90: 217–242, https://doi.org/10.1007/698_2019_383, © The Author(s) 2020, corrected publication 2020, Published online: 3 June 2020

3.4   Contribution of EfDOM from Seine-Aval WWTP to the Copper-Binding Site Flux
        in the Seine River Downstream of Paris 232
3.5   Modelling of Trace Metal Speciation in the Seine River Upstream and Downstream
        of the Paris Conurbation 233
3.6   Evaluation of the Role of DOM on Copper Bioavailability Toxicity to *Daphnia
        magna* 235
3.7   Effect of EfDOM on Trace Metal Sorption by Mineral Particles in Aquatic Systems
        Subjected to Strong Urban Pressure 236
3.8   Effect of EfDOM on PAH Adsorption by Mineral Particles in Aquatic Systems
        Subjected to Strong Urban Pressure 236
4   Conclusions and Perspectives 237
References 239

Abstract This research has been conducted over the last 10 years to characterise the spatio-temporal variability of aquatic organic matter (OM) composition in the Seine River watershed upstream and downstream of Paris Megacity and its effect on micro-pollutants. For this purpose, a large number of samples were collected under different hydrological conditions, and, over 1 year, three representative sites were monitored monthly. Furthermore, the evolution of the OM composition along an urbanisation gradient, from upstream to downstream of the Paris agglomeration, was characterised, highlighting the very strong impact of urban discharges, especially during low-water periods. Substantial differences in the chemical composition are emphasised relative to the urban or natural origin of the organic matter. Dissolved organic matter (DOM) interactions with metallic and organic micro-pollutants were studied, allowing us to (1) identify the key role of DOM on their speciation and bioavailability in aquatic systems and (2) demonstrate that these interactions depend on DOM composition and origin. The essential role of urban DOM on the speciation of trace metals in the Seine River downstream of the Paris agglomeration is also shown.

Keywords Adsorption, Complexation, Dissolved organic matter, EEM fluorescence spectroscopy, eLTER, Monitoring campaigns, PAHs, Particulate organic matter, PIREN-Seine, Snapshot campaigns, Trace metals, Water, Zone Atelier Seine

1   Introduction

In freshwater environments, organic matter (OM), including particulate (POM) and dissolved (DOM) forms, is a highly complex and dynamic mixture of organic compounds derived from both natural sources and anthropogenic inputs [1–3]. The allochthonous (terrestrial plant detritus and soils) or autochthonous (in situ aquatic production) origin of natural organic matter (NOM) determines its composition and properties driving microbial processing, carbon cycling and its reactivity in aquatic ecosystems [4, 5]. NOM is mainly formed by biogeochemical processes such as photosynthesis, excretion or secretion by organisms, biomass decay or diagenesis [1]. The concentration and composition of both DOM, including colloidal OM, and

POM depend on the surrounding watershed and on the hydrological connectivity between the main channel, backwaters and groundwater, which control the transfer of terrestrial and aquatic-derived OM in freshwater ecosystems [4]. Despite the conventional agreement that terrestrial organic carbon is recalcitrant and transported conservatively along streams and rivers with little contribution to aquatic metabolism, recent studies indicate that it is much less refractory and has a greater bioavailability than previously thought [4, 6, 7].

OM plays an essential role in shaping aquatic ecosystems [8] because of the number of processes in which it becomes involved. Whatever its sources, it has a pivotal role in the autotrophy/heterotrophy balance in river systems, as described in detail in Garnier et al. [9]. OM is also well known to influence the speciation, solubility, toxicity and transport of organic and inorganic pollutants [10, 11]. Furthermore, DOM is also involved in aqueous photochemical reactions, nutrient cycling and availability [12, 13]. The mechanisms involved in all these processes are strongly dependent not only on the overall concentration of OM but also on its chemical nature, physicochemical properties and composition in the aquatic environment. Moreover, human activities alter fluvial NOM properties with OM inputs from agricultural and forest practices as well as urban, domestic and industrial sewage [1, 2]. OM is also a major concern in wastewater and drinking water treatments [14–16], affecting the efficiency of water treatment processes, the colour, odour and taste in water and resulting in the formation of disinfection by-products [15–17].

Despite their key role in environmental processes, OM composition and reactivity are still poorly understood. The characterisation of OM and its influence on the speciation of pollutants have been investigated, since 2010, in the Seine River system, as described in this chapter, and contribute to new insights into the role of OM in freshwater ecosystem functioning, its spatial and seasonal variability and its interaction with contaminants. This chapter focuses most particularly on the distinction between different OM sources, including urban point sources, interactions with watershed soils and in-stream processes.

The discharge of wastewater (treated or not) is a significant source of organic matter for the Seine River. A major trend in the trajectory of the Seine River system over the past 50 years has been the gradual improvement of wastewater treatment and the reduction in the loading of point sources of contamination to the river. The long-term trends in water quality and especially the changes in organic pollution from urban point sources are fully described in the Chapter 8. Globally, the quality of wastewater treatment has been very strongly and gradually improved since the 1970s [18]. Between 1970 and 1980, the flow of biochemical oxygen demand (BOD_5) discharged into the river decreased from 70,000 tonnes year^{-1} to 45,000 tonnes year^{-1}. This flow remains constant until 1995 and then decreases sharply to reach 20,000 tonnes year^{-1} in 2005 and 10,000 tonnes year^{-1} in 2015. The decrease occurred later for Kjeldahl nitrogen, i.e. from 30,000 tonnes year^{-1} between 1987 and 2007 to less than 10,000 tonnes year^{-1} since 2007, following the commissioning of the nitrification unit at the Seine-Aval wastewater treatment plant (WWTP) [18]. Despite the very significant improvement in wastewater treatment and thus in water quality in the Seine River over the past 50 years, questions still remain on the role of OM in the transport of nutrients and contaminants from land to sea and on the impact of OM characteristics on the ecological status and ecosystem services of the river. Since the analytical methods for characterising OM and its interaction with contaminants were not available in the 1970s, when wastewater contamination was much higher, it is thus not possible to identify any trend regarding how the influence of the type of OM on water quality and river system functions has evolved in comparison to the situation in the 1970s. The studies described in this chapter nevertheless present the current state of the system and will allow to study in the future potential changes in OM characteristics and reactivity.

2 Sources and Variability of Organic Matter in the Seine River Watershed

2.1 Context and Objectives

OM in aquatic systems plays an important role in water quality and biogeochemical processes [1]. The biogeochemical functioning of the Seine River basin has been largely studied over the last few decades within the PIREN-Seine multidisciplinary scientific programme [19]. The water quality and ecological functioning of the Seine River watershed have been mainly investigated through material budget approaches applied to biogenic elements [20, 21] as well as to chemical and microbiological contaminants [22, 23]. Previous investigations concerning carbon in the Seine River basin mainly focused on organic carbon fluxes [24], biogeochemical mass balances (C, N, P, Si) [20], methane emissions [25] or production and respiration [21]. However, not only dissolved organic carbon (DOC) quantity but also its quality reflect the dynamic interaction between OM sources and biogeochemical processes [26] and are likely to have ecological repercussions [27, 28].

To achieve a better understanding of OM environmental role in aquatic ecosystems, it is essential to improve the knowledge on its physicochemical properties and composition in the aquatic environment. The main goals of the studies conducted since 2010 on the biogeochemistry of the Seine River within the PIREN-Seine programme were to characterise the sources and properties of OM and their spatio-temporal variability in the Seine River watershed.

2.2 Methods for Characterising Organic Matter in River Water

To assess the variability of OM characteristics from both quantitative and qualitative points of view, surface water samples were collected in 23 and 39 sites at low water (November 2011 and September 2012) and 40 sites at flood levels (Fevrier 2013), from the Seine River, the Oise River, the Marne River and its main tributary the

Grand Morin River upstream of the Paris urban area and in the Seine River downstream of Paris urban area and downstream of the largest discharge point of wastewater occurring through the effluents of the Seine-Aval WWTP at Achères (Fig. 1).

Monthly monitoring was carried out over a 1-year period upstream and downstream of the Paris agglomeration in order to highlight the impact of the urban discharges on the DOM of the Seine River in terms of both quality and quantity [29]. There were nine sampling campaigns from October 2010 to September 2011, and three sites were sampled:

- Ussy-sur-Marne (labelled "Upstream I"; open blue circle on Marne River in Fig. 1) and Fontaine-le-Port (labelled "Upstream II"; open blue circle on Seine River upstream Paris in Fig. 1) are located upstream of the conurbation on the Marne and Seine Rivers, respectively. Due to the low population density in their catchment, these two sites were chosen as reference sites given that they were exposed to a low impact from urban discharges.
- Andresy (labelled "Downstream"; open blue circle on Seine River downstream Paris in Fig. 1) is located downstream of Paris approximately 9 km from the Seine-Aval WWTP outlet, ensuring a good level of mixing in Seine River water. This site was selected as being under high urban pressure.

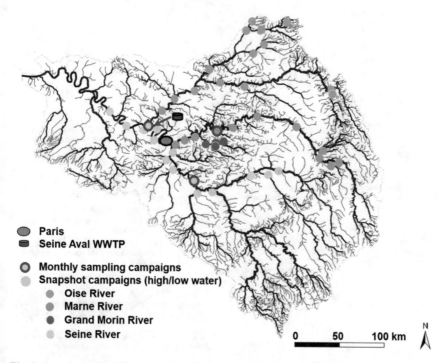

Fig. 1 Location of the different sampling sites in the Seine River and its main tributaries: the Oise River, the Marne River and the Grand Morin River. Filled circles, snapshot campaigns; open blue circles, monthly campaigns upstream and downstream from the Paris conurbation

In addition, more sampling campaigns were carried out in low-water periods (eight campaigns between 2011 and 2014) and flood in February 2013 to determine longitudinal profiles of particulate organic matter (POC), DOC and biodegradable DOC (BDOC) concentrations.

The mean discharge observed at the Paris gauging station (Austerlitz) was around 100 m^3 s^{-1} for both low-water campaigns. The contribution of the large reservoirs in the upstream basin regulating the hydrology of the Seine River was about 50% in August/September 2012 compared to only 20% in November 2011. The high-water campaign occurred in February 2013 with flows amounting to 700 m^3 s^{-1} for the Seine River at Alfortville – the confluence of the Seine and Marne Rivers – with a contribution of 350 m^3 s^{-1} by the Marne River at Gournay, and 280 m^3 s^{-1} at Creil for the Oise River, a downstream tributary of the Seine.

DOM, including colloidal organic matter, and POM were separated immediately after sampling by filtration on precombusted (450°C) glass fibre filter (Whatman GF/F 0.70 µm). As shown in Fig. 2, both DOM and POM fractions could then be characterised in terms of biodegradability, hydrophobicity and other chemical properties owing to a wide range of analytical methods.

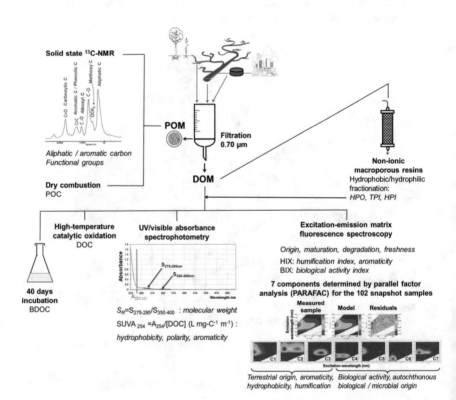

Fig. 2 Different analytical techniques used for the separation or the characterisation of particulate (POM) and dissolved (DOM) organic matter

The organic carbon content was determined by POC and DOC high-temperature combustion measurements, and BDOC was assessed through incubation experiments [30].

Nuclear magnetic resonance (NMR) spectroscopy has been largely used to investigate NOM structure [31]. To identify the main functional groups of OM in the Seine River catchment, the solid-state ^{13}C NMR spectra were recorded for POM isolated on glass fibre filters after filtration of 3–10 L of surface waters collected from selected sampling sites during the low water (2012) and flooding (2013) campaigns.

For selected sampling sites, about 10 L of surface waters collected in low water (2012) and flooding (2013) periods were fractionated on non-ionic macroporous DAX-8 and XAD-4 resins so as to separate DOM into different fractions according to polarity criteria. The hydrophobic (HPO), transphilic (TPI) and hydrophilic (HPI) fractions were thus isolated according to a protocol described elsewhere [29].

The application of optical properties as a proxy for DOM composition and reactivity in aquatic environments has been extensively applied [28]. To gain information on its origin, dynamics and degree of transformation, UV/visible absorbance and excitation emission matrix (EEM) fluorescence spectroscopy combined with parallel factor analysis (PARAFAC) were used to characterise DOM in the Seine River watershed [32]. UV/visible absorbance spectroscopy was used to determine DOM properties at natural pH. The spectral slope ratio S_R ($S_R = S_{275-295\ nm}/S_{350-400\ nm}$) was used to estimate the variation of the molecular weight of DOM; the molecular weight decreases when S_R increases [33]. The specific UV absorbance at 254 nm ($SUVA_{254}$) is strongly correlated with the hydrophobic organic acid fraction of DOM and is a useful proxy for DOM aromatic content and molecular weight [34]. Information on the origin and the transformation degree of DOM can be obtained through the calculation of the humification index (HIX) [35] and the biological index (BIX) [36].

2.3 Sources and Spatio-Temporal Variability of OM in the Seine River Basin

For the three snapshot campaigns, DOC concentrations were not significantly different among the Seine, Marne and Oise subbasins, whether it was in low or high waters (about 3 mgC L^{-1}), with, however, slightly higher values observed for the Oise subbasin. For the Seine River, DOC concentrations were higher downstream of Paris, and this increase of DOC amounts between the upstream and downstream of the Paris conurbation was more pronounced at low water. The biodegradable fraction (BDOC), however, differed by a factor >4 (28% vs. 6% in low water vs. high water, respectively). The Oise River was clearly distinguished by its higher POC content in suspended solids during the high flow in February 2013, much higher than the values of the other subbasins.

The HPO, TPI and HPI fractions were isolated, according to DOM polarity, from 14 surface water samples collected in the low-water period in 2012 and 17 during flooding in 2013. At low-water period, the proportion of DOC in the different fractions showed little variation between sampling sites with the HPO, TPI and HPI fractions averaging 31%, 27% and 42% of DOC, respectively. The HPI proportion was slightly higher in the subbasins of the Marne River and the Seine River, however. These results point out the higher presence of fresh autochthonous DOM at this low-water period. During flooding, the distribution of DOC was also homogenous between the three fractions, with 43%, 29% and 28% for the HPO, TPI and HPI fractions, respectively. A strong increase of the hydrophobic fraction (+40%) and a decrease of the hydrophilic fraction highlighted greater terrestrial inputs for this hydrological condition.

The structural characterisation of POM was investigated by ^{13}C NMR analysis of a few selected samples, representative of the whole Seine River watershed. The analysis of the ^{13}C NMR spectra demonstrated great spatial variability of POM characteristics. Samples from the Oise River and the upstream Seine River collected in 2012, during a low-water period, were characterised by fresh OM, mainly polysaccharides, with also a high terrestrial contribution revealed by the presence of lignin. Spectra of samples collected in the upstream Oise River and in the Seine River downstream of Paris at the same period showed a decrease of aliphatic carbon and carbohydrate signals with a relative increase of the aromatic carbon signal, highlighting the presence of more humified organic material. Except for one location in the upstream Oise River, differences were observed between spectra recorded for all sampling sites in low-water and flood conditions. During flooding, POM for samples in the Oise River and the upstream Seine River exhibited more aromatic characteristics, while fresh material was observed in the upstream Oise River, in contrast to results for the low-water stage. For the two samples in the Oise River and the upstream Seine River, the lower freshness character of OM observed with a higher DOC content could suggest an input of OM from groundwater. Conversely, soil surface leaching could explain the presence of more recent OM and higher DOC levels for the upstream Oise River sample. A high spatial and temporal variability of POM characteristics depending on hydrological connectivity and conditions was thus highlighted by ^{13}C NMR analysis.

The optical properties of DOM were determined, after filtration, for all surface water samples collected during the three snapshot campaigns.

The highest values of S_R were observed for samples collected from the Marne River in low-water periods (2011 and 2012), reflecting the presence of lower molecular weight OM in this area, whereas we noted, for all the sampling sites in 2013 (flooding), the lowest S_R values, corresponding to higher-molecular-weight OM. Three of the four samples collected from the Grand Morin River in 2011 were also characterised by organic components of high molecular weight (HMW; low S_R values) as in the 2013 flood. As for the samples of the Seine River in 2012, they were characterised by low-molecular-weight (LMW) DOM and especially upstream of Paris [32].

It is considered that natural waters with high SUVA$_{254}$ values ($\geq$4 L mg^{-1}C m^{-1}) have relatively high contents of hydrophobic, aromatic and high-molecular-weight DOM, whereas waters with $SUVA_{254}$ values ≤ 2–3 L $mg^{-1}C$ m^{-1} contain a mixture of hydrophobic and hydrophilic DOM from various sources with a range of molecular weights. The highest $SUVA_{254}$ value was determined for one sample from the Oise River in 2011 (5.3 L $mg^{-1}C$ m^{-1}), reflecting a relatively high content of complex heterogeneous macromolecular organic compounds rich in aromatics, while DOM in all other samples exhibited lower $SUVA_{254}$ values (1.0–3.8 L $mg^{-1}C$ m^{-1}), implying a wide range of DOM characteristics and molecular weights [32].

All surface waters were then analysed using EEM fluorescence spectroscopy as previously described [36, 37]. Spatial and temporal variations of DOM quality were highlighted according to hydrological conditions [32]. The higher concentrations of fluorescent materials were found for waters collected in 2011 (low water) in comparison to those sampled in 2012 (low water) and in 2013 (flooding). The HIX [35] and BIX [36] indices were calculated to gain information on DOM origin and transformation. The highest HIX values, corresponding to aromatic and mature organic material, were observed in the Oise basin with maxima in the forest zones (≥ 20). A relatively high biological activity (BIX > 0.6) characterised all the samples, the highest values (BIX > 0.75) being observed for the Seine River in 2011 and 2012 and in the Marne River basin in 2012. The flood period (2013) was characterised by high HIX values (mainly ≥ 10), while the lowest values were observed during the low-water stage (2012), mainly associated with high BIX values, indicating a strong biological activity for these samples [32].

A seven-component model was determined by PARAFAC analysis, explaining 99.8% of the total data set (102 surface water samples) variability [32]. The EEM spectra contour plots of the seven components are given Fig. 3 (top). The seven components determined showed similarities, with peaks identified in previous studies and related in the literature [32]. Components 4, 5, 6 and 7 were related to biological activity, while components 1, 2 and 3 to aromatic and mature terrestrial material (C1 and C2 being mainly linked to aromaticity).

The distributions of these components within the different subbasins and according to the sampling periods are given in Fig. 3 (bottom). Three types of OM were distinguished, the Seine River sub-catchment being mainly characterised by the strongest biological activity, the Oise River subbasin by more terrestrial signatures (even more pronounced for samples taken from forest areas) and a part of the Marne River basin by a third specific type of OM, with a higher contribution of component 7 in particular.

The highest biological signature determined in the Seine River was for samples collected downstream from Paris, pointing out the impact of urban discharges on DOM properties [32]. Moreover, mature DOM of terrestrial origin characterised the samples collected during the flood, while samples collected in the Seine River and the Marne River subbasins during low-water periods (2011 and 2012) were composed of constituents derived from biological activity. However, at low water, a terrestrial signature still described the Oise River sub-catchment. Furthermore, during the flood period, a similar distribution of the seven components was observed for all the samples, whatever the river subbasin, demonstrating a predominant source of terrigenous DOM in the entire watershed.

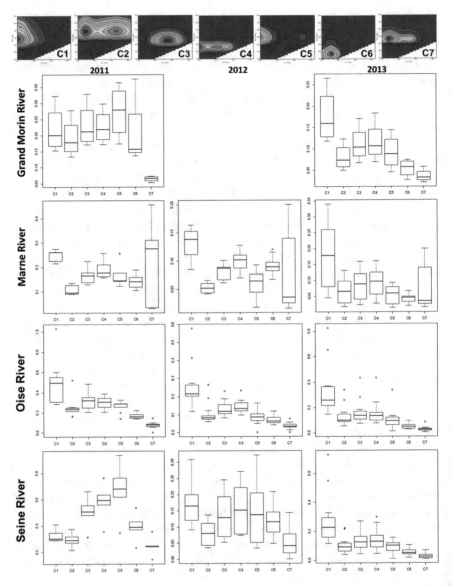

Fig. 3 (Top) EEM contour plots of the seven components identified by PARAFAC C1, C2 and C3, from terrestrial origin, humification process; C4–C7, from biological activity, autochthonous biological/microbial origin. (Bottom) Distribution of the components horizontally, river basin, and vertically, sampling period (2011 and 2012 low water, 2013 high water). Fluorescence intensities in Raman units

Both qualitative and quantitative variations of DOM were thus related to hydro-logical conditions (flood, low water) and specific geographical locations. The results obtained with the various DOM characterisation tools used in this study are

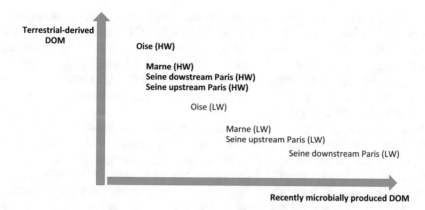

Fig. 4 Origin of DOM in the various watersheds at **high water (HW)** and low water (LW) according to the different DOM characterisation tools used in this study

consistent and highlight that, overall, during high water, DOM presents a greater contribution of DOM from terrigenous origin, especially for Oise River. In low-water periods, the proportion of DOM from terrestrial origin is significantly lower with a higher relative contribution of recently microbially produced dissolved organic matter, which seems to indicate a recent production of autochthonous DOM and/or a contribution from effluent DOM (EfDOM) especially in the Seine River downstream Paris (Fig. 4). At low water, greater differences in the properties of DOM were observed between the three river subbasins.

2.4 Impact of the Paris Conurbation Discharges on DOM Quantity and Quality in the Seine River

Monthly monitoring was carried out over a 1-year period upstream (Upstream I and II sites) and downstream ("Downstream" site) of the Paris agglomeration in order to highlight the impact of the urban discharges on the DOM of the Seine River in terms of both quality and quantity. The water quality parameters measured are presented elsewhere [38]. Throughout the campaign, DOC concentrations at the Upstream I and II sites are very close and follow the same trend: they range from 1.5 to 3.3 mgC L^{-1}, and in both cases, the maximum DOC value is reached during the high-water period in November and January and is correlated with the flow increase. This suggests that this OM originates in part from runoff on the soil and/or erosion from the degrading litter, which is abundant at this period on the banks.

At the downstream site, DOC concentrations were higher than those observed upstream of the agglomeration and varied between 2.5 and 4.5 mgC L^{-1}. This is most likely due to urban discharges from the urban area. It can be highlighted that the DOC concentration differences between downstream and upstream sites are related to river flow and therefore to the dilution of Paris metropolitan area WWTP discharges (23 m^3 s^{-1} for SIAAP stations). During the high-flow period

$(950 \text{ m}^3 \text{ s}^{-1})$, the DOC increase at the downstream site compared to the upstream sites was less than 0.5 mgC L^{-1}, while it reached 1.5 mgC L^{-1} during periods of low water $(100–200 \text{ m}^3 \text{ s}^{-1})$. These values are fully consistent with DOC concentrations in the range of 10–15 mgC L^{-1} in treated effluent from the Seine-Aval WWTP.

Regarding the DOC distribution between HPI, TPI and HPO fractions at upstream and downstream sites, a high temporal variability can be observed. Indeed, the hydrophobic fraction (HPO) exceeded 60% in November when soil runoff and leaching started again and then decreased during winter floods (45% of the DOC). The lowest values, less than 30%, were observed during low-water periods due to autochthonous production as well as the lower terrigenous input. The spatial variability between the upstream and downstream sites is quite clear: the HPI fraction accounts for about 50% of DOC in low water (between March and September 2011) at the downstream site and only around 40% at upstream sites. This increase of the HPI fraction at the downstream site is probably due to effluent dissolved organic matter (EfDOM) from WWTPs.

From these data, instantaneous DOC fluxes and DOC flux distribution according to the HPO, TPI and HPI fractions were determined. For each sampling campaign, the DOC flux increased across the Paris conurbation, but the higher increase was noted during low-water periods (from March to September): the DOC flux increase was on average 100%, 140% for the HPI fraction and 75% for both HPO and TPI fractions. This means that in the receiving environment downstream of the agglomeration, during low-water periods, 50% of the DOC came from EfDOM, and this proportion reached nearly 60% for HPI DOC.

This increase in DOC flux through the Paris conurbation area was greater during low-water periods due to the lower dilution of EfDOM. This flux increase was particularly high for the HPI fraction, which is fully consistent with the distribution of DOC observed in EfDOM [29]. These results therefore demonstrate that, even with the very significant improvement in wastewater treatment over the past 50 years, EfDOM of the Paris conurbation remains a very important DOM source for the Seine River. It has a very strong impact on DOM present downstream, both in terms of quantity and quality, causing a significant increase in hydrophilicity downstream of the agglomeration.

3 Influence of Organic Matter on the Fate of Micro-pollutants and the Role Played by Urban Organic Matter

3.1 Context

In receiving waters, DOM is a key component of both trace metal speciation [10] and bioavailability [11, 39]. Many studies have highlighted the trace metal-binding ability of DOM in surface waters [10, 11, 40], yet the majority of published works generally concern either the DOM collected from surface waters not affected by urbanisation or humic substances (HS), as isolated from natural waters. HS represent the hydrophobic acid fraction of DOM [41] and typically constitute 40–60% of DOC in most natural surface waters [42–44]. However, in rivers under strong urban pressure such as the Seine River in the Paris conurbation, the proportion of the HPO fraction of DOM decreases as a result of various urban discharges, and the HPI fraction can increase to 50% of DOC, as was shown in the previous section.

At the beginning of this study, very little information focusing specifically on trace metal binding by HPI DOM in freshwater was available in the literature, most likely due to the difficulty of its isolation and purification. The aim of this study was to determine trace metal-binding parameters of EfDOM from WWTPs, most particularly its HPI fraction, in order to improve the knowledge of metal speciation and bioavailability within aquatic systems under strong urban pressure.

The study of metal complexation by EfDOM and by DOM from receiving water under high urban pressure was conducted in two phases. In the first phase, DOM was fractionated according to its polarity into HPO, TPI and HPI fractions [45, 46]. In this first phase, lead, mercury and zinc [47–49] complexation by the different DOM fractions was studied. Unfortunately, this very time-consuming method can only be applied to a small number of samples, which makes it difficult to characterise DOM spatio-temporal variability in receiving waters.

For this reason, in the second phase, we decided to no longer fractionate DOM so that the complexing properties of the urban DOM could be investigated for a larger number of samples and then the DOM spatio-temporal variability taken into account. Only copper complexation was studied [29, 38] in this second phase, focusing on the role of EfDOM from WWTPs in copper speciation and bioavailability in receiving waters across the Paris conurbation.

Furthermore, DOM can play an important role in the sorption/desorption of pollutants on suspended solids in aquatic environments. Indeed, there may be (1) competition between DOM and pollutants for adsorption on the particle surface [50], (2) DOM can also clog the particles pores and thus limit pollutant adsorption [51–53], (3) by adsorbing on particles, DOM can modulate their pollutant binding ability [54] and (4) DOM can also bind pollutants and therefore maintain them in solution or, on the other hand, make it easier for them to adsorb as they are associated with DOM [55, 56]. These interactions may depend on the quality of the DOM. EfDOM and DOM of terrigenous origin could therefore have a different effect on the pollutant sorption onto particles. In this study [57], the effect of EfDOM or fulvic acid onto adsorption of PAHs and trace metals was studied.

3.2   Material and Methods

3.2.1   Sampling Points

To collect EfDOM, seven campaigns were conducted at the Seine-Aval WWTP (see Fig. 1), which collected, in 2012, over 70% of dry weather flows from the Paris conurbation (six million inhabitants). These effluents accounted for more than 80%

of the effluent dissolved organic carbon discharged into the Seine River from the entire conurbation, thus offering highly representative samples. Treatment consisted of primary settling for suspended solid removal, aerobic activated sludge for carbon removal and biofilters for nitrogen removal (nitrification and denitrification).

To characterise the spatial evolution of DOM trace metal complexation parameters in receiving waters throughout the Paris conurbation, three sampling sites (labelled Downstream and Upstream I and II) were selected (see Sect. 2.2 and Fig. 1). Five sampling campaigns were carried out at these three sites during both the high-flow period (November 2010 to January 2011) and low-flow period (March 2011, June 2011 and September 2011).

A standard fulvic acid (Suwannee River Fulvic Acid, SRFA 2S101F), purchased from the International Humic Substances Society, was used as a "natural organic matter" reference for follow-up investigations relative to DOM influence on copper bioavailability and pollutant adsorption onto particles.

3.2.2 Experimental Procedures

Experimental procedures concerning (1) sample treatment; (2) characterisation of copper-, lead-, zinc- and mercury-binding abilities; and (3) characterisation of DOM influence on copper bioavailability as well as the methodology used for trace metal speciation modelling are given in several publications [29, 38, 47–49].

Concerning the influence of the DOM type on pollutant adsorption, experimental procedures are described elsewhere [57]. Briefly, EfDOM or SRFA (10 mgC L^{-1}) and particles (0–5,000 mg L^{-1} of montmorillonite, goethite or quartz) were mixed for 72 h. Seven metals (Cd, Co, Cu, Ni, Pb, V and Zn) as well as As were then added, and the whole was left in equilibrium for 72 h. Trace metal concentrations were around 1 μmol L^{-1} except for cadmium (0.1 μmol L^{-1}). Tubes were centrifuged at a relative centrifugal field of 4,000 $\times$ g for 60 min. Supernatant solutions were filtered through a 0.45-μm polypropylene filter, and the total dissolved metal concentration was determined by ICP-AES in order to deduce the proportion of adsorbed metals. Another aliquot was filtered through a chelating disc (Chelex 100®) to distinguish between labile complexes (mainly mineral) and inert complexes (mainly organic) [58]. The control was processed without DOM.

Adsorption of six PAHs – pyrene (Pyr), benzo[a]anthracene (B[a]A), chrysene (Chry), benzo[b]fluoranthene (B[b]F), benzo[k]fluoranthene (B[k]F) and benzo[a] pyrene (B[a]P) – were studied with the same procedure.

3.3 Characterisation of DOM Trace Metal-Binding Ability

3.3.1 Binding Site Modal Distribution

For the sake of simplicity, in the bimodal distribution of binding site stability constants, the modal distribution with a lower stability constant was labelled "L_1", and the modal distribution with the higher stability constant was labelled "L_2". L_1 sites could correspond to carboxylic-type groups and L_2 sites to phenolic-type groups, although the reality is much more complex [59].

For copper and lead complexation by DOM from receiving waters, the trace metal complexation was modelled with a bimodal distribution, in accordance with the literature [60]. Whereas for their complexation by EfDOM, a trimodal binding site distribution proved to be more suitable [29, 47]. In the case of copper, there is a very high-affinity site type (L_3), in addition to the typical low- (L_1) and high-affinity sites (L_2). For lead, this is a third modal distribution corresponding to sites with an intermediate stability constant between stability constants of low- and high-affinity sites. This third family of binding sites could be associated with proteinic structures, amines and amide groups present in the hydrophilic fraction of EfDOM [61]. Some of these binding sites might also be associated with amino polycarboxylates, phosphonates and hydroxycarboxylates [62].

In contrast, for zinc, a bimodal distribution was suitable for all DOM fractions including EfDOM, whereas for mercury only one site type was obtained for all DOM fractions. This single site is usually associated with reduced N (amide, amine) and especially S (thiol) groups of proteinic structures and, to a lesser extent, with carboxylic groups.

3.3.2 Me-DOM Stability Constants and Binding Site Density

In the case of copper complexation at pH 8, for the five campaigns in receiving waters and seven campaigns for EfDOM, the conditional stability constants ranged from $10^{7.3}$ M^{-1} to $10^{7.9}$ M^{-1} for low-affinity sites (L_1) and from $10^{10.2}$ M^{-1} to $10^{11.2}$ M^{-1} for high-affinity sites (L_2), and no significant difference was found across the various sampling points; moreover, no temporal variations in stability constant were observed.

The L_1 copper-binding site density values were similar between upstream sites in receiving waters (≈ 1 mol of binding sites per kg of organic carbon) and higher than for EfDOM (≈ 0.5 mol kg^{-1}). The L_2 binding site density values were lower than for L_1 binding sites with equivalent values for receiving waters and EfDOM, ranging from 0.1 to 0.3 mol kg^{-1}. EfDOM presented by far the lowest (L_1/L_2) ratio, with an average value, equal to 2.7, much smaller than those obtained for receiving water DOM, whose average ratio varied between 7.3 and 7.9. In the case of lead, the L_1/L_2 ratio value was 7.3 for receiving waters and reached only 0.7 for the EfDOM fractions. Even more than in the case of copper, a higher presence of high-affinity binding sites was noted in EfDOM fractions compared to receiving water DOM. The total lead-binding site density was much higher for the HPI EfDOM fraction compared to other EfDOM and downstream fractions.

In the case of very high-affinity (L_3) copper-binding sites that were highlighted only for EfDOM, conditional stability constants ranged from $10^{12.2}$ M^{-1} to $10^{13.6}$ M^{-1}. The L_3 site density (mean, 5.7×10^{-2} mol kg^{-1}) was less than L_2 and L_1 densities; it amounted to between 25% and 33% of the L_2 density. The L_3

binding site concentration was slightly less than 10^{-6} mol L^{-1} in Seine-Aval WWTP effluent. Even if it was lower than those of L$_1$ and L$_2$ sites, it was environmentally relevant given that trace metal concentrations in receiving waters were considerably lower, ranging from 10^{-10} to 10^{-7} mol L^{-1}.

For zinc, there was not a very-high-affinity site but, regardless of the sampling site (receiving waters or EfDOM), the HPI fraction could systematically be characterised by higher stability constants (K$_1$ and K$_2$) and a lower L$_1$/L$_2$ ratio compared to the other fractions [49].

Regarding mercury, stability constants for EfDOM fractions ($10^{25.2}$ M^{-1}) were significantly higher than for the SRFA standard ($10^{22.8}$ M^{-1}) but roughly equal to those observed for the Upstream I site DOM fractions ($10^{24.6}$ M^{-1}) [48]. In both cases, no significant differences were observed depending on the different fractions. Furthermore, it has been shown that stability constants are positively correlated with total organic N and S. N- and S-enriched organic materials, such as EfDOM, had stronger Hg-complexing ligands [48].

Overall, these results highlight the strong binding capacity of EfDOM. The high binding ability of EfDOM and especially its HPI fraction could be explained by their nitrogen and sulphur group contents [47–49]. It could be noted that EfDOM revealed a high trace metal-binding ability despite its low aromaticity, as indicated by its low SUVA [29].

3.4   Contribution of EfDOM from Seine-Aval WWTP to the Copper-Binding Site Flux in the Seine River Downstream of Paris

During the low-water period, the copper-binding site molar concentrations in the Seine River evolved significantly from upstream to downstream of Paris, with an increase of 173% for L$_1$ sites and 157% for L$_2$ sites [29]. This increase was probably due to EfDOM discharges in the receiving waters. L$_1$ and L$_2$ site concentrations in EfDOM were, respectively, three times and ten times higher than in upstream waters [29]. Based on the experiments carried out in the Piren-Seine programme, the contribution of EfDOM from the Seine-Aval WWTP to the binding site fluxes in receiving water downstream of the Paris conurbation was assessed [29].

The ratio of binding sites originating from EfDOM to total binding sites measured downstream of Paris was calculated as described in [29]. This instantaneous flux ratio was calculated for all sampling dates (Fig. 5).

During high-flow periods, the contribution of EfDOM from the WWTP Seine-Aval to the total binding site concentration at the downstream site was around 5% and 23%, respectively, for low-affinity (L$_1$) and high-affinity (L$_2$ + L$_3$) sites. Due to the lower EfDOM dilution, during the low-flow period, these averaged values were two to five times higher, i.e. reaching 19% and 58%, respectively. Downstream of Paris, during the low-water period, the largest proportion of high-affinity binding

Fig. 5 Seine River flow at Paris Austerlitz and temporal variations in the percentage of the Cu-binding site flux (L1, top; and L2 + L3, bottom) at the downstream site originating from EfDOM. Figure reproduced from Fig. 3 in Ref. [29]

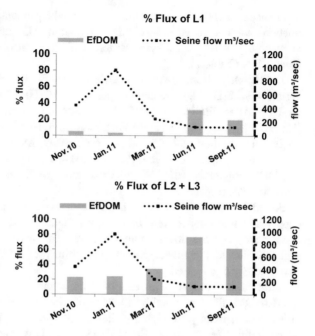

sites stems from EfDOM; the other smaller portion originates from upstream of Paris. Downstream of Paris, the percentages of high-affinity binding sites ($L_2 + L_3$) originating from EfDOM are much greater than those of low-affinity binding sites (L_1), which underscores the strong impact, especially during low-flow periods, of urban discharges on the high-affinity binding site concentration downstream of Paris.

3.5 Modelling of Trace Metal Speciation in the Seine River Upstream and Downstream of the Paris Conurbation

To assess the environmental implications relative to the strong metal-binding ability of EfDOM, metal speciation was computed in the Seine River upstream and downstream of the Seine-Aval WWTP discharge point. Binding parameters experimentally determined in works carried out within the Piren-Seine programme [29, 47–49] were used to conduct the computation. The competition of metal with other trace metals or major elements for DOM complexation was not taken into account. The pH was either set at 8, which is the known pH value for the Seine River, or 6.8, which accounts for the DOM-induced pH value in Hg-DOM stability constant assessment tests. The complete methodology for modelling is presented elsewhere [29, 47–49].

In the case of Hg, the distribution of dissolved Hg among the three different DOM fractions was calculated in receiving waters at the Upstream I site, in EfDOM and in receiving waters considering different dilution factors (α) of EfDOM in the Seine River (Table 1).

The Hg-DOM from Upstream I consisted of 94% Hg-HPO, 5% Hg-TPI and 0.7% Hg-HPI. For EfDOM, these percentages were 25% Hg-HPO, 22% Hg-TPI and 53% Hg-HPI. After dilution of EfDOM in Upstream I water, the [Hg-DOM]$_{EfDOM}$: [Hg-DOM]$_{Upstream\ I}$ ratio values decreased from 19 down to 2.3 and 0.8, respectively, for a dilution of 4, 25 and 80, respectively. As regards the [Hg-HPI]$_{EfDOM}$: [Hg-HPO]$_{Upstream\ I}$, ratios were 10, 1.4 and 0.4. The contribution of urban ligands to [Hg-DOM] of mixed EfDOM and Upstream I waters decreased from 95%, 70% down to 43%, respectively.

For other trace metals, we obtained the same type of results. For copper, inorganic copper and free copper were both more abundant upstream of Paris, where free copper exceeded the downstream value by a factor of 2–4 [29]. This result was probably due to the higher complexing affinity of DOM downstream of Paris, as previously highlighted.

In the case of zinc, downstream of Paris, at the low-water period and therefore low EfDOM dilution, roughly 49–94% of total zinc was bound to the EfDOM HPI fraction, mainly through L_2 sites at 10 and 1 μg L^{-1} total dissolved zinc concentration, respectively. EfDOM and especially the HPI fraction therefore controlled zinc speciation in the Seine downstream of Paris to a large extent. It can be highlighted that the presence of EfDOM downstream of Paris decreases the percentage of free zinc by a factor ranging from 1.8 at $[Zn]_{total} = 10$ μg L^{-1} to 12 at $[Zn]_{total} = 1$ μg L^{-1}.

Table 1 Modelled dissolved Hg partition between Upstream I site DOM and EfDOM considering different dilution factors of EfDOM in the Seine River

Type of DOM	DOM fraction and total contribution	Dissolved Hg distribution among DOM fractions and DOM types under various water dilution conditions			
		$\alpha = 1$ (prior to dilution)	$\alpha = 4$ (low waters)	$\alpha = 25$ (mean dilution)	$\alpha = 80$ (high waters)
Upstream I site DOM	HPO	94 ± 7	5 ± 1	28 ± 2	54 ± 4
	TPI	5 ± 1	0.3 ± 0.2	1.0 ± 0.7	3 ± 1
	HPI	0.7 ± 0.5	<0.1	0.2 ± 0.1	0.4 ± 0.2
	Total	100	5 ± 2	30 ± 3	57 ± 6
EfDOM	HPO	25 ± 3	24 ± 2	18 ± 1	11 ± 1
	TPI	22 ± 2	20 ± 1	15 ± 1	9 ± 1
	HPI	53 ± 4	51 ± 3	38 ± 3	23 ± 2
	Total	100	95 ± 7	70 ± 6	43 ± 5

Data were obtained using laboratory conditions with experimentally determined stability constants and binding site density values

Concerning lead, the results highlighted that between 82% and 90% of Pb is bound to the EfDOM HPI fraction, overcoming by a wide margin the complexation to the river HPO fraction (humic substances) [47].

Despite the uncertainties, it has been demonstrated in this work that dissolved organic matter contained in WWTP-treated effluents and particularly its hydrophilic fraction, due to their very high complexing properties, can control a large proportion of the dissolved trace metal speciation downstream of the Paris conurbation, particularly in low-water periods because of low EfDOM dilution.

3.6 Evaluation of the Role of DOM on Copper Bioavailability Toxicity to Daphnia magna

The median effective concentration (EC50) values are the toxic concentration for 50% of a population of *Daphnia magna*. They are expressed in total dissolved copper concentration ($EC50_{tot}$) for each DOM studied and for the inorganic matrix (Fig. 6). Compared with tests conducted in the inorganic matrix, the $EC50_{tot}$ values were systematically higher in the presence of DOM. By binding copper, DOM actually reduced its bioavailability and therefore caused an EC50 increase to higher concentrations. This outcome was very consistent with the literature, which describes the role of organic matter in reducing the toxicity of metals [63–67].

Statistically significant differences (by a factor of 2) in $EC50_{tot}$, depending on the type of DOM, were observed. DOM originating upstream of Paris displayed significant differences with the lower $EC50_{tot}$ value for Upstream II when compared to Upstream I. The Upstream I value was comparable to those obtained for the Suwannee River Fulvic Acid (SRFA) and for downstream site. A slightly higher $EC50_{tot}$ was found for EfDOM, which confirms the influence of EfDOM in reducing copper toxicity [29, 39]. As opposed to what has been observed in natural waters [65], no positive correlation could be derived either between $EC50_{tot}$ and SUVA of

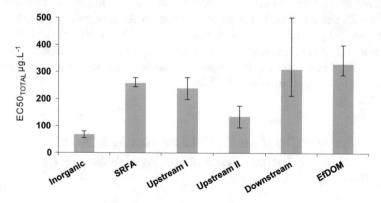

Fig. 6 EC50 expressed in dissolved total copper concentrations, for each DOM studied, SRFA and the inorganic matrix. Figure reproduced from Fig. 4 in Ref. [29]

DOM, which acts as a good proxy for DOM aromaticity, or between EC50$_{tot}$ and the DOM aromaticity percentage [29, 49]. The highest EC50$_{tot}$ values, however, were observed for EfDOM with very low aromaticity and a low UV absorbance. Compared to DOM from Upstream I and II, EfDOM and Downstream DOM presented higher densities of high-affinity binding sites (L$_2$) and even very high-affinity binding sites (L$_3$) for EfDOM, which may be responsible for the greater protective role played by these DOM samples.

Logically, given the strong affinity of EfDOM for copper, these results prove that EfDOM has a considerable influence on copper toxicity on *Daphnia magna* with a slightly greater protecting effect than both SRFA and DOM from upstream of Paris.

3.7 Effect of EfDOM on Trace Metal Sorption by Mineral Particles in Aquatic Systems Subjected to Strong Urban Pressure

The results obtained showed that in many cases trace metal adsorption onto montmorillonite (MMT) and goethite was strongly modified by DOM compared to the control without DOM [57]. In the case of adsorption onto MMT, there was no influence of the DOM type on copper adsorption, with a 65% decrease with both EfDOM and SRFA. The adsorption decrease for Zn was 17% with SRFA and reached 60% with EfDOM. Given the very large increase of inert complexes in the dissolved fraction (+200%) with EfDOM compared to SRFA, it would appear that the substantial Zn adsorption decrease with EfDOM is due to maintaining Zn in solution by complexation with EfDOM. As for Zn, Cd was weakly adsorbed on MMT in the presence of EfDOM due to its retention in solution as inert complexes with EfDOM.

In the case of adsorption onto goethite, the Cd adsorption with SFRA was higher compared to the control (+162%), while it was lower in the presence of EfDOM (−20%). Zn showed a similar behaviour, while for As, adsorption was lower with DOM, especially with EfDOM (−65%). For Cu and Ni, adsorption with SRFA was similar to that of the control, while it decreased by 54% and 77%, respectively, with EfDOM. For Cd, Ni and Zn, the sorption decrease is probably due to maintaining in solution by complexation with EfDOM.

3.8 Effect of EfDOM on PAH Adsorption by Mineral Particles in Aquatic Systems Subjected to Strong Urban Pressure

Adsorption of six PAHs, pyrene (Pyr), benzo[a]anthracene (B[a]A), chrysene (Chry), benzo[b]fluoranthene (B[b]F), benzo[k]fluoranthene (B[k]F) and benzo[a]

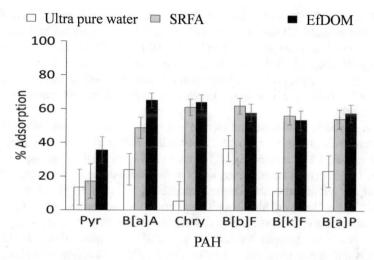

Fig. 7 Adsorption of PAH (250 ng L^{-1}) classified from lowest to highest molecular weight onto quartz (500 mg L^{-1}) in ultrapure water and with EfDOM (10 mgC L^{-1}) or SRFA (10 mgC L^{-1}). Figure reproduced from Fig. 75 in Ref. [57]

pyrene (B[a]P) onto MMT, goethite and quartz were studied in ultrapure water and with SRFA or EfDOM [57].

In ultrapure water, the adsorption of PAH onto MMT is higher than for other particles in the order MMT > goethite > quartz. With DOM, the adsorption varied according to the nature of DOM, the mineral phase and the properties of the PAHs (Fig. 7).

Concerning MMT, PAH adsorption was lower with DOM, especially in the presence of SRFA (−25% to −70%) compared to EfDOM (−10 to −50%).

For goethite, the presence of SRFA mainly reduced the adsorption of light PAHs (Pyr and B[a]A) by about 60%, while the absorption decrease is only about 20% for heavier PAHs (Chry, B[b]F, B[k]F, B[a]P). With EfDOM, the adsorption decrease is about 40–70% regardless of PAH molecular weight.

For quartz, unlike goethite and MMT, adsorption increased as DOM increased. The very low PAH adsorption in ultrapure water was increased by about 50% for B [b]F and pyrene to a factor of 5 for Chry and B[k]F. The differences in PAH adsorption on quartz depending on the type of DOM were not statistically significant, and therefore the sorption increase did not seem to depend here on the DOM quality.

4 Conclusions and Perspectives

The spatio-temporal variability of aquatic OM composition in the Seine River watershed upstream and downstream of Paris Megacity and its effect on micropollutant speciation and fate have been widely studied, especially since 2010, within the PIREN-Seine programme. Both qualitative and quantitative variations of DOM have been related to hydrological conditions and specific geographical locations. Furthermore, it has been demonstrated that EfDOM of the Paris conurbation, despite the very significant improvement in wastewater treatment over the past 50 years, had a very strong impact on DOM present downstream, both in terms of quantity and quality, causing a significant increase in hydrophilicity downstream of the agglomeration.

With regard to speciation and bioavailability of dissolved metals, these studies have demonstrated the strong trace metal-binding ability of EfDOM in spite of its low aromaticity. The remarkably high binding ability of EfDOM and especially of its HPI fraction could be explained by its proteinic nature as well as high nitrogen and sulphur group contents. Through modelling, it has been demonstrated that EfDOM, due to their very high complexing properties, can control, to a large degree, the dissolved trace metal speciation downstream of the Paris conurbation, particularly in low-water periods because of low EfDOM dilution. These results also highlight that the trace metal speciation computation in surface water subjected to important urban discharges should include not only the humic and/or fulvic part of DOM but also other more hydrophilic fractions as depicted by EfDOM. Logically, given the strong affinity of EfDOM for trace metals, these studies have pointed out that EfDOM had a high influence on copper toxicity to *Daphnia magna*, with a slightly greater protective effect than one fulvic acid standard and DOM from the upstream Paris conurbation. Much research is still needed, however, in order to fully understand how the shape and molecular structures constitutive of EfDOM molecules favour the interactions between binding moieties and metallic pollutants.

With regard to pollutant-suspended particle interactions, our results highlight that in many cases trace metal and PAH adsorption onto particles was strongly modified in presence of DOM. Furthermore, the DOM influence on pollutant adsorption depends on the DOM type. Hence, in addition to playing a central role in dissolved metal speciation, EfDOM could also influence the pollutant transport in aquatic systems as well as their transfers among aquatic compartments.

The combination of all these results indicates that it is essential to take into account the great spatio-temporal variability of DOM properties to investigate water quality and river system functioning. It also highlights that hydrophilic organic matter from major WWTPs could play an important role in the biogeochemical cycle of pollutants within aquatic systems under strong urban pressure.

Acknowledgements We sincerely thank Dr. Benoit Pernet-Coudrier, assistant professor at the University of Western Brittany, who passed away accidentally in 2016. His PhD work has strongly contributed to the study of interactions between DOM and metals discussed in this chapter. This work was conducted in the framework of the PIREN-Seine research programme (www. piren-seine.fr), a component of the Zone Atelier Seine within the international Long-Term Socio Ecological Research (LTSER) network.

References

1. Artifon V, Zanardi-Lamardo E, Fillmann G (2019) Aquatic organic matter: classification and interaction with organic microcontaminants. Sci Total Environ 649:1620–1635
2. Cawley KM, Butler KD, Aiken GR, Larsen LG, Huntington TG, McKnight DM (2012) Identifying fluorescent pulp mill effluent in the Gulf of Maine and its watershed. Mar Pollut Bull 64(8):1678–1687
3. Bauer JE, Bianchi TS (2011) Dissolved organic carbon cycling and transformation. In: Wolanski E, McLusky DS (eds) Treatise on estuarine and coastal science, vol 5. Academic Press, Waltham, pp 7–67
4. Besemer K, Luef B, Preiner S, Eichberger B, Agis M, Peduzzi P (2009) Sources and composition of organic matter for bacterial growth in a large European river floodplain system (Danube, Austria). Org Geochem 40(3):321–331
5. Lambert T, Bouillon S, Darchambeau F, Morana C, Roland FAE, Descy J-P, Borges AV (2017) Effects of human land use on the terrestrial and aquatic sources of fluvial organic matter in a temperate river basin (The Meuse River, Belgium). Biogeochemistry 136:191. https://doi.org/10.1007/s10533-017-0387-9
6. Battin TJ, Kaplan LA, Findlay S, Hopkinson CS, Marti E, Packman AI, Newbold JD, Sabater F (2009) Biophysical controls on organic carbon fluxes in fluvial networks. Nat Geosci 1(8):95–100. https://doi.org/10.1038/ngeo101
7. Wiegner TN, Tubal RL, MacKenzie RA (2009) Bioavailability and export of dissolved organic matter from a tropical river during base-and stormflow conditions. Limnol Oceanogr 54(4):1233–1242
8. Stedmon CA, Markager S, Bro R (2003) Tracing dissolved organic matter in aquatic environments using a new approach to fluorescence spectroscopy. Mar Chem 82(3–4):239–254. https://doi.org/10.1016/S0304-4203(03)00072-0
9. Garnier J, Marescaux A, Guillon S et al (2020) Ecological functioning of the Seine River: from long term modelling approaches to high frequency data analysis. In: Flipo N, Labadie P, Lestel L (eds) The Seine River basin. Handbook of environmental chemistry. Springer, Cham. https://doi.org/10.1007/698_2019_379
10. Buffle J (1988) Complexation reactions in aquatic system: an analytical approach. Ellis Horwood, New York, p 692
11. Campbell PGC (1995) Interactions between trace metals and aquatic organisms: a critique of the free-ion activity model. In: Tessier A, Turner DR (eds) Metal speciation and bioavailability in aquatic systems. Wiley, Chichester, pp 45–102
12. Bormann FH, Likens GE (1967) Nutrient cycling. Science 155(3761):424–429. https://doi.org/10.1126/science.155.3761.424
13. Nebbioso A, Piccolo A (2013) Molecular characterization of dissolved organic matter (DOM): a critical review. Anal Bioanal Chem 405(1):109–124. https://doi.org/10.1007/s00216-012-6363-2
14. Amy G, Drewes J (2007) Soil aquifer treatment (SAT) as a natural and sustainable wastewater reclamation/reuse technology: fate of wastewater effluent organic matter (EfOM) and trace organic compounds. Environ Monit Assess 129:19–26. https://doi.org/10.1007/s10661-006-9421-4
15. Baghoth SA, Sharma SK, Amy GL (2011) Tracking natural organic matter (NOM) in a drinking water treatment plant using fluorescence excitation emission matrices and PARAFAC. Water Res 45:797–809. https://doi.org/10.1016/j.watres.2010.09.005
16. Lidén A, Keucken A, Persson KM (2017) Uses of fluorescence excitation-emissions indices in predicting water treatment efficiency. J Water Process Eng 16:249–257. https://doi.org/10.1016/j.jwpe.2017.02.003
17. Bieroza MZ, Bridgeman J, Baker A (2010) Fluorescence spectroscopy as a tool for determination of organic matter removal efficiency at water treatment works. Drinking Water Eng Sci 3:63–70. https://doi.org/10.5194/dwes-3-63-2010

18. Rocher V, Azimi S (2017) Evolution de la qualité de la Seine en lien avec les progrès de l'assainissement de 1970 à 2015. Editions Johanet, Paris, p 76
19. Billen G, Garnier J, Mouchel J-M, Silvestre M (2007) The Seine system: introduction to a multidisciplinary approach of the functioning of a regional river system. Sci Total Environ 375:1–12. https://doi.org/10.1016/j.scitotenv.2006.12.001
20. Garnier J, Leporcq B, Sanchez N, Philippon (1999) Biogeochemical mass-balances (C, N, P, Si) in three large reservoirs of the Seine Basin (France). Biogeochemistry 47:119–146
21. Garnier J, Billen G (2007) Production vs. respiration in river systems: an indicator of an "ecological status". Sci Total Environ 375:110–124. https://doi.org/10.1016/j.scitotenv. 2006.12.006
22. Thévenot DR, Moilleron R, Lestel L, Gromaire M-C, Rocher V, Cambier P, Bonté P, Colin J-L, de Pontevès C, Meybeck M (2007) Critical budget of metal sources and pathways in the Seine River basin (1994–2003) for Cd, Cr, Cu, Hg, Ni, Pb and Zn. Sci Total Environ 375:180–203. https://doi.org/10.1016/j.scitotenv.2006.12.008
23. Servais P, Garcia-Armisen T, George I, Billen G (2007) Fecal bacteria in the rivers of the Seine drainage network (France): sources, fate and modelling. Sci Total Environ 375:152–167. https://doi.org/10.1016/j.scitotenv.2006.12.010
24. Vilmin L, Flipo N, Escoffier N, Rocher V, Groleau A (2016) Carbon fate in a large temperate human-impacted river system: focus on benthic dynamics. Global Biogeochem Cycles 30:1086–1104. https://doi.org/10.1002/2015GB005271
25. Garnier J, Vilain G, Silvestre M, Billen G, Jehanno S, Poirier D, Martinez A, Decuq C, Cellier P, Abril G (2013) Budget of methane emissions from soils, livestock and the river network at the regional scale of the Seine basin (France). Biogeochemistry 116:199. https://doi. org/10.1007/s10533-013-9845-1
26. Jaffé R, McKnight D, Maie N, Cory R, McDowell WH, Campbell JL (2008) Spatial and temporal variations in DOM composition in ecosystems: the importance of long-term monitoring of optical properties. J Geophys Res 113:G04032. https://doi.org/10.1029/2008JG000683
27. Cory RM, Boyer EW, McKnight DM (2011) Spectral methods to advance understanding of dissolved organic carbon dynamics in forested catchments. In: Levia DF et al (eds) Forest hydrology and biogeochemistry: synthesis of past research and future directions, ecological studies, vol 216. Springer, Cham. https://doi.org/10.1007/978-94-007-1363-5_6
28. Jaffé R, Cawley KM, Yamashita Y (2014) Applications of excitation emission matrix fluorescence with parallel factor analysis (EEM-PARAFAC) in assessing environmental dynamics of natural dissolved organic matter (DOM) in aquatic environments: a review. Chapter 3. Advances in the physicochemical characterization of dissolved organic matter: impact on natural and engineered systems; Rosario-Ortiz, ACS symposium series. American Chemical Society, Washington. https://doi.org/10.1021/bk-2014-1160.ch003
29. Matar Z (2015) Influence of effluent organic matter on copper speciation and bioavailability in rivers under strong urban pressure. Environ Sci Pollut Res 22:19461–19472. https://doi.org/ 10.1007/s11356-015-5110-6
30. Servais P, Barillier A, Garnier J (1995) Determination of the biodegradable fraction of dissolved and particulate organic carbon in waters. Ann Limnol 31(1):75–80
31. Mao J, Cao X, Olk DC, Chu W, Schmidt-Rohr K (2017) Advanced solid-state NMR spectroscopy of natural organic matter. Prog Nucl Magn Reson Spectrosc 100:17–51. https://doi.org/10. 1016/j.pnmrs.2016.11.003
32. Nguyen PT (2014) Study of the aquatic dissolved organic matter from the Seine River catchment (France) by optical spectroscopy combined to asymmetrical flow field-flow fractionation. Analytical chemistry. Université de Bordeaux
33. Helms JR, Stubbins A, Ritchie JD, Minor EC, Kieber DJ, Mopper K (2008) Absorption spectral slopes and slope ratios as indicators of molecular weight, source, and photobleaching of Chromophoric dissolved organic matter. Limnol Oceanogr 53(3):955

34. Hansen AM, Kraus TEC, Pellerin BA, Fleck JA, Downing BD, Bergamaschi BA (2016) Optical properties of dissolved organic matter (DOM): effects of biological and photolytic degradation. Limnol Oceanogr 61(2016):1015–1032. https://doi.org/10.1002/lno.10270

35. Zsolnay A, Baigar E, Jimenez M, Steinweg B, Saccomandi F (1999) Differentiating with fluorescence spectroscopy the sources of dissolved organic matter in soils subjected to drying. Chemosphere 38:45–50

36. Huguet A, Vacher L, Relexans S, Saubusse S, Froidefond JM, Parlanti E (2009) Properties of fluorescent dissolved organic matter in the Gironde estuary. Org Geochem 40:706–719

37. Parlanti E, Wörz K, Geoffroy L, Lamotte M (2000) Dissolved organic matter fluorescence spectroscopy as a tool to estimate biological activity in a coastal zone submitted to anthropogenic inputs. Org Geochem 31(12):1765–1781. https://doi.org/10.1016/S0146-6380(00)00124-8

38. Matar Z (2012) Influence de la matière organique dissoute d'origine urbaine sur la spéciation et la biodisponibilité des métaux dans les milieux récepteurs anthropisés, Thèse Université Paris-Est, Champs-sur-Marne, 258 p

39. Pernet-coudrier B, Clouzot L, Varrault G, Tusseau-vuillemin MH, Verger A, Mouchel JM (2008) Dissolved organic matter from treated effluent of a major wastewater treatment plant: characterization and influence on copper toxicity. Chemosphere 73:593–599

40. Town RM, Filella M (2000) A comprehensive systematic compilation of complexation parameters reported for trace metals in natural waters. Aquat Sci 62:252–295

41. Thurman EM, Malcolm RL (1981) Preparative isolation of aquatic humic substances. Environ Sci Technol 15:463–466

42. Martin-Mousset B, Croue JP, Lefebvre E, Legube B (1997) Distribution and characterization of dissolved organic matter of surface waters. Water Res 31:541–553

43. Perdue EM, Ritchie JD (2003) Dissolved organic matter in freshwaters. Treatise on Geochem 5:273–318

44. McDonald S, Bishop AG, Prenzler PD, Robards K (2004) Analytical chemistry of freshwater humic substances. Anal Chim Acta 527:105–124

45. Leenheer JA et al (2000) Comprehensive isolation of natural organic matter from water for spectral characterizations and reactivity testing. American Chemical Society, Washington, pp 68–83

46. Croue JP (2004) Isolation of humic and non-humic NOM fractions: structural characterization. Environ Monit Assess 92:193–207

47. Pernet-Coudrier B, Companys E, Galceran J, Morey M, Mouchel JM, Puy J, Ruiz N, Varrault G (2011) Pb-binding to various dissolved organic matter in urban aquatic systems: key role of the most hydrophilic fraction. Geochim Cosmochim Acta 75:4005–4019

48. Muresan B, Pernet-Coudrier B, Cossa D, Varrault G (2011) Measurement and modeling of mercury complexation by dissolved organic matter isolates from freshwater and effluents of a major wastewater treatment plant. Appl Geochem 26:2057–2063

49. Louis Y, Pernet-Coudrier B, Varrault G (2014) Implications of effluent organic matter and its hydrophilic fraction on zinc(II) complexation in rivers under strong urban pressure: aromaticity as an inaccurate indicator of DOM–metal binding. Sci Total Environ 490:830–837. https://doi.org/10.1016/j.scitotenv.2014.04.1230048-9697/

50. Moon J-W, Goltz MN, Ahn K-H, Park J-W (2003) Dissolved organic matter effects on the performance of a barrier to polycyclic aromatic hydrocarbon transport by groundwater. J Contam Hydrol 60:307–326

51. Newcombe G, Morrison J, Hepplewhite C (2002a) Simultaneous adsorption of MIB and NOM onto activated carbon. I. Characterisation of the system and NOM adsorption. Carbon 40:2135–2146

52. Newcombe G, Morrison J, Hepplewhite C, Knappe DRU (2002b) Simultaneous adsorption of MIB and NOM onto activated carbon: II. Competitive effects. Carbon 40:2147–2156

53. Li Q, Snoeyink VL, Mariñas BJ, Campos C (2003) Pore blockage effect of NOM on atrazine adsorption kinetics of PAC: the roles of PAC pore size distribution and NOM molecular weight. Water Res 37:4863–4872

54. Saada A, Breeze D, Crouzet C, Cornu S, Baranger P (2003) Adsorption of arsenic (V) on kaolinite and on kaolinite–humic acid complexes: role of humic acid nitrogen groups. Chemosphere 51:757–763
55. Wu P, Tang Y, Wang W, Zhu N, Li P, Wu J, Dang Z, Wang X (2011a) Effect of dissolved organic matter from Guangzhou landfill leachate on sorption of phenanthrene by Montmorillonite. J Colloid Interface Sci 361:618–627
56. Wu P, Zhang Q, Dai Y, Zhu N, Dang Z, Li P, Wu J, Wang X (2011b) Adsorption of Cu (II), Cd (II) and Cr (III) ions from aqueous solutions on humic acid modified Camontmorillonite. Geoderma 164:215–219
57. Soares Pereira-Derome C (2016) Influence de la matière organique dissoute d'origine urbaine sur la spéciation des micropolluants: de la station d'épuration au milieu récepteur. Thèse, Université Paris-Est, Champs-sur-Marne, 297 p
58. Chaminda GGT, Nakajima F, Furumai H (2008) Heavy metal (Zn and Cu) complexation and molecular size distribution in wastewater treatment plant effluent. Water Sci Technol 58:1207–1213
59. Benedetti MF, Van Riemsdijk WH, Koopal LK, Kinniburgh DG, Gooddy DC, Milne CJ (1996) Metal ion binding by natural organic matter: from the model to the field. Geochim Cosmochim Acta 60(14):2503–2513
60. Milne CJ, Kinniburgh DG, Tipping E (2001) Generic NICA-Donnan model parameters for proton binding by humic substances. Environ Sci Technol 35:2049–2059
61. Croue JP, Benedetti MF, Violleau D, Leenheer JA (2003) Characterization and copper binding of humic and nonhumic organic matter isolated from the South Platte River: evidence for the presence of nitrogenous binding site. Environ Sci Technol 37(2):328–336
62. Knepper T (2003) Synthetic chelating agents and compounds exhibiting complexing properties in the aquatic environment. Trac-Trends Anal Chem 22:708–724
63. Markich SJ, Brown PL, Jeffree RA, Lim RP (2003) The effects of pH and dissolved organic carbon on the toxicity of cadmium and copper to a freshwater bivalve: further support for the extended free ion activity model. Arch Environ Contam Toxicol 45:479–491
64. Schwartz ML, Curtis PJ, Playle RC (2004) Influence of natural organic matter source on acute copper, lead, and cadmium toxicity to rainbow trout (Oncorhynchus mykiss). Environ Toxicol Chem 23:2889–2899
65. De Schamphelaere KAC, Vasconcelos FM, Tack FMG, Allen HE, Janssen CR (2004) Effect of dissolved organic matter source on acute copper toxicity to Daphnia magna. Environ Toxicol Chem 23:1248–1255
66. DePalma S-G-S, Arnold W-R, McGeer JC, Dixon DG, Smith DS (2011) Effects of dissolved organic matter and reduced sulphur on copper bioavailability in coastal marine environments. Ecotoxicol Environ Saf 74:230–237
67. Trenfield MA, McDonald S, Kovacs K, Lesher EK, Pringle JM, Markich SJ, Ng JC, Noller B, Brown PL, van Dam RA (2011) Dissolved organic carbon reduces uranium bioavailability and toxicity. 1. Characterization of an aquatic fulvic acid and its complexation with uranium VI. Environ Sci Technol 45:3075–3081

Experience Gained from Ecotoxicological Studies in the Seine River and Its Drainage Basin Over the Last Decade: Applicative Examples and Research Perspectives

M. Bonnard, I. Barjhoux, O. Dedourge-Geffard, A. Goutte, L. Oziol, M. Palos-Ladeiro, and A. Geffard

Contents

1   Introduction 244
2   Value of In Vitro Bioassays for an Ex Situ Evaluation of Water Contamination: The Case
    of Endocrine Disruptors 245
    2.1   In Vitro Endocrine Disruptor Bioassays in Water Contamination Monitoring 246
    2.2   Comparison of In Vitro Bioassays to Chemical Analysis in Water Contamination
          Monitoring 246
    2.3   In Vitro Bioassays Applied to Sediment Matrix in Aquatic Environment Quality
          Monitoring 248
    2.4   Conclusion 250
3   Biomarkers for the In Situ Biomonitoring of Water Contamination 250
    3.1   Presentation of Sentinel Species and Their Advantages in Ecotoxicological Studies 250
    3.2   Monitoring of Oxidative Stress in Fish Models 252
    3.3   Value of Digestive Activities as Biomarkers in Two Aquatic Invertebrates:
          Application to the Seine River Basin 253
    3.4   Value of Haemocytes of Zebra Mussels for the Development of Immune
          and Genotoxic Biomarkers 256
4   Combining Multidisciplinary Data for a Global Environmental Quality Diagnosis:
    Development of an Integrated Approach 259
    4.1   Introduction 259
    4.2   The WOE Methodology: Towards an Index of Ecological Disturbance 260

The copyright year of the original version of this chapter was corrected from 2019 to 2020. A correction to this chapter can be found at https://doi.org/10.1007/698_2020_667

M. Bonnard (✉), I. Barjhoux, O. Dedourge-Geffard, M. Palos-Ladeiro, and A. Geffard
Université de Reims Champagne-Ardenne (URCA), UMR-I 02 SEBIO, Reims, France
e-mail: marc.bonnard@univ-reims.fr

A. Goutte
PSL University, UMR 7619 METIS, SU/CNRS/EPHE, Paris, France

L. Oziol
Université Paris-Sud, CNRS, AgroParisTech, UMR 8079 ESE, Orsay, France

Nicolas Flipo, Pierre Labadie, and Laurence Lestel (eds.), *The Seine River Basin*,   243
Hdb Env Chem (2021) 90: 243–268, https://doi.org/10.1007/698_2019_384,
© The Author(s) 2020, corrected publication 2020, Published online: 3 June 2020

4.3 Application of the Integrative WOE Approach to a 1-Year Case Study on the Seine River 260
4.4 Conclusion 263
5 Conclusions and Perspectives 264
References 265

Abstract The Seine River and its drainage basin are recognised as one of the most urbanised water systems in France. This chapter gathers typical applications of complementary ecotoxicological tools that were used in PIREN-Seine programmes for a decade to reflect the Seine River contamination as well as its biological repercussions on organisms. Ecotoxicological studies focused on both (1) specific bioassays and (2) (sub)-individual biological responses (i.e. biomarkers) measured in diverse taxa (i.e. crustaceans, mussels and fishes) representative of the trophic network. Experience gained from these studies made it possible to establish reference and threshold values for numerous biological endpoints. They now can be combined with chemical measurements within integrated models (i.e. the Weight of Evidence [WOE] approach) generating a global index of waterbody pollution. These biological endpoints today appear sufficiently relevant and mature to be proposed to water stakeholders as efficient tools to support environmental management strategies.

Keywords Bioassays, Biomarkers, Biomonitoring, eLTER, Freshwater ecotoxicology, PIREN-Seine, WOE approach, Zone Atelier Seine

1 Introduction

The Seine basin is subjected to multiple anthropogenic pressures including intensive agricultural activities and some heavy urbanisation and industrialisation within the Paris conurbation and its surrounding areas [1]. This wide variety of anthropogenic pressures affecting the Seine watershed makes this area an ideal area to study contemporary environmental issues. The short- and long-term repercussions of such diffuse and chronic pressure on ecosystem health remain difficult to evaluate, considering the complexity and the diversity of these exogenous inputs. The contamination level can be determined through chemical analysis as defined by the presence of substances that would not normally occur or at concentrations above the natural background. However, the pollution status assessment additionally integrates chemical bioavailability and the biological impacts of contaminants [2]. Consequently, integrated and multidisciplinary strategies are clearly recommended and even required by the European Water Framework Directive 2000/60/CE [3]. One of the main strengths of the PIREN-Seine programme is that it is based on a substantial interdisciplinary scientific network, whose partners work jointly contributing their own expertise to improve the understanding of the ecological and biogeochemical functioning of the whole Seine watershed, including human-induced modifications.

Two main strategies have been used in the PIREN-Seine programme for a decade in order to characterise the ecotoxicological status of the Seine waterbodies. The first one has consisted in an ex situ strategy with the development of complementary bioassays performed on cell cultures and/or laboratory model organisms to investigate the (eco)toxicity of environmental samples. In particular, the study of the oestrogenic potential of water and sediment samples using specific in vitro bioassays is illustrated in this chapter (see Sect. 2). The second strategy consisted in in situ measurements of various biological responses, called biomarkers, to reflect the impacts on sentinel species. This approach is illustrated here with examples of studies conducted on (1) field populations for fish species, i.e. passive biomonitoring, and (2) transplanted organisms for aquatic invertebrates such as mussels and crustaceans, i.e. active biomonitoring (see Sect. 3). Most of these ecotoxicological studies investigated three pilot sites of the Seine River, upstream and downstream of the Paris agglomeration (i.e. Marnay, Bougival and Triel-sur-Seine) and, more episodically, sites on the Orge River, an important tributary of the Seine River. Experience gained from all these studies allowed establishing reference and threshold values for many biological endpoints. They now can be combined with chemical measurements within integrated models (i.e. the Weight of Evidence [WOE] approach; see Sect. 4) generating a global index of waterbody pollution.

2 Value of In Vitro Bioassays for an Ex Situ Evaluation of Water Contamination: The Case of Endocrine Disruptors

Among pollutants discharged into the environment, many belong to the family of endocrine disruptors (EDs), chemicals able to lead to endocrine disorders in both humans [4] and wildlife [5], even at trace levels. Polychlorinated biphenyls (PCBs), polybrominated diphenyl ethers (PBDEs) and phthalates are such chemicals, which have been quantified both in water and the roach *Rutilus rutilus* in the Seine basin [6]. The main ED disorders may affect the oestrogenic pathway of aquatic species and induce the 'feminisation' of male fishes in some cases [7]. However, aquatic contaminants may target other endocrine axes. Among the endocrine-disrupting effects affecting oestrogenic, androgenic or thyroid endpoints, the most pronounced effects of aquatic contamination were oestrogenicity, as brought to light by a European research programme [8]. In this context, in vitro bioassays applied for endocrine disruptors [9, 10] are a good compromise between in situ bioassays, which can be expensive and time- and animal-consuming, and hence not suitable as high-throughput screening tools, and chemical analyses, which are limited to a restricted number of quantified pollutants and do not allow one to consider their harmful effects in mixtures representative of an in situ exposure. In vitro bioassays can therefore be considered as intermediate sensitive and global screening tools for monitoring aquatic ED contamination, to assess its intrinsic dangerousness because they integrate interaction effects (additive, antagonistic or synergistic) between low doses of chemicals (cocktail effects). Such in vitro cellular bioassays were used for the monitoring of organic pollution by endocrine disruptors in European surface waters [8] and were also implemented in the PIREN-Seine programme for the evaluation of the Seine River quality. Only data of oestrogenic activities are presented, given their ubiquity and levels in both the water column and sediments.

2.1 In Vitro Endocrine Disruptor Bioassays in Water Contamination Monitoring

The oestrogenicity of the organic contamination of surface waters collected over nine campaigns between February and November 2010 at Marnay-sur-Seine, Bougival and Triel-sur-Seine was studied using the MELN reporter gene assay according to [11]. Regardless of the sampling location, surface waters showed significant oestrogenic-mimetic activities, on the order of 1 ng E2-eq/L[1]: median EEQ[1] values of 0.27, 0.40 and 0.51 ng E2-eq/L were measured for Marnay-sur-Seine, Bougival and Triel-sur-Seine, respectively [12]. Similar oestrogenic activity levels were reported under the PIREN-Seine programme for organic contamination of rivers sampled in the Orge catchment area near Paris during September 2007 [13]. Both mean and median EEQ values were the lowest for Marnay-sur-Seine and the highest for Triel-sur-Seine. The data tended to highlight an upstream–downstream gradient of increasing oestrogenic activity in the Seine River: Marnay-sur-Seine < Bougival < Triel-sur-Seine, even if temporal variations due to hydrologic conditions may mask the increasing gradient when the results are averaged over the year. Indeed, the strongest mean oestrogenic activity was observed for Bougival water sampled in July [12], corresponding to the low-flow period. The same general pattern was shown for Triel-sur-Seine. In contrast, the lowest oestrogenic activities were detected in winter, highlighting a dilution of water contamination during high-flow periods. Overall, the cellular bioassay implemented made it possible to monitor the Seine River contamination by bioactive chemicals at the oestrogenic axis, in accordance with both seasonal and spatial expected distribution of pollution.

2.2 Comparison of In Vitro Bioassays to Chemical Analysis in Water Contamination Monitoring

The in vitro bioassays may be used alone or in combination with chemical analysis to monitor the aquatic contamination by endocrine disruptors. In the framework of the

---

[1]The transcriptional activities measured using in vitro ED bioassays are generally expressed as 17β-estradiol (E2) equivalent (eq) quantity (EEQ), in ng E2-eq/L of water or in ng E2-eq/g of sediment, by comparing them with the activity of the specific hormonal ligand E2.

PIREN-Seine programme, several chemicals were analysed in water samples. Among them, the following oestrogens were especially analysed, the natural oestrogens, E2 or its isoform α, oestrone (E1) and oestriol (E3) and the synthetic one, ethinylestradiol (EE2), and their contribution to the oestrogenic potential of the water column was studied. For this purpose, the oestrogenic potency of water samples collected between February and November 2010 in Marnay-sur-Seine, Bougival and Triel-sur-Seine was expressed in two ways: as biological EEQ in ng E2-eq/L derived from the bioassay approach and as chemical EEQ in ng E2-eq/L, corresponding to the sum of the relative biological potency to E2 of each oestrogen multiplied by their respective concentration in the dissolved phase of water samples [13]. Thus, the biological EEQs measured were compared to the chemical EEQ estimated according to E2 (isoforms α and β), E1, E3 and EE2 concentrations weighted by their respective oestrogenicity on MELN cells [12]. A good correlation was highlighted between chemical and biological EEQs on all the three Seine axis pilot sites, the correlation coefficient R values ranging from 0.84 for Marnay-sur-Seine to 0.99 for Bougival.

Other contaminants, known or suspected as endocrine disruptors, were monitored during these water-sampling campaigns: phthalates, alkylphenols, polycyclic aromatic hydrocarbons (PAHs), PCBs, PBDEs and bisphenol A [12]. Among these chemicals, only 4-nonylphenol and bisphenol A are proven endocrine disruptors, with reference toxicological values established regarding their ED effects [14, 15]. For all water samples, the correlation between biological and chemical EEQs was much weaker for 4-nonylphenol ($R = 0.57$) and bisphenol A ($R = 0.73$) than for oestrogens ($R = 0.96$), with E2 as the largest contributor by far to the oestrogenicity of surface waters from the Seine River, followed by E1, in accordance with most studies reviewed by [16]. E3 and EE2 were not detected in the water samples collected in 2010, unlike in previous samples from the Orge catchment in 2007 [13]. Comparatively, bisphenol A and 4-nonylphenol contributed to a minor extent (~2%) to the ED potential of Seine River organic contamination. A safe oestrogenic equivalent value of 0.3 ng E2-eq/L, regarding steroid oestrogens, was based on the assumption that these chemicals are responsible for more than 90% of in vitro oestrogenicity of municipal wastewater treatment plant effluents [16]. The above-mentioned results are in line with this consideration, and the proposed provisional trigger value was exceeded by mean EEQ values measured for water samples from downstream Seine River sites: Bougival (0.84 ng E2-eq/L) and Triel-sur-Seine (0.55 ng E2-eq/L). In the latter campaigns (2011–2012), the trigger value was exceeded only for water samples from Triel-sur-Seine, especially in June 2012, showing an EEQ of 2.1 ng E2-eq/L.

2.3 In Vitro Bioassays Applied to Sediment Matrix in Aquatic Environment Quality Monitoring

Since biological activity of dissolved phase samples may reflect only an intermittent contamination of surface waters, in vitro bioassays were also applied to sediment samples. Indeed, sediments act as a pollution storage tank concentrating many persistent or pseudo-persistent compounds for long periods (e.g. PAHs, PCBs, sex steroids, alkylphenols, plasticisers) as shown for French sediments [17–19]. In this context, the oestrogenic potential was simultaneously measured on sediment and water samples collected in the three pilot sites of the Seine axis during three sampling periods in 2011–2012, using the MELN bioassay [20]. The spatial and temporal oestrogenicity of sediment organic extracts are shown in Fig. 1a, b, respectively.

The oestrogenicity dose–response curves showed a clear upstream–downstream gradient for sediments collected during the low-flow period (June 2012) (Fig. 1a), with EEQ values ranging from 1.95, 0.867 to 0.0005 ng E2-eq/g, for Triel, Bougival and Marnay, respectively. These EEQ levels were in accordance with those observed between small good-quality French rivers (Aisne, Vallon du Vivier and Lézarde Rivers with oestrogenicity at 0.20–0.83 ng E2-eq/g sediment) and poor-quality rivers (Rhonelle and Réveillon, with oestrogenicity at 1.69 and 6.43 ng E2-eq/g sediment, respectively) according to French water agencies [18]. Triel-sur-Seine sediments collected in the low-flow period showed a higher activity (EEQ, 1.95 ng E2-eq/g in June 2012) than those sampled during other periods (EEQ, approximately 0.3 ng E2-eq/g for September 2011 and December 2012) (Fig. 1b). These spatial and temporal variations in the oestrogenic potential of Seine River sediment were in accordance with those evidenced for water samples (Fig. 1). Furthermore, the sediments showed differences in oestrogenic activity in accordance with their contamination levels in various hydrophobic and persistent organic pollutants belonging to ED compound families (PAHs, PCBs, organochlorine pesticides-OCPs, PBDE) [20].

Contrary to the water column, today there are no trigger values for sediment oestrogenicity regarding the environmental risks for benthic organisms [21]. Over the two matrices studied, sediment appears, however, as the most suitable for highlighting the ubiquitous oestrogenicity of the aquatic contamination and the time–space variations of its biological activity. Compared to intermittently collected water samples, the sediment extracts are more concentrated (while being non-cytotoxic in the MELN bioassay) and integrate long-term contaminations. As a result, the fold inductions of oestrogenic activity obtained with sediment extracts (from ×4 to ×30 in comparison to blank samples; Fig. 1) were greater than those obtained with dissolved phase extracts (from ×3 to ×12), ensuring quantifiable transcriptional activities in each sediment sample, contrary to water samples.

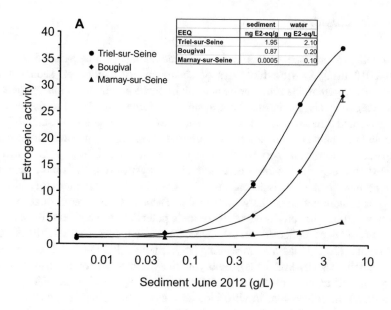

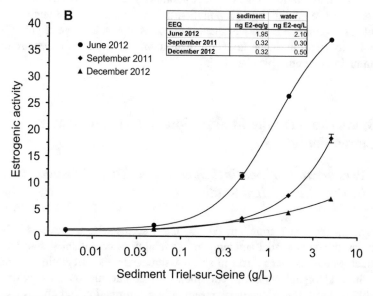

Fig. 1 Spatial (**a**, for June 2012 samples) and temporal (**b**, for Triel-sur-Seine samples) variation of the oestrogenic potential in sediments (in g/L of culture medium in the MELN bioassay) collected from the Seine River in September 2011, June 2012 and December 2012 at Marnay-sur-Seine, Bougival and Triel-sur-Seine. Graph data correspond to mean ± SEM of one experiment repeated three times. Insert data represent EEQ values in ng E2-eq/g and in ng E2-eq/L of the dissolved phase of water samples collected at the same time

2.4   Conclusion

The quality of surface waters was monitored alongside the Seine River over 2010–2012 using in vitro bioassays of oestrogenicity applied on sediment and water matrix contamination. The results highlighted a contamination gradient by bioactive chemicals, increasing from upstream (Marnay-sur-Seine) to downstream (Bougival and Triel-sur-Seine). In addition, a lower oestrogenic potential of both water and sediment contamination was observed under high-flow conditions (winter) compared to low-flow periods (summer). The in vitro bioassay implemented revealed an acceptable quality of the Seine River upstream of Paris but a degraded quality downstream of Paris according to the provisional trigger values for endocrine disruptor contamination of the water column. These data are overall in accordance with those from other bioassays of ED or toxic potential measurements on sediment or water matrices collected in the same three pilot sites in the PIREN-Seine programme [20]: bioassays measuring thyroidicity (PC-DR-LUC cells), anti-androgenicity (MDA-kb2 cells), embryotoxicity/teratogenicity on fish (Medaka embryo-larval assay), cytotoxicity (Microtox® assay) and genotoxicity (SOS chromotest). In conclusion, in vitro bioassays stood out as being suitable tools to monitor the quality of the Seine River. In addition, sediment may be considered as a suitable matrix to monitor aquatic pollution using a bioassay-based approach, since it is able to concentrate bioactive aquatic pollutants of a wide range of polarities and especially to act as an integrative adsorbent matrix of the ongoing environmental contamination by anthropic chemicals.

3   Biomarkers for the In Situ Biomonitoring of Water Contamination

3.1   Presentation of Sentinel Species and Their Advantages in Ecotoxicological Studies

To highlight the water contamination status, it seems wise to develop indicators integrating the daily spatial and temporal variability of contaminant concentrations in surface waters. Besides direct in situ measurements, the integration of contamination in wild organisms could contribute further information on environmental hazard [22]. Moreover, since the exposure, biotransformation and effects pathways of contaminants may differ between vertebrates and invertebrates; biomarkers need to be studied in diverse taxa for an ecological relevance.

Biomonitoring of wild fish provides a spatially and temporally integrated view of pollutant exposure and ecotoxicological effects, because of their relatively high mobility, long life span (bioaccumulation over time [23]), high trophic levels (i.e. biomagnification [24]) and slow genetic adaptation due to a slow turn over [25]. The European chub (*Squalius cephalus* L.) is extensively used as a bioindicator species in field studies [26–28]. This fish species is common and widely distributed throughout Europe, in both clean and polluted freshwaters [29]. The abundance growth rate is stable in France, and feral chubs in the Seine River occur in satisfactory numbers to allow ecotoxicological field studies. Previous studies have highlighted the high sensitivity of chubs to pollutants, with high levels of DNA damage and high EROD (ethoxyresorufin-O-deethylase) activity [30–32] in polluted freshwaters.

Aquatic invertebrates such as filter feeders (bivalves) or detritivorous species (amphipods) are good sentinel species. Bivalves can concentrate and retain contaminants in their tissues for a long time, which represents a valuable integrative characteristic. Among continental bivalves, studies are mainly based on a model organism, the zebra mussel *Dreissena polymorpha*, which has proved its value [33]. *D. polymorpha*, originating from the Ponto–Caspian region, is characterised by its abundance and widespread distribution in European and North American rivers, great filtration capacities leading to high levels of xenobiotic accumulation and good tolerance to environmental stressors [33]. As regards amphipod crustaceans, the genus *Gammarus* plays a key role in food webs such as the common shredder, playing a major role in leaf-litter breakdown. The reduction in its feeding activity is directly correlated with the reduction in the processing of leaf litter [34]. Furthermore, in *Gammarus* spp., the ingestion of leaves leads to the assimilation of energy and the production of faecal pellets. Gammarids and their faeces are a food resource for many aquatic species. Since *Gammarus fossarum* is widespread in Central and Eastern Europe, this common species has been defined as a very good candidate to study the impact of chemical stressors.

In accordance with their sedentary lifestyle, individual responses of the invertebrates used as biomarkers may be correlated with the quality of the sampling site, thereby allowing for the development of early warning and sensitive diagnosis tools exhibiting high ecological relevance [35]. Biomarkers related to energy metabolism [36] and immune functions could be useful prognostic tools, bearing in mind the close relationships existing between the energy balance and the immunity status in the health of organisms. Concerning population preservation, biomarkers relative to reproductive processes are undoubtedly relevant and promising. Regarding field studies, *D. polymorpha* and *G. fossarum* could be used not only in passive biomonitoring but also in active strategies. In fact, the caging of individuals from the same population (and a fortiori the same species) in different sites (1) limits or avoids the influence of intrinsic parameters and (2) exposes organisms for a controlled time period, so that recent contamination can be evidenced. As for the contamination level or associated biological effects, several studies have underlined the value of the active biomonitoring approach using *D. polymorpha* [37–39] and the amphipod *G. fossarum* [40, 41]. Therefore, they constitute valuable bioindicator species, largely used as a freshwater biomonitoring tool [35, 42, 43], as illustrated in the next sections by studies undertaken in the PIREN-Seine programme.

3.2   Monitoring of Oxidative Stress in Fish Models

3.2.1   Introduction

In aerobic organisms, the use of molecular oxygen for the metabolism of organic carbon leads to the generation of reactive oxygen species (ROSs) that refer to oxygen free radicals and nonradical reactive species. The deleterious effects of ROSs are prevented by enzymatic or non-enzymatic antioxidant defence mechanisms. However, an excessive generation of ROS can overwhelm the organism's antioxidant capacity, thus inducing oxidative stress, protein and DNA damage, lipid peroxidation and cellular ageing. The exposure to environmental pollutants also induces oxidative stress, through the formation of ROS during xenobiotic metabolism and especially during the hydroxylation phase catalysed by the cytochrome P-450 monooxygenase system [44, 45]. In addition, antioxidant efficiency can also be reduced by micropollutants.

3.2.2   Field Applications

A common approach is based on the comparison of oxidative stress levels between fish populations at clean and polluted sites. Here, our aim was to explore the influence of pollutant load on the ROS levels and antioxidant capacity of chubs at an individual level in the Seine River basin. Feral chubs are expected to be exposed to numerous aquatic pollutants, and it was shown that the muscle pollutant load of chubs from the Seine River followed this pattern: phthalate diesters > pyrethroid pesticides > polychlorinated biphenyls (PCBs) > PAHs > organochlorine pesticides > PBDEs [46]. Plasmatic levels of antioxidant efficiency significantly decreased with increasing levels of phthalate metabolites in chub liver [47]. This result corroborates a previous study in *Carassius auratus* that highlighted an inhibitory effect of phthalates on antioxidant enzyme activity and expression [48]. Similarly, a study in humans has pointed out that phthalate monoester metabolites in urine samples are positively correlated with increased plasmatic markers of oxidative stress [49]. Hence, the biotransformation of these widespread plasticisers may reduce the antioxidant efficiency in wild chub from the Seine River. Thiobarbituric acid reactive substances (TBARS) are good proxies of oxidative damage that are formed as a byproduct of lipid peroxidation [45]. Our results pointed out a significant increase of TBARS with increasing PCB burden in chub muscle [47], as experimentally shown in goodeid fish (*Girardinichthys viviparous*) exposed to sublethal concentrations of PCBs [50]. Despite bans since 1975 (use in open systems) and 1987 (use in closed systems), PCBs are still of environmental concern, because these historically persistent organic pollutants bioaccumulate over time in fish [23] and are biomagnified across aquatic food webs [24].

3.3 Value of Digestive Activities as Biomarkers in Two Aquatic Invertebrates: Application to the Seine River Basin

3.3.1 Introduction

All of the biochemical and physiological processes involved in the life cycle of organisms are closely related to energy metabolism. The energy contributions of individuals are determined, among other processes, by their ability to transform raw energy ingested in the form of food into assimilable energy in the form of nutrients. This ability depends in particular on digestive enzyme activities, notoriously sensitive to various chemical stressors [51]. Disruption of digestive processes may affect individual energy balance and result in energy allocation disturbances with possible consequences on individual fitness. Moreover, as part of an environmental quality assessment, variations in enzymatic activities are often one of the first responses to stress; they therefore represent useful points to confirm the occurrence of toxic effects before they are perceptible at higher levels of biological organisation. The PIREN-Seine programme provided the opportunity to study the activity of digestive enzymes in two aquatic organisms largely used in ecotoxicology, the zebra mussel (*D. polymorpha*) and the gammarid (*G. fossarum*). This work involved both the characterisation of the studied response variability with respect to biotic and abiotic factors and the application of these tools in the Seine River watershed.

3.3.2 Characterisation of Digestive Enzyme Activities

The response of biomarkers might be influenced by environmental factors (both biotic and abiotic), potentially confounding factors that may complicate interpretation of the results. It thus appears necessary to determine the sources of variations, the amplitude of the response modulations related to these confounding factors with respect to the contamination factors, and to specify the measurement conditions of these biomarkers.

In the gammarid, previous experiments conducted under controlled conditions made it possible to study the effects of (1) temperature and conductivity; (2) breeding and moulting cycles in the female; and (3) the quantity of available food [52]. The results indicate that conductivity is not a key parameter influencing the digestive enzyme activities during short-term bioassays. Conversely, temperature must be taken into account for reliable data interpretation, especially in environments with temperatures below 12°C [52]. In addition, amylase, cellulase and trypsin activities in females decreased significantly during the reproductive cycle, particularly during the amplexus. In contrast, gammarid males retain higher levels of digestive activity during this breeding season and therefore appear to be unaffected by the development cycle. Finally, the application of trophic stress decreases the activity of amylase, which is observed in both females and males [53].

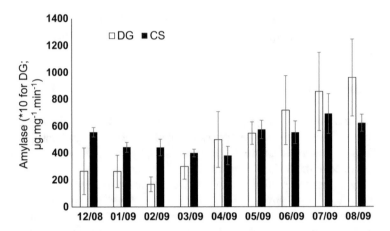

Fig. 2 Seasonal variation in digestive enzyme activities (μg maltose mg^{-1} protein min^{-1}; mean $\pm$ SD, $n = 7$–8 organs) recorded in the crystalline style (CS) and the digestive gland (DG, concentrations $*10$) of mussels from Vesle over the monitoring period. Monthly sampling dates: i.e. 12/08 corresponds to December 2008; 04/09 corresponds to April 2009

In the case of *D. polymorpha*, a 12-month active biomonitoring study was conducted within the PIREN-Seine programme on the Vesle basin [39]. A clear seasonal trend was observed in the study of the digestive gland (DG), with mussels exhibiting especially high digestive enzyme activities in spring and early summer, competing with the development and maturation of gametes (Fig. 2, illustration for amylase). These digestive patterns were obviously related to the interaction between the reproductive and nutritional status of mussels. It appears that the level of the digestive enzyme activity closely follows temperature variations. While seasonal modulations of amylase and cellulase activities were recorded in the DG of mussels, this was not the case for these activities when recorded in their crystalline style (CS) (Fig. 2, illustration for amylase). The absence of seasonal variations in the CS could be related to the innate nature of the style, i.e. a secretion organ whose weight does not vary according to season. On the contrary, the DG is an organ whose weight exhibits variations over the year, with a potential influence on digestive enzyme activities. When using the DG, the temperature must be taken into consideration during field evaluation [39].

3.3.3 Consequences of Digestive Activity Impairment

This aspect was particularly considered in the gammarid *G. fossarum*. Along with the measurements taken on the activity of digestive enzymes, the energy reserves and responses related to fecundity (number of oocytes) and fertility (number of embryos) in females were quantified in controlled conditions. Concomitantly with the inhibition of the digestive activities studied, a decrease in the available energy (lipids + glycogen + proteins) is observed in females, as well as a decrease in the fecundity and fertility of the females subjected to a trophic stress [54]. Thus, the disruption of the digestive capacity of animal organisms could, after 15 days of exposure to trophic stress, induce a risk of reduction of the energy available for reproduction. In the case of exposure to chemical contaminants, in addition to a disruption of the digestive capacity of organisms, it should be noted that individuals must allocate some of their energy to defend mechanisms, potentially to the detriment of reproduction. Therefore, the digestive enzymes studied show their relevance as biomarkers in ecological risk assessment (ERA).

3.3.4 Field Applications

Following studies carried out under controlled conditions, a methodology was established for the in situ studies of water quality assessment. The strategy adopted consisted in an active approach to organisms, i.e. transplantation of organisms. Such an approach makes it possible, using calibrated organisms, to limit as much as possible the influence of both endogen (sex, age, etc.) and abiotic (temperature, conductivity, etc.) parameters. To better assess the biological impact of the multi-contamination of a small urban river, transplanted zebra mussels were exposed for 2 months at two sites in the Orge River basin, a tributary of the Seine River: one at the outlet of the basin (Athis-Mons, downstream site, highly urbanised) and the other in its upper part (Villeconin, upstream site, rural) [38]. Amylolytic and cellulolytic activities measured in the CS displayed similar response patterns, both activities being always 20–30% lower downstream than upstream (Fig. 3, illustration for amylase). Two major characteristics of the upstream–downstream gradient may

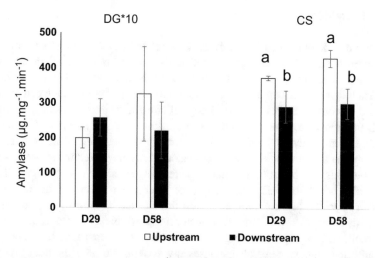

Fig. 3 Amylase activities (μg maltose mg^{-1} protein min^{-1}) measured in digestive gland (DG) and crystalline style (CS) of mussels after 29 days (D29) and 58 days (D58) at both sites (Upstream and Downstream). Different letters upon bars indicate that significant differences were observed (mean $\pm$ SD, t-test, $P < 0.05$)

explain the differences in CS enzyme activities: the trophic status and the exposure to micropollutants [38]. Contrary to CS, no significant difference between the two sites could be observed for DG enzymes (Fig. 3 illustration for amylase). This contrast between the two organs may be attributed to higher availability of enzymes that may interact with pollutants in extracellular spaces (such as the style sac and stomach lumen) than in intracellular spaces (e.g. the digestive cells of the DG). The water quality along the Seine River (comparison between Marnay-sur-Seine, Bougival and Triel-sur-Seine) also impacted digestive capacity in transplanted gammarids [20]. As with *Dreissena*, the value of digestive enzymes has also been demonstrated in another model species in ecotoxicology.

3.3.5   Conclusion

The objective of the various actions carried out in the PIREN-Seine programme was to validate the use of digestive enzymes in invertebrates as biomarkers of the quality of aquatic environments. The digestive activities were therefore characterised with respect to certain parameters of the environment; their ecological relevance was demonstrated; and, finally, their application in biomonitoring proved to be relevant.

3.4   Value of Haemocytes of Zebra Mussels for the Development of Immune and Genotoxic Biomarkers

3.4.1   Introduction

Among biomarkers studied in bivalves, those related to the immune system are of great interest since they can be indicative of the health status of an organism. Considered as one of the first lines of defence of the body after the physical barrier of the shell, haemocytes are key cells of individual homeostasis, which can either be activated to counter a chemical or biological stressor or increase their turnover in organisms exposed to contaminated sites to guarantee mussel survival. A functional immune system indicates favourable living conditions for an organism; conversely, a weak immune system could reveal a degraded environment requiring organisms to adapt. For this reason, immune markers are developed to qualify not only the individual's health but also the quality of their environment. With an opened circulatory system, haemolymph circulates through the body, bringing nutrients to the various organs. Drawn from the adductor muscle of bivalves after a small breach in the shell, haemolymph sampling is not lethal for organisms, a clear advantage for biomonitoring studies. Several responses involved in the physiological processes may be measured on haemocytes. For this purpose, a multiple biomarker approach was adopted on haemocytes of *D. polymorpha*, including biomarkers of cytotoxicity (cellular distribution and mortality), genotoxicity (primary DNA strand breaks) and immunotoxicity (capacity and efficiency of the phagocytosis process); all of them have shown their relevance in ERA. This strategy has been employed in the PIREN-Seine programme to assess the water quality at the three pilot sites along the Seine River following the upstream–downstream gradient: Marnay-sur-Seine, Bougival and Triel-sur-Seine.

3.4.2 Immune Marker Response

Two sampling experiments were conducted in different hydrologic conditions: one during the zebra mussel 'breeding' period between May and July 2016 and another one during the 'resting' period between November and December 2016. All mussels originated from the same population in a control site (Sainte Marie du Lac Nuisement, lac du Der-Chantecoq, 48°36′22.02″N, 4°46′34.0″E). The day after sampling, haemolymph was withdrawn and cellular parameters were measured. Haemocyte distribution, cell mortality and responses relative to the phagocytosis process were measured with flow cytometry. Using this technique, the following data can be acquired: (1) phagocytosis capacity, defined as the percentage of cells which have engulfed at least one bead, quantifies how many haemocytes in the entire cell population are able to ingest beads; (2) phagocytosis efficiency, corresponding to the percentage of cells that have engulfed at least three beads, indicates how much phagocytic cells have efficient phagocytosis activity and are not just randomly bound to beads; and (3) the intensity of phagocytosis activity quantified by the haemocyte avidity, represented by the mean number of beads engulfed per cell. During the first campaign (May–July 2016), the response of haemocyte distribution demonstrated a site effect. In mussels caged at Triel-sur-Seine, 71% of haemocytes showed an efficient phagocytic activity for both sampling times, whereas it was 10–20% lower for the two other sites. We did not observe this site effect during the second campaign, but the results were time-dependent. In fact, the mean percentage of phagocytic cells was 79% in November and 59% in December. This result was also confirmed by another marker of phagocytic effectiveness: the avidity of haemocytes. In November, the mean avidity of haemocytes was 10 ± 2 beads per cell, whereas in December, the average avidity decreased to only 6 ± 2 beads per cell. These results would suggest stimulated immune functions in November in comparison with December. Globally, necrosis affected 10–28% of haemocytes over the year. Thus, the biological endpoints measured with the flow cytometry technique clearly show a site effect on the first campaign with the stimulation of haemocyte functions associated with lower mortality from Marnay-sur-Seine to Triel-sur-Seine. Full interpretation of these results is ongoing. In particular, these results still have to be compared with environmental concentrations of contaminants to confirm their utility as efficient biomarkers in ERA.

3.4.3 Genotoxic Marker Response

In the same field study, primary DNA strand breaks were measured by the comet assay (SCGE, single cell gel electrophoresis assay) in the haemocytes of caged zebra mussels. This assay has gained importance in environmental genotoxicity studies because it is a rapid, relatively inexpensive (compared to other methods) and sensitive technique for highlighting a variety of DNA damage to individual cells, i.e. single-strand breaks (SSBs), double-strand breaks (DSBs) or alkaline-sensitive sites (ALSs) [55]. The results obtained in terms of levels of DNA strand breaks in haemocytes and the sensitivity of zebra mussels to a genotoxic stress were (1) in accordance with immune marker responses (see above) and (2) in keeping with genotoxicity results obtained on other organs (i.e. gills, DG cells) of caged zebra mussels during previous field studies carried out in the PIREN-Seine programme. For example, the upstream–downstream effect related to the gradient of contamination between Marnay-sur-Seine and Triel-sur-Seine was evidenced with the levels of DNA strand breaks in gill cells of zebra mussels. Levels of damage varied according to the duration or the season of transplantation [56]. In the same study, the levels of irreversible chromosomal damage such as micronuclei measured in gills were also positively correlated with PAH bioaccumulation and related to the gradient of water contamination between the three pilot sites. In another study [57], specific DNA adduct patterns with levels varying between 3 and 18 adducts/10^6 nucleotides were detected both in the gills and the DG of mussels, after 1 or 2 months of transplantation at the three sites on the Seine River. Conversely, this was not the case for mussels in the control site. The DNA adduct formation could be correlated with the modulation or the decrease of gene expression implicated in detoxification processes. Further work is necessary to fully understand the relationship between the responses measured at different genomic scales (from DNA to chromosome), as well as the incidence of structural and functional damage (phagocytosis, DNA repair, apoptosis or necrosis, etc.) on cells and, more globally, on individual fitness. Genotoxicity responses may then be regarded not only as biomarkers of exposure but also as biomarkers of effects, which would reinforce their relevance as predictive tools of ecological dysfunctions.

3.4.4 Conclusion

In situ experiments demonstrate the relevance of haemocytes as target cells for immune and genotoxic biomarker measurements in *D. polymorpha* because they reflect the contrasted levels of impact of the sites studied as well as the physiological conditions of mussels in relation to the stage of sexual maturity. Research is continuing for a better characterisation of the seasonal variability of biomarker responses as well as the possible incidence of (a)biotic confounding factors. These aspects are critical to define both thresholds and reference values useful for a robust diagnosis of the environment's quality. Because of their central functions on organism physiology, the development of a panel of markers analysed in haemocytes, including cytotoxic, immunotoxic and genotoxic responses and their integration into a 'cellular index', would provide (1) a better understanding of the overall toxicity on the mussel's physiology and (2) new data for integrative tools designed for the diagnosis of environmental quality.

4 Combining Multidisciplinary Data for a Global Environmental Quality Diagnosis: Development of an Integrated Approach

4.1 Introduction

The large multidisciplinary scientific network involved in the PIREN-Seine programme has been particularly solicited in the field of ecotoxicology for several years. As a matter of example, between 2011 and 2012, a dozen research groups collaborated on a synchronous and integrative multi-marker approach to assess the chemical and ecological/ecotoxicological status of the three pilot sites, situated upstream and downstream of the Paris agglomeration along the Seine River. This 1-year monitoring study generated about 550 variables per site for each of the four sampling periods (corresponding to distinct seasons). These variables characterised in particular (1) the metallic and organic contamination levels in the water column and in the sediment, (2) the bioaccumulation levels of pollutants of concern and (3) the biological responses (biomarkers) in transplanted gammarids, as well as (4) the (eco)toxicity of water and sediment samples using laboratory bioassays.

Embedding such a multidisciplinary data set in a 'Weight of Evidence' (WOE) approach is widely recommended and even required by regulatory authorities for efficient environmental risk assessment (ERA) [3, 58]. The principle of the WOE approach is based on integrating and combining complementary data from different packages, called 'Lines of Evidence' (LOEs), in order to draw an overall conclusion on the ecological status of the aquatic ecosystem considered [59]. The environmental diagnosis thus results from the combination of contamination levels, bioavailability analyses and biological responses in key species or model organisms at different levels of organisation [60, 61]. A helpful model converting a conceptual WOE strategy to evaluate sediment hazard into logical flow charts and calculations was proposed by [62], which was adapted to the data set acquired within the PIREN-Seine programme [20].

4.2 The WOE Methodology: Towards an Index of Ecological Disturbance

The WOE methodology consisted in gathering the selected parameters into four LOEs: sediment and water chemistry (contamination level analysis), contaminant bioavailability (bioaccumulation levels in caged gammarids), biomarker responses (in gammarids) and bioassay responses (on sediment and water samples). A hazard quotient (HQ) was then calculated for each LOE and reported in an evaluation grid for a rapid and clear hazard classification (i.e. ranging from absent to severe). An overall risk assessment was finally obtained by compiling the HQs of the four LOEs into a single index (WOE index) associated with a global hazard class. In addition to the quality and the relevance of the data, there are some prerequisites for parameter integration in the model: the availability of a reference value for each endpoint and, for biological endpoints, an additional threshold value defining the basal/natural variations of the response considered is needed. Accordingly, around 160 endpoints acquired between 2011 and 2012 were selected to be injected into the model. They included 27 and 60 substances of concern for the water column and sediment contamination, respectively (LOE chemistry); 37 compounds bioaccumulated in gammarids (LOE bioavailability); 9 biological responses characterising neurotoxicity, energy acquisition, feeding rate, reproduction and survival in caged gammarids (LOE biomarkers); and 11 and 14 endpoints assessing survival, development, teratogenicity, growth, endocrine disruption, cytotoxicity and genotoxicity using bioassays on water and sediment samples, respectively (LOE bioassays) [20]. Data integration was performed for the three sites and the four sampling campaigns, as well as for an annual average for each site. One of the strengths of the model stems from the differential weighting of each endpoint. For instance, each compound integrated into the chemical and the bioavailability LOEs is weighted according to its status in the Water Framework Directive [63], giving higher weighting factors to priority hazardous substances and priority substances. In the biomarker and bioassay LOEs, each endpoint is weighted against the biological significance of the response considered according to an arbitrary scale of 0–3. For bioassays, the weighting factors also take into account the test matrix and the exposure duration [20, 62]. Moreover, the procedure was improved (1) using external reference values for biomarkers and bioaccumulation levels established for the same population of gammarids during previous studies and (2) integrating natural variations of the responses as well as the effects of confounding factors.

4.3 Application of the Integrative WOE Approach to a 1-Year Case Study on the Seine River

The HQs generated for the annual average clearly reflected the well-known anthropogenic gradient along the Seine River, with the lowest values observed for

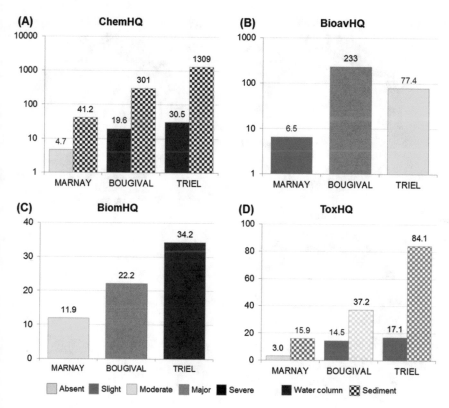

Fig. 4 Annual hazard quotients generated from the LOEs chemistry (**a**), bioavailability (**b**), bio-markers (**c**) and bioassays (**d**) for the three sites of the Seine River: Marnay-sur-Seine (upstream of Paris), Bougival (about 40 km downstream of Paris) and Triel-sur-Seine (about 80 km downstream of Paris and the Oise River confluence). Data taken from [20]

Marnay-sur-Seine (upstream of Paris) and the highest for the two downstream sites (Bougival and Triel-sur-Seine) (Fig. 4). Three out of four LOEs identified Triel-sur-Seine as the most heavily impacted site. However, the highest HQ value was observed for Bougival within the LOE bioavailability (Fig. 4b). The hazard classi-fication was also widely dependent on the LOE: it varied, for instance, from absent (ToxHQ for water samples; Fig. 4d) to severe (ChemHQ for sediment samples; Fig. 4a) for Marnay-sur-Seine and from slight (ToxHQ for water samples; Fig. 4d) to severe (ChemHQ for water and sediment samples and BiomHQ; Fig. 4a, c) for Triel-sur-Seine. The ChemHQs calculated for sediment evaluated as severe the risk associated with the chemical contamination for all sites (Fig. 4a). While the ChemHQ values increased along the downstream–upstream gradient, the hazard classes attributed to each site based on contaminant analysis in sediments failed to discriminate the three stations, contrary to the other LOEs. Conversely, the toxicity testing of water samples did not reveal a significant risk at any of the three sites. This shows that the results, if interpreted independently, can alternatively over- or underestimate the environmental risk depending on the LOE considered.

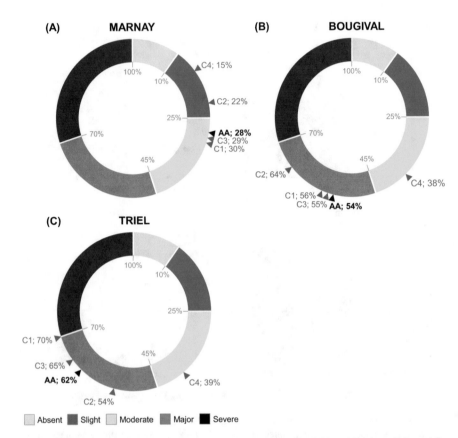

Fig. 5 WOE indexes generated for Marnay-sur-Seine (**a**), Bougival (**b**) and Triel-sur-Seine (**c**) for the four sampling periods (C1–C4) and the annual average (AA). C1, autumn 2011; C2, spring 2012; C3, summer 2012; C4, winter 2012. Note that only data on water contamination and bioavailability were evaluated in the C2 campaign. Data taken from [20]

This again highlights the importance of a multidisciplinary approach to evaluate the global environmental risk in aquatic ecosystems. Indeed, a comprehensive and efficient ERA must jointly evaluate a site's contamination (i.e. level of contaminants in the environment) and pollution status (i.e. including the incidence and the severity of the biological impacts of the contamination) [2, 64].

In this study, the overall risk assessment was reached with the final WOE integration (Fig. 5). Annual WOE indexes reflected the anthropogenic gradient with a moderate hazard in Marnay-sur-Seine and a major hazard for the Bougival and Triel-sur-Seine stations. Seasonal trends were also highlighted with the winter campaign (C4: high-flow period) associated with slight to moderate hazard, whereas the autumn campaign (C1: low-flow period) resulted in the highest risk (moderate to major, with the Triel-sur-Seine value very close to the severe hazard class; Fig. 5). These observations are in line with the seasonal flow rate variations in the Seine

River, generating a dilution or a concentration of the contaminants during the respective sampling periods [65].

The final integration coherently synthesised the outputs of each LOE. Despite its high integrative dimension, this method did not result in a loss of information. The identification of the (class of) contaminants or biological endpoints responsive for the HQ values is possible by calculating endpoint contribution within each LOE. This procedure was able, for instance, to identify perfluorooctanesulfonic acid (PFOS) as an ubiquitous contaminant along the segment of the Seine River studied, with measured concentrations in the water column systematically above the environmental quality standard of 0.65 ng/L. This data examination also pinpointed the herbicide metazachlor as a specific contaminant for the winter period. The contributions calculated within the LOE bioavailability indicated that PAHs were the main problematic class of compounds regarding their bioaccumulation in gammarids at the two downstream stations. In particular, pyrene and fluoranthene exceeded more than 30- and 10-fold the gammarid basal impregnation levels in organisms transplanted in Bougival and Triel-sur-Seine, respectively. Within the LOE biomarker, contribution examinations resulted in differential response spectra depending on the site. For the annual results, only a slight decrease in gammarid survival was responsive for the BiomHQ calculated for Marnay-sur-Seine, whereas in Triel-sur-Seine the response spectrum was more complex with noticeable contributions of reproduction, feeding, energy acquisition and survival markers. In a more general way, contributions of HQ values to the annual WOE indexes were particularly high for the ChemHQs (from 80% for Marnay-sur-Seine to 54% for Triel-sur-Seine), indicating a noticeable contamination of the sites, including the upstream station usually considered as a reference site. These contributions decreased for the downstream sites as the annual ToxHQs and BiomHQs increased their contributions to around 30% for Bougival and up to more than 40% at Triel-sur-Seine. This traced the emergence of toxic responses in caged gammarids downstream of Paris in relation to the contamination levels, thus classifying these sites as polluted.

4.4 Conclusion

One of the main challenges of the study was to propose a way to compile, synthesise and interpret the countless results of the multi-marker study conducted between 2011 and 2012. The WOE approach developed was successfully applied in this context and resulted in an integrative ERA based on the compilation of more than 150 endpoints to characterise the contamination/pollution status of each site. The results of this method are easily interpreted and transferable to environmental stakeholders through the generation of a global index of perturbation (WOE index) associated with a hazard class. However, several biological responses studied during the programme could not be integrated into the model due to the lack of available references. In particular, modifications of bacterial communities and tolerance acquisition in biofilms (pollution-induced community tolerance 'PICT' method) gave results that highly agreed with those generated by the model. Such endpoints are particularly useful in an ERA because they reflect alterations at higher levels of biological organisation (i.e. community level). The development of robust baselines for this type of response would be very valuable for their integration into the WOE approach, widening the type of biological responses included in the model.

The versatility of the WOE model allows its application in a wide variety of contexts. For instance, it can be considered to monitor the environmental repercussions of installations, industries and river development or rehabilitation projects: the model would be helpful to characterise the global ecological impact by comparing the results before and after such modifications of the environment. This tool could also be applicable in the context of temporal and comparative monitoring of study sites, e.g. along the Seine River continuum.

5 Conclusions and Perspectives

The different examples given in this chapter illustrate the application of a wide array of bioassays or biomarkers in the PIREN-Seine programme, which aimed to investigate the ecotoxicological quality of the Seine River and some of its tributaries (e.g. Oise and Orge). With the development of new biotechnologies, bioassays and biomarkers will certainly evolve in the future to (1) improve our knowledge of toxic pathways of water contaminants and (2) increase their sensitivity in the detection of adverse effects on biota linked to water contamination. For instance, the recent development of 'omics' appears promising to identify new biological targets of emerging contaminants. In the near future, there is no doubt that the establishment of both reference and threshold values of biomarkers will authorise their integration into ERA for monitoring the efficiency of water treatment plants or restoration schemes for aquatic ecosystems, as may be requested by national or European directives. However, biomonitoring water quality may not be restricted to a limited geographic area and must cover the entire ecological continuum, from continental waters to the estuary as for the Seine River, for example, or must be carried out jointly between European members for cross-border rivers. For this type of biomonitoring survey, we need to go beyond various technical and scientific barriers, implying the choice of several species belonging to the same taxa along the salinity gradient (from continental to brackish/marine waters) representative of aquatic ecosystems and various trophic levels; a single species could not live along the entire continuum. Likewise, the development of reference and threshold values for biomarkers will allow responses between species to be compared. Therefore, the cartography of the overall ecotoxicity representative of the different human pressures on watersheds could be established regardless of the target species. This is an exciting challenge that will guide our research strategies on the Seine River basin for the coming years.

Acknowledgements This work was conducted in the framework of the PIREN-Seine research programme (www.piren-seine.fr), a component of the Zone Atelier Seine within the international Long Term Socio Ecological Research (LTSER) network.

References

1. Billen G, Garnier J, Mouchel JM et al (2007) The Seine system: introduction to a multidisciplinary approach of the functioning of a regional river system. Sci Total Environ 375:1–12. https://doi.org/10.1016/j.scitotenv.2006.12.001
2. Chapman PM (2007) Determining when contamination is pollution – weight of evidence de terminations for sediments and effluents. Environ Int 33:492–501
3. European Commission (EC) (2000) Directive 2000/60/EC of the European Parliament and of the Council of 23 October 2000 establishing a framework for Community action in the field of water policy. Off J Eur Union 327:1–73
4. Kabir ER, Rahman MS, Rahman I (2015) A review on endocrine disruptors and their possible impacts on human health. Environ Toxicol Pharmacol 40:241–258
5. Wagner M, Kienle C, Vermeirssen ELM et al (2017) Endocrine disruption and in vitro eco toxicology: recent advances and approaches. Adv Biochem Eng Biotechnol 157:1–58
6. Teil MJ, Tlili K, Blanchard M et al (2014) Polychlorinated biphenyls, polybrominated diphenyl ethers, and phthalates in roach from the Seine River basin (France): impact of densely urbanized areas. Arch Environ Contam Toxicol 66:41–57
7. Sumpter J, Jobling S (1995) Vitellogenesis as a biomarker for estrogenic contamination of the aquatic environment. Environ Health Perspect 103:173–178
8. Tousova Z, Oswald P, Slobodnik J et al (2017) European demonstration program on the effect-based and chemical identification and monitoring of organic pollutants in European surface waters. Sci Total Environ 601–602:1849–1868
9. Leusch FDL, Neale PA, Hebert A et al (2017) Analysis of the sensitivity of in vitro bioassays for androgenic, progestagenic, glucocorticoid, thyroid and estrogenic activity: suitability for drinking and environmental waters. Environ Int 99:120–130
10. Wangmo C, Jarque S, Hilscherová K et al (2018) In vitro assessment of sex steroids and related compounds in water and sediments – a critical review. Environ Sci Process Impacts 20:270–287
11. Jugan ML, Oziol L, Bimbot M et al (2009) *In vitro* assessment of thyroid and estrogenic endocrine disruptors in wastewater treatment plants, rivers and drinking water supplies in the greater Paris area (France). Sci Total Environ 407:3579–3587
12. Gasperi J, Moreau-Guigon E, Labadie P et al (2011) Contamination de la Seine par les micropolluants organiques: Effet des conditions hydriques et de l'urbanisation. PIREN-Seine, Phase V, Rapport de synthèse 2007–2010
13. Miège C, Karolak S, Gabet V et al (2009) Evaluation of estrogenic disrupting potency in aquatic environments and urban wastewaters by combining chemical and biological analysis. Trends Anal Chem 28:186–195
14. EFSA (2015) Scientific opinion on the risks to public health related to the presence of bisphenol A (BPA) in foodstuffs: opinion on BPA. EFSA J 13:3978
15. ANSES (2017) Élaboration de VTR par voie orale fondée sur des effets reprotoxiques pour les nonylphénols. Saisine n°2017-SA-0211
16. Jarošová B, Bláha L, Giesy JP et al (2014) What level of estrogenic activity determined by in vitro assays in municipal waste waters can be considered as safe? Environ Int 64:98–109
17. Fenet H, Gomez E, Pillon A et al (2003) Estrogenic activity in water and sediments of a French river: contribution of alkylphenols. Arch Environ Contam Toxicol 44:1–6

18. Kinani S, Bouchonnet S, Creusot N et al (2010) Bioanalytical characterisation of multiple endocrine- and dioxin-like activities in sediments from reference and impacted small rivers. Environ Pollut 158(1):74–83

19. Creusot N, Dévier MH, Budzinski H et al (2016) Evaluation of an extraction method for a mixture of endocrine disrupters in sediment using chemical and in vitro biological analyses. Environ Sci Pollut Res 23(11):10349–10360. https://doi.org/10.1007/s11356-016-6062-1

20. Barjhoux I, Fechner LC, Lebrun JD et al (2018) Application of a multidisciplinary and integrative weight-of-evidence approach to a 1-year monitoring survey of the Seine River. Environ Sci Pollut Res 25:23404–23429. https://doi.org/10.1007/s11356-016-6993-6

21. Brack W, Dulio V, Ågerstrand M et al (2017) Towards the review of the European Union Water Framework management of chemical contamination in European surface water resources. Sci Total Environ 576(2017):720–737

22. Zuykov M, Pelletier E, Harper D (2013) Bivalve mollusks in metal pollution studies: from bioaccumulation to biomonitoring. Chemosphere 93:201–208

23. Vives I, Grimalt JO, Ventura M et al (2005) Age dependence of the accumulation of organo-chlorine pollutants in brown trout (Salmo trutta) from a remote high mountain lake (Redo, Pyrenees). Environ Pollut 133(2):343–350

24. Kelly BC, Ikonomou MG, Blair JD et al (2007) Food web–specific biomagnification of persistent organic pollutants. Science 317(5835):236–239

25. Rowe CL (2008) The calamity of so long live: life histories, contaminants, and potential emerging threats to long-lived vertebrates. Bioscience 58(7):623–631. https://doi.org/10.1641/B580709

26. Flammarion P, Devaux A, Nehls S et al (2002) Multibiomarker responses in fish from the Moselle River (France). Ecotoxicol Environ Saf 51(2):145–153

27. Winter MJ, Verweij F, Garofalo E et al (2005) Tissue levels and biomarkers of organic contaminants in feral and caged chub (Leuciscus cephalus) from rivers in the West Midlands, UK. Aquat Toxicol 73:394–405

28. Krča S, Žaja R, Calic V et al (2007) Hepatic biomarker responses to organic contaminants in feral chub (Leuciscus cephalus) – laboratory characterization and field study in the Sava river, Croatia. Environ Toxicol Chem 26:2620–2633

29. Durand JD, Persat H, Bouve Y (1999) Phylogeography and postglacial dispersion of the chub Leuciscus cephalus in Europe. Mol Ecol 8:989–997

30. Devaux A, Flammarion P, Bernardon V et al (1998) Monitoring of the chemical pollution of the River Rhone through measurement of DNA damage and cytochrome P4501A induction in chub (Leuciscus cephalus). Mar Environ Res 46:257–262

31. Larno V, Laroche J, Launey S et al (2001) Responses in chub (Leuciscus cephalus) populations to chemical stress assessed by genetic markers, DNA damage and cytochrome P4501A induction. Ecotoxicology 10:145–158

32. Viganò L, Camoirano A, Izzotti A et al (2002) Mutagenicity of sediments along the Po River and genotoxicity biomarkers in fish from polluted areas. Mutat Res Genet Toxicol 515 (1):125–134

33. Binelli A, Della Torre C, Magni S et al (2015) Does zebra mussel (Dreissena polymorpha) represent the freshwater counterpart of Mytilus in ecotoxicological studies? A critical review. Environ Pollut 196:386–403

34. Forrow DM, Maltby L (2000) Toward a mechanistic understanding of contaminant-induced changes in detritus processing in streams: direct and indirect effects on detritivore feeding. Environ Toxicol Chem 19:2100–2106

35. Palos Ladeiro M, Barjhoux I, Bigot-Clivot A et al (2017) Mussel as a tool to define continental watershed. Quality in organismal and molecular malacology. InTech, London. https://doi.org/10.5772/67995. Available via https://www.intechopen.com/books/organismal-and-molecular-malacology/mussel-as-a-tool-to-define-continental-watershed-quality

36. Dedourge-Geffard O, Palais F, Biagianti-Risbourg S, Geffard O, Geffard A (2009) Effects of metals on feeding rate and digestive enzymes in *Gammarus fossarum*: an *in situ* experiment. Chemosphere 77:1569–1576
37. Bervoets L, Voets J, Chu S et al (2004) Comparison of accumulation of micropollutants between indigenous and transplanted zebra mussels (Dreissena polymorpha). Environ Toxicol Chem 23(8):1973–1983
38. Bourgeault A, Gourlay-Francé C, Vincent-Hubert F et al (2010) Lessons from a transplantation of zebra mussels in a small urban river: an integrated ecotoxicological assessment. Environ Toxicol 25:468–478
39. Palais F, Dedourge-Geffard O, Beaudon A et al (2012) One-year monitoring of core biomarker and digestive enzyme responses in transplanted zebra mussels (Dreissena polymorpha). Ecotoxicology 21(3):888–905
40. Coulaud R, Geffard O, Xuereb B et al (2011) *In situ* feeding assay with *Gammarus fossarum* (Crustacea): modelling the influence of confounding factors to improve water quality biomonitoring. Water Res 45:6417–6429
41. Lacaze E, Devaux A, Mons R et al (2011) DNA damage in caged *Gammarus fossarum* amphipods: a tool for freshwater genotoxicity assessment. Environ Pollut 159:1682–1691
42. Borcherding J (2010) Steps from ecological and ecotoxicological research to the monitoring for water quality using the zebra mussel in a biological warning system. In: van der Velde G, Rajagopal S, bij de Vaate A (eds) The zebra mussel in Europe. Backhuys Publishers/Margraf Publishers, Leiden/Welkersheim, pp 279–283
43. Voets J, Bervoets L, Smolders R et al (2010) Biomonitoring environmental pollution in freshwater ecosystems using Dreissena polymorpha. In: van der Velde G, Rajagopal S, bij de Vaate A (eds) The zebra mussel in Europe. Backhuys Publishers/Margraf Publishers, Leiden/Welkersheim, pp 301–321
44. Valavanidis A, Vlahogianni T, Dassenakis M et al (2006) Molecular biomarkers of oxidative stress in aquatic organisms in relation to toxic environmental pollutants. Ecotoxicol Environ Saf 64(2):178–189
45. Lushchak V (2011) Environmentally induced oxidative stress in aquatic animals. Aquat Toxicol 101:13–30
46. Goutte A, Alliot F, Azimi S et al (2018) Influence du niveau trophique sur l'imprégnation des chesvesnes par les micropolluants organiques en agglomération parisienne. Rapport PIREN-Seine
47. Goutte A (2018) Métabolisation des micropolluants: Imprégnation et dommages potentiels chez les poissons d'eau douce de la Marne et de ses affluents. Rapport Agence de l'Eau Seine Normandie
48. Qu R, Feng M, Sun P et al (2015) A comparative study on antioxidant status combined with integrated biomarker response in *Carassius Auratus* fish exposed to nine phthalates. Environ Toxicol 30(10):1125–1134. https://doi.org/10.1002/tox.21985
49. Ferguson KK, McElrath TF, Chen YH et al (2014) Urinary phthalate metabolites and biomarkers of oxidative stress in pregnant women: a repeated measures analysis. Environ Health Perspect 123(3):210–216
50. Vega-López A, Galar-Martínez M, Jiménez-Orozco FA et al (2007) Gender related differences in the oxidative stress response to PCB exposure in an endangered goodeid fish (Girardinichthys viviparus). Comp Biochem Physiol A Mol Integr Physiol 146(4):672–678
51. Dedourge-Geffard O, Palais F, Geffard A et al (2013) Origin of energy metabolism impairments. In: Amiard-Triquet C, Amiard JC, Rainbow PS (eds) Ecological biomarkers – indicators of Ecotoxicological effects. CRC Press, Taylor & Francis Group, Boca Raton, pp 279–306
52. Charron L, Geffard O, Chaumot A et al (2013) Effect of water quality and confounding factors on digestive enzyme activities in *Gammarus fossarum*. Environ Sci Pollut Res 20:9044–9056
53. Charron L, Geffard O, Chaumot A et al (2014) Influence of molting and starvation on digestive enzyme activities and energy storage in *Gammarus fossarum*. PLoS One 9(4):e96393. https://doi.org/10.1371/journal.pone.0096393

54. Charron L, Geffard O, Chaumot A et al (2015) Consequences of lower food intake on the digestive enzymes activities, the energy reserves and the reproductive outcome in *Gammarus fossarum*. PLoS One 10(4):e0125154. https://doi.org/10.1371/journal.pone.0125154

55. de Lapuente J, Lourenço J, Mendo SA et al (2015) The comet assay and its applications in the field of ecotoxicology: a mature tool that continues to expand its perspectives. Front Genet 6:180. https://doi.org/10.3389/fgene.2015.00180

56. Michel C, Bourgeault A, Gourlay-Francé C et al (2013) Seasonal and PAH impact on DNA strand-break levels in gills of transplanted zebra mussels. Ecotoxicol Environ Saf 92:18–26

57. Chatel A, Faucet-Marquis V, Gourlay-Francé C et al (2015) Genotoxicity and activation of cellular defenses in transplanted zebra mussels Dreissena polymorpha along the Seine River. Ecotoxicol Environ Saf 114:241–249

58. US EPA (2016) Weight of evidence in ecological assessment. Epa/100/R-16/001

59. Burton GA, Chapman PM, Smith EP (2002) Weight-of-evidence approaches for assessing ecosystem impairment. Hum Ecol Risk Assess 8:1657–1673. https://doi.org/10.1080/20028091057547

60. Chapman PM, Hollert H (2006) Should the sediment quality triad become a tetrad, a pentad, or possibly even a hexad? J Soils Sediments 6:4–8. https://doi.org/10.1065/jss2006.01.152

61. Dagnino A, Sforzini S, Dondero F et al (2008) A "Weight-of-Evidence" approach for the integration of environmental "Triad" data to assess ecological risk and biological vulnerability. Integr Environ Assess Manag 4:314–326. https://doi.org/10.1897/IEAM_2007-067.1

62. Piva F, Ciaprini F, Onorati F et al (2011) Assessing sediment hazard through a weight of evidence approach with bioindicator organisms: a practical model to elaborate data from sediment chemistry, bioavailability, biomarkers and ecotoxicological bioassays. Chemosphere 83:475–485. https://doi.org/10.1016/j.chemosphere.2010.12.064

63. European Commission (EC) (2013) Directive 2013/39/EU of the European Parliament and of the Council of 12 August 2013 amending Directives 2000/60/EC and 2008/105/EC as regards priority substances in the field of water policy

64. Chapman PM, Wang F, Janssen CR et al (2003) Conducting ecological risk assessments of inorganic metals and metalloids: current status. Hum Ecol Risk Assess 9:641–697

65. Labadie P, Munoz G, Peluhet L et al (2013) Les polluants organiques persistants dans la Seine: dynamique spatio-temporelle et transfert vers le compartiment biologique, Rapport d'activité PIREN-Seine

Sedimentary Archives Reveal the Concealed History of Micropollutant Contamination in the Seine River Basin

Sophie Ayrault, Michel Meybeck, Jean-Marie Mouchel, Johnny Gaspéri, Laurence Lestel, Catherine Lorgeoux, and Dominique Boust

Contents

1 Introduction 271
2 Long-Term Reconstruction of Past Contamination Trajectories Based on River
  Sedimentary Archives 272
  2.1 Conceptual Steps for the Study of Contaminants in Cores 272
  2.2 Contamination Fluxes Estimated from Floodplain Cores 274
  2.3 Circulation of Contaminants Within River Basins 275
3 The Seine River Basin and Its Contamination by Micropollutants 276
  3.1 The Seine River Basin: A Sensitive Territory Under High Pressures 276
  3.2 Regulatory Surveys Failed to Assess Micropollutants (1971–2006) 277
  3.3 Spatial Position of Sedimentary Archives Within the Seine River Basin 278
  3.4 Core Dating 280
4 The Evolution of the Seine River Basin Contamination (1910–2015) Unveiled from
  Sediment Cores 280
  4.1 Temporal Evolution of Persistent Organic Pollutants at the Basin Outlet
      and in the Estuary 280

The copyright year of the original version of this chapter was corrected from 2019 to 2020. A correction to this chapter can be found at https://doi.org/10.1007/698_2020_667

S. Ayrault (✉)
Université Paris-Saclay, Laboratoire LSCE, CEA-CNRS-UVSQ, Gif-sur-Yvette, France
e-mail: sophie.ayrault@lsce.ipsl.fr

M. Meybeck, J.-M. Mouchel, and L. Lestel
Sorbonne Université, CNRS, EPHE, UMR Metis, Paris, France

J. Gaspéri
Université Paris-Est, Laboratoire LEESU, Champs-sur-Marne, France

C. Lorgeoux
Univ Lorraine, CNRS, GeoRessources, Nancy, France

D. Boust
IRSN, Institut de Radioprotection et de Sûreté Nucléaire, Laboratoire de Radioécologie de Cherbourg-Octeville (LRC), Cherbourg, France

Nicolas Flipo, Pierre Labadie, and Laurence Lestel (eds.), *The Seine River Basin*, Hdb Env Chem (2021) 90: 269–300, https://doi.org/10.1007/698_2019_386, © The Author(s) 2020, corrected publication 2020, Published online: 3 June 2020

    4.2   Metals (Ag, Cd, Cr, Cu, Hg, Ni, Pb, Sb, Zn) and Arsenic (As) at the Seine River
           Outlet (1935–2005) 283
    4.3   Antibiotics at the Seine River Outlet (1954–2004) 284
5   Intra-basin Comparisons in the Seine River Basin Reveal Contrasted Trends
    in Subbasins 285
    5.1   Compared Metal Contamination History: The Cadmium Example 285
    5.2   Compared POP Contamination History: The PAH Example 286
6   Interbasin Comparison of Metal Contamination Trends in Western European Rivers
    Using Sedimentary Archives 287
7   Circulation of Material Within Basins and Its Impact on River Fluxes 290
    7.1   Leakage Ratio of Metals Within the Seine River Basin (1950–2005) 290
    7.2   Intercomparison of River Contaminant Fluxes in Relation to Their Population
           in Western European Rivers 291
8   Trajectories of the Environmental Issues Affecting River Basins and Society: Example
    of PCB Contamination in the Seine River Basin 293
9   Conclusions 295
References 296

Abstract Sedimentary archives provide long-term records of particulate-bound pollutants (e.g. trace metal elements, PAHs). We present the results obtained on a set of selected cores from alluvial deposits within the Seine River basin, integrating the entire area's land uses upstream of the core location, collected upstream and downstream of Paris megacity and in the estuary. Some of these cores go back to the 1910s. These records are complemented by in-depth studies of the related pollution emissions, their regulation and other environmental regulations, thereby establishing contaminant trajectories. They are representative of a wide range of contamination intensities resulting from industrial, urban and agricultural activities and their temporal evolution over a 75,000 km^2 territory. A wide set of contaminants, including metals, radionuclides, pharmaceuticals and up to 50 persistent organic pollutants, have been analysed based on the Seine River sediment archives. Altogether, more than 70 particulate contaminants, most of them regulated or banned (OSPAR convention, European Water Framework Directive (WFD 2000/60/EC)), were measured in dated cores collected at 7 sites, resulting in a large data set.

After drawing a picture of the literature devoted to sedimentary archives, the findings resulting from several decades of research devoted to the Seine River basin will be used, together with other studies on other French and foreign rivers, to illustrate the outstanding potential of sedimentary archives. The limitations of using sedimentary archives for inter-site comparison and the approaches developed in the PIREN-Seine to overcome such limitations such as selecting pertinent indicators (specific fluxes, per capita release, leakage rate, etc.) will be described. The very complex interactions between humans and their environment will be addressed through questions such as the impact on the spatial and temporal trajectories of contaminants of factors such as wastewater management, deindustrialisation within the Seine River basin, implementation of national and EU environmental regulations, etc. This chapter will show how such studies can reveal the persistence of the contamination and the emergence of new pollutants, e.g. antibiotics. It will propose indicators for the evaluation of the environment resilience and the efficiency of environmental policies.

Keywords Contaminant trajectories, Emerging pollutants, Metals, Persistent organic pollutants, Sediment archives

1 Introduction

Sedimentary archives have been used since the early 1970s to reconstruct the past contaminant contents of river particles settled in the river continuum, particularly in the United States. They were used to reconstruct the past contamination of the Chesapeake Bay [1], the progressive eutrophication of the Great Lakes and their gradual contamination with mercury [2] and polychlorinated biphenyls (PCBs) [3]. Cores taken in the Mississippi River Delta integrate the contamination over 3.2 M km^2 [4, 5] for trace metals, polycyclic aromatic hydrocarbons (PAHs) and chlorinated organic compounds. Their analysis demonstrated that the lead (Pb) profile showed a doubling of concentrations from 1900 to 1970, followed by a significant decline of Pb contamination over this wide territory. These trends could not be demonstrated by the regulatory monitoring performed on filtered and unfiltered (whole-water) water sample pairs [6]: the analysis of a single core provided more valuable results on river contamination than 15 years of water quality surveys over the entire United States. To overcome the difficulties raised by this protocol, between 1991 and 2015, the United States Geological Survey (USGS) added the analysis of fine deposited river sediments to regulatory monitoring (Horowitz, pers. comm.) The sedimentary archive approach was then recommended by the National Oceanic and Atmospheric Administration (NOAA) [7].

In Europe, sedimentary archives were also considered in the 1970s in Alpine lakes [8] and the Rhine River delta [9]. In this basin, contamination for As, Cd, Cr, Cu, Hg, Ni, Pb and Zn started before 1920 and peaked in the 1970s, a trend confirmed later [10]. Most of these studies focused on the evolution of contaminant levels, generally at the river mouth or in deltas, therefore integrating the whole river basin.

In France, sediment archive analysis started with the metal contamination in the Garonne estuary (Gironde) [11] and then in the Lot reservoirs, a tributary of the Garonne, downstream of a former zinc mining site, which was the major source of Zn and Cd for this system [12, 13]. More recently, sediment cores have been analysed in the Seine basin [14, 15], the Loire basin [16] and the Rhône basin [17, 18]. In addition to trace elements and persistent organic pollutants (POPs), pharmaceutical products have been analysed more recently [19, 20]. This brief and non-comprehensive review does not cover the many lake cores analysed in France [21]. At all French sites, the sediment archives reveal the evolution of micropollutant contamination for the period when their regulatory analyses in water were limited or insufficient (<1990s) and long before, over periods that may exceed 100 years.

Apart from some notable exceptions, the archive-based studies, conducted by geochemists, have rarely documented the river basin history, which is needed to correlate the sediment archives with the evolution of the river basin upstream of the coring site, and its complex interactions with societies.

The interdisciplinary approach, developed by the PIREN-Seine to go further than a geochemical description of the contamination trends, is presented in Sect. 2. Section 3 briefly presents the Seine River basin, our coring strategy and core analysis. Greater detail on the basin and its pollutant sources can be found in other chapters of this volume and in original papers for coring, dating and analysis details. Sections 4–7 present the different steps of the historical reconstruction of the basin contamination to allow for intra- and interbasin comparisons, as conceptualised in Fig. 1, illustrated with different particulate contaminants. The final section links the sediment archives with the basin history (past pressures, awareness, conflicts, societal responses and regulations, creation of institutions), i.e. defining the contamination trajectory, taking PCB as an example.

2 Long-Term Reconstruction of Past Contamination Trajectories Based on River Sedimentary Archives

2.1 Conceptual Steps for the Study of Contaminants in Cores

These studies can be schematised by a succession of steps, from the most common ones to those developed on the Seine River by the PIREN-Seine programme, as follows.

Step A. *Depth contaminant profile*: cored fine sediments – clay and silt deposits – are sliced, and then targeted contaminants are analysed on each slice. To minimise the effect of grain size on the contents of pollutants, samples can be sieved at the clay–silt fraction, typically <63 µm. Al, Th or Sc should also be measured as quantitative tracers of this fraction. Particulate organic pollutant contents are often normalised to the particulate organic carbon (POC). When the sedimentation rate is regular, the slices are averaged temporal windows, e.g. yearly averages. The best cores (e.g. lakes and reservoirs) provide an annual resolution. Cores taken in alluvial plains record the deposits corresponding to high flows and floods: the temporal record is more fragmented. Sometimes a single flood event corresponds to a multicentimetre layer.

Step B. *Core dating*: generally using the measurement of Cs-137, an artificial radioisotope which provides three temporal markers: (1) the Cs-137 emergence in 1945, (2) a first maximum in 1962 originating from atmospheric nuclear test bombs and (3) a second and sharper maximum generated by the Chernobyl nuclear accident in 1986; Pb-210 decay can be used as well [22]. The evolution of the sedimentation rate is then used to convert the depth profile into a temporal evolution of contamination. Reliable geochemical tracers for dating before 1945 are lacking.

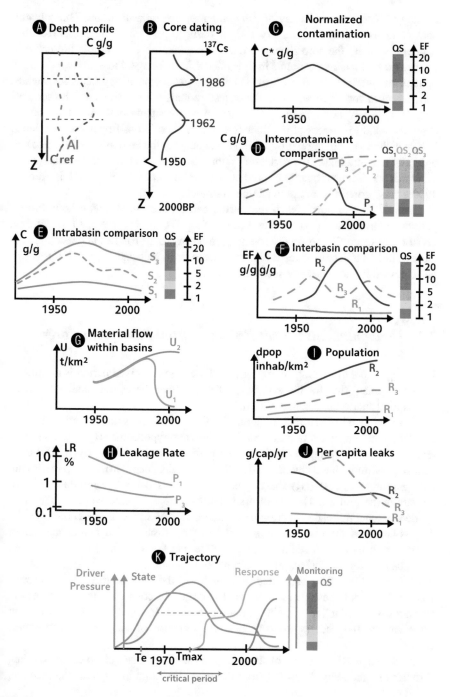

Fig. 1 Environmental contamination history based on multistep analysis of sedimentary archives, from state to environmental indicators and trajectories

Step C. *Temporal evolution of contamination and normalisation*. The contaminant contents expressed on a dry weight basis ($\mu g\ g^{-1}$ dw) are presented on a temporal scale. If the core slices are relatively thick, this evolution is presented as a succession of average contents in time slices. At this stage, there are two different options to interpret these profiles: (1) in the geochemical approach for metals, enrichment factors (EF) represent the excess contamination, which are the contents at time t doubly normalised to the background contents determined in natural conditions and to a tracer of the finer fraction. A variation of EFs is the Igeo index [23], which has a geometric progression to account for the increase of EFs, up to two orders of magnitude; (2) the regulatory approach is based on environmental quality standards for sediment quality or predicted no-effect concentrations [24]. Before the 1990s, these criteria were not available in France.

Once dated and normalised, the results can be used to compare how various pollutants have evolved at a given station (Step D), to compare stations within a given basin (Step E) and finally to compare basins (Step F). The understanding of these pollution trajectories needs historical socio-economic data (Steps G–K), as will be discussed in the following sections.

2.2   Contamination Fluxes Estimated from Floodplain Cores

In this approach, the average composition of the core at a given period is multiplied by the sediment flux for the same period to generate the contaminant flux. Then excess river fluxes are calculated [25].

This approach has several limitations, however. The extension of core analysis to river contaminant fluxes is based on the following hypothesis: (H_1) these deposits represent the flux-averaged quality of the particles carried by the river; (H_2) once deposited (without hiatus, which makes the age model uncertain), these particles are not (bio)turbated; and (H_3) the targeted compounds do not decompose (H_4) or migrate in the profiles. This set of assumptions is applied only over a long period, i.e. decades and more, considering that the contaminant fluxes are first controlled by the evolution of the chemical composition of river suspended particulate matter (SPM) much more than sediment fluxes. Note that H_1 is discussed for metals and radionuclide trends in [26]. The interpretation should also take into account the possible retention of contaminated particulates in the river basin [27], as in reservoirs, which could result in a decreased leakage not connected to environmental measures (e.g. reduction of point pollution sources). In some basins, the sediment loads can be reduced up to 95% as a consequence of large dam or reservoir cascades [28].

In the Seine River, some of these hypotheses have been checked. Floodplain particles are generally deposited from the peak discharge throughout the receding stage. Cd analysis on river SPM during Seine River floods has shown that actually the maximum content is observed at the receding stage, i.e. when particulates settle in the floodplain [29], so that the sediment contamination is not underestimated and hypothesis H_1 is satisfied. For the assessment of excess metal fluxes, the effect of interannual variations of suspended loads (H_2) is minimised by taking a 2 to 5 cm core slice, homogenised and then analysed (corresponding to an average period of 4–6 years, depending on the sedimentation rate), to which an average Seine River interannual sediment load (700,000 t year^{-1}; [30]) has been attributed throughout the core. In some cores of the Seine River (e.g. [14]), the Cs-137 peak maximum is very high, an indication of the limited mobility of this element and the absence of post-depositional bioturbation (H_3, H_4).

Finally, it must be remembered that sediment transport in fluvial systems results in a mixing of particulates of various ages of mobilisation/remobilisation, ranging from a few hours (remobilised bottom sediments) to a few hundred years and more (erosion of floodplain banks). Thus, the comparison of contemporary river output of polluted particles to the present-day circulation of material is an approach that is mostly used to compare the life cycle of different chemicals (they are affected by the same bias) or to assess the general trend of exported pollutants. These reservations explain why modelling particulate pollutant transfer in the anthroposphere is so difficult.

2.3   Circulation of Contaminants Within River Basins

The river's excess fluxes of particulate contaminants are then compared to the circulation of products and materials within the anthroposphere, i.e. the material flow [31] such as imports and exports, transformation and use of products and recycling (e.g. for metals). In the case of metals, one should also take into account mining and smelting, which are not always present within the river basin intercepted by the core. The anthroposphere also gradually stores contaminants in contaminated soils, infrastructures, etc. In large basins, only a very small proportion of this enormous cycling of material actually leaks into the aquatic environment. In small urban streams, it may be greater. Such material flow analysis has been done for Paris for the metal circulation [32, 33] and for the whole basin for specific elements such as nitrogen and phosphorus [34]. Fluvial sediment archives are key elements in the validation of contaminant circulation in river basins. They concern environmental transfers between the major Earth system components – atmosphere, biosphere, hydrosphere, soils and sediments – which intercept different diffuse and point sources of contamination, which are considered as leaks to the river system: contaminated atmospheric fallout, agricultural runoff (sometimes impacted by reuse of urban sewage sludge on cropland), sewage outfalls and leaks from waste dumps.

The ratio of the exported river pollutant fluxes to the circulation of the related material within the intercepted basin generates the *leakage ratio* (Fig. 1, Step H), a dimensionless factor, which is a performance indicator for regulation measures (pressures/driver). Another indicator of environmental pressure, rated to the basin population, is the per capita *excess load* (Fig. 1, Steps I and J), i.e. the excess flux divided by the basin's population upstream of the coring station (ELcap), expressed in g cap^{-1} year^{-1}. Both indicators can be used to compare the efficiency of environmental regulations in river basins and their trajectories, provided that past metal contents are reconstructed on the basis of dated sediment archives and past metal demands are recovered by an analysis of the area's economic history [25, 35, 36].

3   The Seine River Basin and Its Contamination by Micropollutants

The Seine River basin has one of the highest ratios of pressures to river sediment fluxes, which results in a very high contamination level in the lower Seine River, downstream of Paris. This contamination by metals and by numerous organic pollutants was not assessed before 2006 when new monitoring strategies were applied, following the EU Water Framework Directive (WFD). At that time, the first cores taken in the Seine River basin revealed the unknown long-term history of contamination, which reached very severe levels, when compared to current sediment quality guidelines.

3.1   The Seine River Basin: A Sensitive Territory Under High Pressures

In addition to the pressures (e.g. climate, morphology, land use, etc.) described earlier in this volume [37], the basin is characterised by a highly biased population distribution: about 70% of the basin's population is concentrated in the Paris megacity (2,750 km^2 with ten million inhabitants), which covers only 4% of the basin area. The upper Seine River and Marne tributary meet at Paris (Fig. 1 [37]), and the lower Seine River receives the other main tributary, the Oise River, downstream of the Paris megacity. The Paris megacity's main wastewater treatment plant (WWTP) at Achères, today called "Seine-Aval", discharges the treated effluent just upstream of the Seine-Oise confluence, some 70 river km downstream of the centre of Paris. Its capacity gradually grew in three decades to 8 M people at the end of the 1980s and then decreased to 6.5 M people because of the construction of additional WWTPs, which still makes it the largest WWTP in Europe.

The lower Seine River is structurally very sensitive to wastewater inputs, due to its minimal sediment transport and, above all, to its limited summer dilution during low flows (110 m^3 s^{-1}).

Most Seine River cores presented here intercept the 1945–2005 time window, which covers the reconstruction period after the World War II, from 1945 to 1974, also termed the Glorious Thirties. During that period, important changes and innovations occurred in the industrial sector. From 1970 to 2000, sewage collection and treatment plants were generalised in the Paris megacity (ten million inhabitants), i.e. more than two-thirds of the contamination point sources. Industries, which used to discharge their wastes directly into river courses, installed recycling processes and were gradually connected to sewage networks after 1970 [25]. Many heavy industries were delocalised outside the Paris area and then outside the Seine basin. During that period, new types of products were consumed, new drugs were used, and some products were legally withdrawn.

Since its beginning, the PIREN-Seine programme launched a series of studies on particulate micropollutants in the basin, particularly on metals [29, 38–41] and PCBs [42]. Other compounds such as PAHs, alkylphenols, PBDEs [15] and pharmaceuticals [19] were later studied. Various strategies have been set up: river profiles upstream and downstream of Paris, temporal variations with river flow and TSS at selected sampling stations, spatial distribution of the contamination using of fresh floodplain deposits [43], in-depth source inventories and circulation of compounds within the anthroposphere including atmospheric inputs. At the same time, the long-term history of human activities on the basin and of the social responses to water quality issues was addressed [33, 44–46]. The selection of core sites, core analysis and interpretation greatly benefited from this accumulated knowledge.

3.2 Regulatory Surveys Failed to Assess Micropollutants (1971–2006)

In France, trace metals and chlorinated insecticides were first surveyed within three national river quality inventories (1971, 1976 and 1981) on unfiltered waters, to comply with World Health Organization (WHO) drinking water criteria. Analyses were carried out 4 to 6 times per year at 25 key stations in the Seine River basin. Due to substantial bias (contamination during field and laboratory operations, excessive quantification limits in untrained laboratories), the first water quality inventory carried out by the newly created Agence de l'Eau Seine-Normandie (AESN) did not provide usable data [47]. The first environmental assessment of the European Environmental Agency in 1994 [48] also failed to produce quantitative figures on micropollutant concentrations.

The pioneering micropollutant studies in the Seine River basin were conducted on river SPM and concerned trace elements such as Cd and Hg [49, 50]. These academic studies failed to assess the state of extreme contamination they observed given the lack of sediment quality criteria and a reference natural background. Therefore, their use for environmental assessment remained very limited. The use of sediment archives had been recommended very early by marine scientists [51] and by the

Ministry of Transport's technical laboratory [52], but this approach was not used by the river basin authorities.

From 1982 to 1992, the six French river basin agencies developed their own monitoring strategies. The AESN turned to deposited sediments and rotating surveys on 204 sediment stations. The bed sediment analyses were carried out on the fraction <2 mm, i.e. including the sand fraction; unfortunately, ancillary data (e.g. % fine fraction, aluminium), which would have allowed for grain-size effect correction, were lacking, limiting data interpretation [53]. The official assessment of the metal contamination (Ministère de l'environnement, 1985) stated: "one finds high concentrations only for copper: metals in sediments are low". This statement ignored the severe Cd and Hg contamination occurring at that time, gradually unveiled by the PIREN-Seine programme in the 1990s using SPM analysis (e.g. [41]). In 2006, the monitoring within the WFD considered the dissolved fraction only. It therefore appears that the shifting strategies of micropollutant monitoring from 1971 to 2006 did not generate a general spatial and temporal assessment for contamination.

The sediment quality guidelines (SQG) that were or could have been used to assess the contamination of micropollutants is another limitation to contamination assessments. From 1971 to 2003, the guidelines concerning metal contamination evolved considerably, depending on the media (e.g. unfiltered waters, agricultural soils, dredged river materials, urban sludges). For a long time, the SQG did not exist in France, until they were finalised in 2003 by the French water quality assessment system, the SEQ-Eau [54]; sediment guidelines were therefore replaced by the WFD criteria on waters in 2006. Metal contamination in rivers has not been assessed on a national scale [55, 56], ignoring this critical issue.

3.3 Spatial Position of Sedimentary Archives Within the Seine River Basin

The core position within river systems should be chosen depending on the expected environmental information. In the Seine River basin, our coring strategy has considered very different positions of the intercepted basin area (Fig. 2). In addition, the coring sites are geographically located in Fig. 1 [37]. Coring sites are listed in Table 1 and greater detail can be found in the related articles. Cores taken directly from the river bed by a diver in 1994 [57], which cannot be dated, presented sharp contamination boundaries, suggesting erosion of previous deposits, are not used here.

Our pristine or sub-pristine site (PRI) is a headwater reservoir, which intercepts forested area. The medium impacts are studied on two floodplain cores, downstream of the city of Troyes (URB), on the Upper Seine River, and in the middle part of the Oise River (IND), a major Seine River tributary impacted by heavy industries. The maximum impact is assessed on a core taken 20 km downstream of the centre of Paris, in a navigated reach blocked by a sluice, which actually acts as a sediment trap

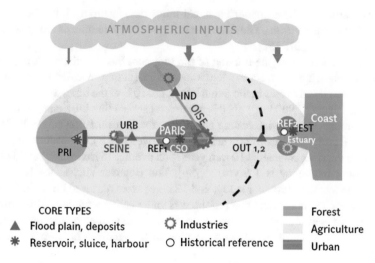

Fig. 2 Schematic positions of sediment archives used in the Seine River basin (see text)

Table 1 Principal coring sites considered in the Seine River basin

	Site (period covered)	Basin area (km^2)	Dpop[a] (cap. km^{-2})	Contaminants[b]	Reference(s)
EST	Rouen harbour (1965–2008)	75,000	208	Metals, major elements, lead isotopes, radionuclides, PAH, PCB, PBDE, OCP	[26, 60, 61]
OUT$_1$	Bouafles (1945–2004)	64,277	277	Metals, major elements, lead isotopes, radionuclides, PAH, PCB, PBDE, alkylphenols, antibiotics	[14, 15, 19, 60–63]
OUT$_2$	Muids (1933–2003)	64,484	276	Metals, major elements, lead isotopes, radionuclides, PAH	[14, 63, 64]
IND	Chauny (1954–2004)	3,280	45	Metals, major elements, lead isotopes, radionuclides	[14, 63]
CSO	Chatou (no datable archive)	44,670	na	Metals (Cd, Cu, Pb, Zn, Fe)	[65, 66]
URB	Troyes (1972–2001)	3,410	47	Metals, major elements, lead isotopes, radionuclides	[14, 63]
PRI	Lac de Pannecière (1931–2008)	917	9	Metals, major elements, lead isotopes, radionuclides, PAH	[64]

na not available

[a]Population density (Dpop)

[b]See related articles for further detail

for the main Paris combined sewer overflows (CSO) at low flows. The Seine River basin outlet is sampled on floodplain cores (OUT$_1$ and OUT$_2$), at Muids and Bouafles, near the outlet (Poses). Finally, an estuarine site on a harbour dock at Rouen (EST) integrates both direct estuarine sources with the upstream river sources, particularly from major chemical industries, including P fertiliser production from phosphogypsum. The atmospheric inputs of contaminants are greater at the core sites URB to EST [58].

Floodplain sites store the fraction of the sediment flux circulating during floods, usually during winter, and act as a temporary sediment trap. The SPM transported by high-flow and flood events in the Seine River, i.e. 10% of the time, was estimated for the Marne River (12,900 km^2) at 60% of the total annual flux [30], and samples from floodplains are therefore interesting candidates for the estimation of yearly fluxes.

At the Bassée floodplain, downstream of Troyes (URB), the sedimentation rate 35 m from the river channel is 10 mm year^{-1} on average (5 mm year^{-1} at 70 m), and the flooding frequency is 1.6 year^{-1} [30]. The Bouafles site, at the basin outlet (OUT$_1$), underwent particular study, and 12 cores were retrieved between 2003 and 2010. It was clear in the field that the site was under active accretion, with large trees buried above their roots. Unfortunately, in 2016, the inundated prairie was ploughed, evidencing the issue of on-site conservation of the archives in strongly anthropic basins.

The sampling strategy is completed by the assessment of the natural backgrounds. For organic pollutants, the PRI core can be considered. For metals, a prehistoric reference sediment (REF$_1$) is considered (Marne–Seine Rivers confluence at Paris-Bercy, dated 5,000 years BP). It is similar to another prehistoric estuarine sediment (REF$_2$) [59].

3.4 Core Dating

With the notable exception of the cores from Chatou for which no age model could be drawn, the cores exhibiting a clear chronology based on the measurement of radioactive tracers, ^{137}Cs and ^{210}Pb (and ^{238}Pu for EST), were selected. All cores were checked for a regular and fine-grain sedimentation profile. Core analyses are detailed in the published papers quoted here.

4 The Evolution of the Seine River Basin Contamination (1910–2015) Unveiled from Sediment Cores

4.1 Temporal Evolution of Persistent Organic Pollutants at the Basin Outlet and in the Estuary

4.1.1 Evolution of the River Basin at Its Outlet

The vertical profiles are represented for a set of different pollutants in Fig. 3. Since analyses provide multielemental contents for a given chemical family for each sample, the individual results are often clustered and given as "sum of". As the

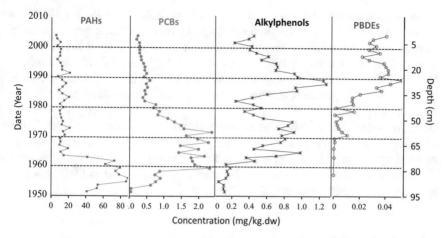

Fig. 3 POP contamination profiles (mg kg^{-1} dry weight) for PAHs, PCBs, alkyl phenols and polybromodiphenylethers (PBDEs) in the Seine River floodplain core OUT$_1$ [15]

conventions may have evolved over the last 50 years, for instance, for the sum of PCBs, one must check if the same basket of molecules is considered.

Four organic pollutant (POPs) families, namely, 13 polycyclic aromatic hydrocarbons (PAHs), 15 polychlorinated biphenyls (PCBs), 3 alkylphenols (APs) and 8 polybromodiphenylethers (PBDEs), were analysed on the Seine River outlet core [15]. The core provides detailed contamination records at a nearly annual scale. Each of them exhibits a specific trend, which can be described by several metrics: (1) the state of contamination at the core bottom, for those already featuring a contamination in 1945, (2) the emergence date (T_e), (3) the period of maximum (T_{max}), (4) the multimodal or monoclinic evolution and (5) the recent – last 10 years – trend and its slope.

The PAH profile reflects an old contamination associated with the use of coal for domestic heating, industries and power plants in the Seine River basin, which had never been impacted by coal mining. Most particularly, the post-war reconstruction period, from 1952 to 1960, is clearly marked with the highest levels (up to 90 mg kg^{-1} for $\sum$PAHs). A marked decrease of PAHs is observed between 1962 and 1965, by a factor of 10. The present-day level is fairly stable at 10 mg g^{-1}. This very rapid decrease of PAHs in flood sediments after 1960 cannot be explained solely by changes in coal consumption for domestic heating but also reflects a major industrial transformation with the closure of several gas plants based on the distillation of coal and the reduction of coal use in the steel industry. After 1968, coal was gradually replaced by fossil fuel in power plants, and contaminated sewage SPM was reduced by urban water management at the basin scale. Pre-industrial levels of PAHs – originating from fires – are not established; they should be a few mg kg^{-1} in the sediments [67].

PCBs, PBDEs and APs have different contamination profiles, reflecting the pattern of use and regulations specific to each family. The starting points (T_e) of contamination correspond to the dates of their first industrial uses: the 1950s for PCBs, 1960s for APs and 1970s for PBDEs. PCBs and PBDEs present a sharp increase, while AP increases are more gradual. The Tmax occurred between 1960 and 1972 for PCBs, between 1965–1975 and 1985–1995 for APs and after 1990 for PBDEs. Finally, each of these peaks is followed by a period of sharp decrease, according to the regulations and the means of available substitution products. For example, a rapid decay of PCBs from 1972 to the 1980s was observed, followed by a slower period of decay. This trend is further discussed in Sect. 8. For APs, a rapid decrease has been observed since 1992 following the gradual restriction of their use, initiated by the OSPAR convention on pollutant inputs into the Atlantic Ocean and due to their substitution in detergents by other compounds such as PBDEs, which show a plateau pattern over the 1995–2005 period.

4.1.2 Evolution of Contamination in the Estuary

The EST core was sampled in a harbour dock that can be considered a perfect sediment trap when considering the ^{137}Cs profile used for dating and the sub-annual definition (1 cm year^{-1}) of the analysed layers, which provides a highly detailed record of the temporal variations of 95 potentially hazardous compounds (Fig. 4) [26]. The contamination pattern is similar to those observed in the OUT$_1$ core for PCBs, with maximum contents in 1970–1975. The comparison of actual PCB levels

Fig. 4 Evolution of POP contamination in the Seine estuary sediment core EST (1970–2008) [26]

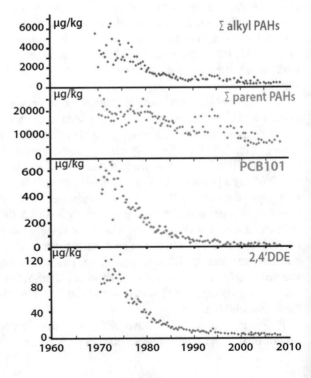

between the two cores is difficult because the analyses do not consider the same compounds. The content of dichlorodiphenyldichloroethane (DDE), a degradation product of dichlorodiphenyltrichloroethane (DDT), a chlorinated insecticide used until 1970, drops sharply after its total ban in the early 1970s.

4.2   Metals (Ag, Cd, Cr, Cu, Hg, Ni, Pb, Sb, Zn) and Arsenic (As) at the Seine River Outlet (1935–2005)

Considering the OUT$_1$ trends, the maximum metal EF occurred before 1965 (Fig. 5). Metal regulation started in the 1970s (Circulaire July 4, 1972, limiting the metal discharges from the surface treatment industry) so that the decontamination started ~10 years before the social responses to contamination. This pattern is first due to the industrial delocalisation outside of the centre of Paris and then from its suburbs; second, to changes in industrial practices, which include the gradual recycling of metals (e.g. in plating, in the 1980s); and third, to gradual use restrictions and then bans on Cd (its use as pigment was banned in 1995), Hg (in thermometers in 1999) and leaded gasoline (2000) [36].

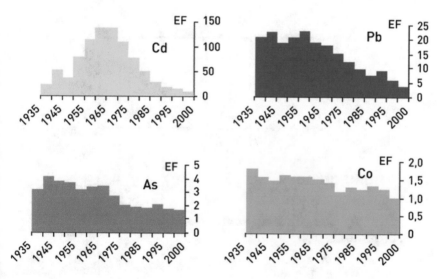

Fig. 5 Evolution of metal enrichment factors – 5-year averages – in the Seine River sediment archives at the basin outlet OUT[1] (1935–2005) [25]

4.3 Antibiotics at the Seine River Outlet (1954–2004)

Several antibiotics were analysed on the Seine River outlet OUT_1 core for a period of 40 years (1954–2004) [19]. The temporal evolution of norfloxacin, flumequine, oxolinic acid, sulfamethoxazole and nalidixic acid is specific to each compound (Fig. 6). Norfloxacin was detected below its quantification limit in two slices out of 45. Flumequine, oxolinic acid and nalidixic acid were accurately recorded: (1) their emergence date in the record is 2–3 years after or before their assumed first use; (2) they present a general maximum near 1988 for flumequine, oxolinic acid and nalidixic acid. The progressive decrease could be due to (1) changes in the

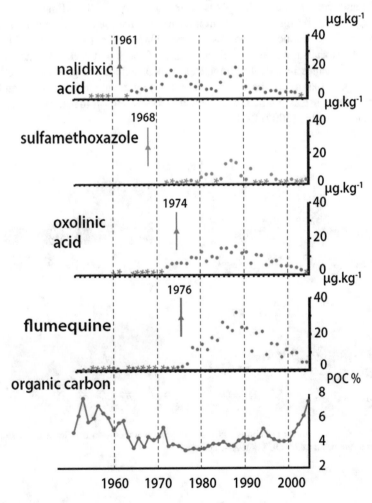

Fig. 6 Evolution of antibiotic contents in sediment archives of the Seine River at its outlet (OUT₁), from [19]. Vertical bars indicate the date of marketing authorisation

prescription strategies of practitioners, (2) a general decrease in antimicrobial consumption between 2000 and 2005 and (3) constant improvements made in the collection and treatment of wastewater coming from Paris and its suburbs during the last 40 years. Sulfamethoxazole has the least regular record and its detection lags 12 years behind its assumed first use. The regularity and the coherence of these profiles are encouraging: this is the first demonstration that an antibiotic's history can be recorded in sedimentary archives, extending this approach to other types of environmental indicators (e.g. [20]).

5 Intra-basin Comparisons in the Seine River Basin Reveal Contrasted Trends in Subbasins

5.1 Compared Metal Contamination History: The Cadmium Example

Cadmium is an iconic contaminant in the Seine River basin. No fewer than seven contamination profiles are compared in Fig. 7 (Step E, Fig. 1).

All enrichment factors were normalised to the thorium content in shale $(12.3 \text{ mg kg}^{-1})$ and to the local background element content of Cd (0.2 mg kg^{-1}), except for the PRI core, located in the Morvan region, a crystalline shield where the background levels are Cd = 0.5 and Th = 18.4 mg kg^{-1} [68]. For the URB and IND cores, the Cd EFs are very low, compared to those observed downstream of Paris and in the estuary (OUT and EST). They clearly evidence that the maximum Cd contamination occurred in 1960–1965 at the outlet [25]. The high Cd levels found in the Lower Seine River (OUT and EST compared to PRI, URB, IND) result from the emissions of numerous plating workshops and industries within Paris, which have been greatly reduced in the last few decades.

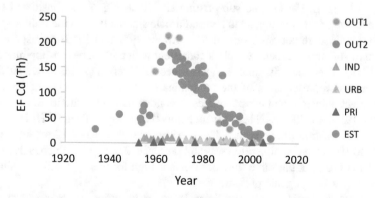

Fig. 7 Comparison of cadmium contamination trends in the Seine River sediment archives, upstream (PRI, URB, IND) and downstream of Paris (OUT₁, OUT₂, EST)

In the estuary site, the main Cd peak is observed around 1970 and a second one is observed in the late 1970s. A third very sharp peak is observed in 1987. These peaks are not observed in the OUT_1 core, suggesting that the 1970, 1975–1980 and 1987 estuary contamination episodes are due to very local events as the direct release of phosphogypsum wastes from a fertiliser industry using Cd-rich ore. There has been continuous dumping of "phosphogypsum residues" from the phosphoric acid industry, treating phosphorus ores at several locations in the estuary. Such waste materials were first released in the upper estuary, over several decades, then in the middle estuary and finally by barges in the outer estuary (Baie de Seine) with substantial overspills. These dumpings were on the order of magnitude of the Seine River sediment load (0.7 Mt. $year^{-1}$ on average): 1974 (0.4 Mt. $year^{-1}$) to 1987 (0.4 Mt. $year^{-1}$) with a maximum between 1980 and 1984 (1.6–2 Mt. $year^{-1}$) [60]. The EST core reveals this long-term impact on many metals and phosphorus (P content up to 5,000 mg kg^{-1} in EST sediment vs 800 mg kg^{-1} for the basin reference) and for Y, La and lanthanides, U, Cd, As, Cr, Mo, Th, Ba, etc. [26]. All show a marked increase of their contents in the estuarine core with the same temporality as phosphorus. This contamination source is estimated to account for 50–70% of these elements' contamination, reaching 90% for Y. This type of generalised contamination by a mineral chemistry industry has rarely been documented.

These contaminations were superimposed to those originating from the upstream river basin: the EFs recorded in the estuarine sediments were extreme in the 1970s, from 100 to 1,000 for many elements (Hg, Pb, Ag, Cd), making this waterbody one of the most polluted in Europe [69].

5.2 Compared POP Contamination History: The PAH Example

The comparison of the PAH contamination trends at five coring stations is illustrated in Fig. 8 (Step E, Fig.1). The sub-pristine site (PRI, Fig. 2) is considered to be representative of the non-impacted levels. Particle size analysis of the sediment core reveals that the core bottom layers (dated between 1952 and 1955) correspond to a mixture of soil and sediment, which is consistent with the reservoir's impoundment date (1949). Therefore, the analyses performed on layers deposited after 1955 are considered as representative of the contamination trend of this area of the basin where direct industrial and urban sources of contamination are minimal.

In the 1950s, the PRI PAH levels were below 0.2 mg kg^{-1}. Then PAHs increased, possibly due to atmospheric inputs. The maximum levels reached 10 mg kg^{-1} ($\sum 13$ PAHs). In 1992, the contamination was still present at 2 mg kg^{-1}, which is five times less than at OUT_1 at the same time but remains high for this type of basin slightly impacted by anthropogenic pressure.

The PAH contamination at the Seine River basin outlet was assessed in three cores taken 10 km apart (two cores at OUT_1 and one core at OUT_2). This allows for a

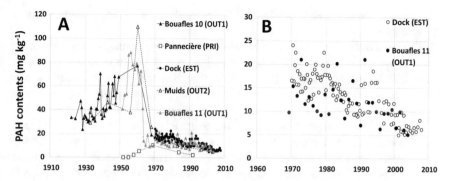

Fig. 8 (**a**) Comparison between the sum of PAH contents (13 PAHs, mg kg^{-1}), in the Seine River sediment archives, downstream of Paris (Bouafles and Muids sediment (black, grey and open triangles)) with the upstream sub-pristine reference (Pannecière reservoir sediment, open squares). (**b**) Enlargement of the 1970–2010 period for basin outlet (Bouafles) and estuary (Dock). These contents are also compared with estuary sediment (dark circles)

validation of the sediment archive approach: there is no significant PAH source between Muids and Bouafles and the two OUT$_1$ cores should be identical. Actually, the contamination pattern is very similar in the three cores (date of maximum contamination, slope of contamination and decontamination), which proves the reliability of the general trend. Small differences (less than 30%) in peak values may stem from local patterns of sedimentation or sediment reworking.

Focusing on the 1970–2000 period, similar PAH decontamination rates are observed in the outlet cores (OUT$_1$ and the EST), but second-order peaks in 1997 and 2004 showed by the EST core suggest a direct PAH source in the estuary. Also, the decontamination rates for OUT$_1$ (3.0 ± 0.3 mg kg^{-1} year^{-1}) and for the upstream site (PRI) (2.8 ± 0.2 mg kg^{-1} year^{-1}) are very close, representative of the general decrease of PAH emissions at the scale of the Seine River basin, probably linked to the decrease in coal consumption over this period (see Sect. 4.1).

6 Interbasin Comparison of Metal Contamination Trends in Western European Rivers Using Sedimentary Archives

Interbasin comparisons can be performed at basin outlets which integrate the whole catchment area (Fig. 1, F). The evolution of metals in nine rivers in Europe (Rhine Volga and Danube, Mersey, Scheldt and Garonne) and in the Mississippi Delta shows various levels of contamination, highly controlled by the river SPM contamination level [70]: the Mississippi had the lowest EFs and the highest dilution power by erosion-derived SPM, while the Rhine and Seine Rivers show opposite characteristics.

The level of contamination of river sediment strongly depends on the ratio of the pressure over the dilution power (sediment load derived from erosion). The Seine River at its mouth has a much higher ratio than the Rhône or the Mississippi. Similar control is observed for PAH contamination when comparing the Orge River, an urban sub-catchment of the Seine River, and other worldwide urban catchments [71].

Compared to the rivers cited above, the Seine River outlet exhibits the highest metal concentrations for the 1940–1960 period, but small rivers impacted by big cities present even higher contamination levels [35]. This is illustrated (Fig. 9) when comparing the metal contamination trends in cores from the Zenne River floodplain (Brussels), a Lambro reservoir downstream of Milan (data courtesy of Luigi Vigano, CNR), the Quentzee Lake on the Havel-Spree River (Berlin) and the Seine River outlet [72]. The maximum contamination reached in the Seine River core is always much lower than for the other rivers (Table 2A). The city impact ranking for the general metal contamination is as follows: Havel/Spree > Lambro > Zenne > Seine rivers. Note that if the CSO core (Seine River), whose sampling site is more similar

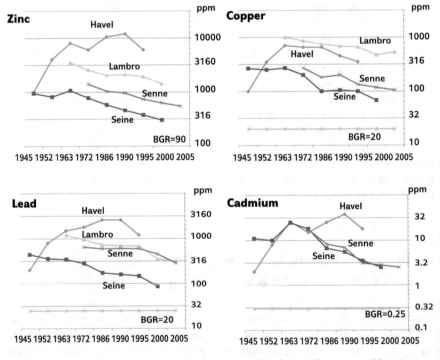

Fig. 9 Comparison of metal contamination trends, resulting from sedimentary archives (ppm or mg kg^{-1}, log scales), in four rivers impacted by major European cities, the Havel-Spree River and Berlin, the Lambro River and Milan, the Zenne River and Brussels and the Seine River and Paris, and background concentration values (BGR) (from [72])

Table 2 Comparisons of metal contaminations in world rivers from highly industrialised countries, using sedimentary archives at their peak contamination stages

	Cd	Hg	Pb	Zn	Basin area (km^2)	Population (10^6 cap.)
A. Maximum contamination (mg kg^{-1}) observed at urban site (year)						
Lambro (Milan) (1)	n.d.	n.d.	1,400 (1964)	3,500 (1963)	2,747	4.2
Zenne (Brussels) (1)	15.5 (1974)		700 (1979)	1,300 (1974)	1,160	1.3
Spree/Havel (Berlin) (1)	38 (1990)		2,600 (1986)	12,000 (1990)	10,105	3.5
Seine (Paris) (1)	28 (1960)		450 (1945)	1,150 (1967)	65,000	10
European and US urban sewage PM (1970–1980) (1)	140		410	3,000		
European rivers background (6)	0.25	0.04	20	90		
B. Per capita in French rivers at outlet positions (g^{-1} cap^{-1} year^{-1})						
Seine (2)	3.4	0.6	50	160	75,000	15
Loire (3)	0.5	0.1	10	40	117,800	7.7
Garonne (4)	1.9	n.d.	50	170	56,198	3.25
Rhône (5)	1.5	0.5	100	190	97,800	14
C. Per capita in other basins (g^{-1} cap^{-1} year^{-1})						
Western European rivers (6) (7)	2.8	0.74	33.7	175	0.53	123
Mississippi (8)	0.58	0.175	30	47.5	3.2	67
Danube (9)	0.6	–	9.5	70	0.8	81
Montreal (10)	0.23	0.05	7.6	64	–	–

(1) [72] Median of 21 analyses, adjusted to Al = 75,000 mg kg^{-1}, (2) [25], (3) derived from [16], (4) derived from [11], (5) derived from [18], (6) derived from [70]. For Montreal: per capita release into St Lawrence River based on sewer analysis. (7) Elbe, Meuse, Garonne, Rhine, Scheldt and Seine river basins [70], (8) [5], (9) [77], (10) [78]
n.d. not determined

to the Spree/Havel, Lambro and Zenne river sampling sites, had been usable as a dated archive, the ranking would have been very different. The Cd content in the CSO core was up to 100 mg kg^{-1} [66]. The decontamination of the Lambro, Zenne and Seine rivers developed around 1960, but the decontamination of the Havel-Spree did not occur before 1990, after the German reunification. Water quality studies in the River Elbe since the German reunification in 1990 revealed the improvement of water quality due to the reduction of industrial wastewater emissions since 1990 [73].

7 Circulation of Material Within Basins and Its Impact on River Fluxes

In a fully sustainable society, the products and goods containing potentially harmful substances should circulate without any leak into the environment (air, soil, water, coastal zone). The previous sections have shown that this is not the case and that river SPMs are a very good indicator of the way a society is handling these environmental issues. Sedimentary archives add a precious time-depth to this question, provided that additional information on the general circulation of targeted materials within the anthroposystem are available. This approach has been applied to establish the metal circulation patterns in Western European rivers (Steps G–K, Fig. 1).

7.1 Leakage Ratio of Metals Within the Seine River Basin (1950–2005)

The national demand for Cd, Cu, Hg, Pb and Zn has been reconstructed since 1900 on the basis of the national consumption of all metals and products containing metals. The metal demand in 1900 is the reference value and is set at 1.0 (Fig. 10, left). The metal demands greatly increased (2.5- to 5-fold) after World War II, and the present-day consumption (year 2000) is much higher than in 1950, except for Hg. This increase is faster than the population increase in the Seine River basin, indicating that the per capita consumption has greatly increased, except for Hg after 1975 and Cd after 1985. The Hg demand has dropped by nearly two orders of magnitude since 1975, and Cd demand has dropped since 1995, both as a result of regulations (gradual Hg ban since the 1970s and Cd use limitation beginning in 1995).

The leakage ratio (in %) is the ratio of exported river pollutant fluxes over circulation of the related material within the intercepted basin. It can be regarded

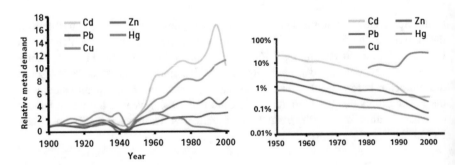

Fig. 10 Left circulation of metals in France, normalised to its value in 1900. Right: leakage rates in % estimated for the Seine River basin [25]

as an efficiency indicator of the capacity of a given society to recycle its metals. It has dramatically decreased for all metals since 1950 (e.g. from 0.8 to 0.05% for Cu), except for Hg, which remains at a high ratio, around 10% of leaks. The metal exported fluxes calculated from core data are therefore powerful tools to observe the general decontamination trend in Europe, which can be linked to the local type of economy and environmental policies, defining the trajectory of a given environmental issue [74]. In the Seine River basin, the location of recycling facilities is a key control factor in contamination by metals [33]. The recycling process (collection of used metal-containing devices and their retreatment outside of the Seine River basin and often outside France, i.e. a kind of pollution delocalisation [36]) seems to have been effective mostly after 1980.

As pointed out in Sect. 2, this leakage estimation is based on several hypotheses: (1) all river particles have the same age; (2) their age is very recent – a few years on average – so that the average export at time t $\pm$ a few years can be compared to the circulation data for the same period. When there is evidence that the river particles have extended age ranges or that they have greater ages, these hypotheses are not valid. In the Seine River, the floodplain deposits are relatively fine (silt), so that their average transit time throughout the river network ranges from a few weeks (large floods) to a few years. The occurrence of antibiotics in core sediment (Fig. 6) within 1 year after first arriving on the market and the decline of DDT metabolite, which started with the ban of this insecticide in France, suggest a rapid reaction of Seine River fine sediment (median grain size in the silt fraction) to source changes. In contrast, coarser material (1 cm) travels in the Seine River at secular rates: gravels contaminated by iron smelting during the eighteenth century have actually been used to study the very slow bedload movement in the Upper Marne (see [75], box 1).

7.2 Intercomparison of River Contaminant Fluxes in Relation to Their Population in Western European Rivers

The excess load of metals and the load of micropollutants can be rated by the population of river basins, to generate the per capita excess load (ELcap), an indicator also used for major ions or nutrients to compare and scale these per capita loads to economic indicators, such as per capita energy consumption [51, 76]. For metals, the comparison of the per capita loads for Western European rivers, the St Lawrence, Mississippi and Danube rivers, at the period of their maximum contamination level near 1970 (Table 2A) showed similar orders of magnitude for each metal, suggesting the existence of a metal metabolism common to these old industrial countries for similar levels of development and environmental regulations [70, 72]. Based on the metal content, the general pressure ranking is Zn $\gg$ Pb $\geq$ Cu > Cr > Ni > Cd = As > Hg, but their maximum EFs in river particulates are completely reversed: Hg > Cd > Pb > Zn. When looking in greater detail, each

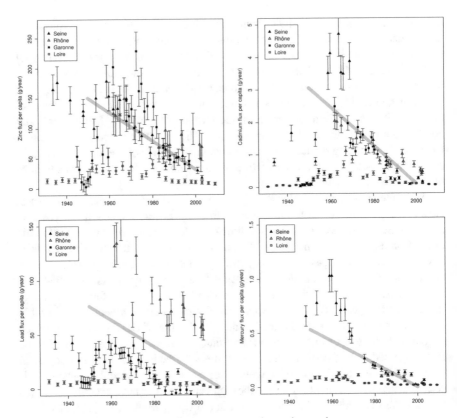

Fig. 11 Long-term evolution of metal fluxes (in g^{-1} cap^{-1} $year^{-1}$) carried by French rivers (Rhône, Seine, Loire and Garonne Rivers). The thick grey line is the per capita linear regression for the French rivers (population-weighted)

basin may have its own maximum period, depending on local environmental and political history [72].

ELcap trends were calculated for Cd, Hg, Pb and Zn based on the sediment archives of the four main French rivers that were studied: the Seine [25], Loire [16], Garonne [11] and Rhône [18] (Fig. 11). All per capita loads in French rivers show a regular decrease after 1960–1965, while the metal demand shows an inverse trend (Fig. 10). This is an outstanding indication of the improved environmental efficiency for metal emissions over the long term in France. It could not have been demonstrated on the basis of the regulatory monitoring of metal contamination in these basins, which is less than 25 years old, and sometimes not relevant for the river contaminant fluxes (e.g. survey of the contamination of aquatic mosses by the Rhône basin authority).

The levels of per capita loads reveal a totally different ranking of the four basins, with regard to their actual river SPM contamination (Table 2B). While the levels of metal concentrations are always lower in the Rhône River and higher in the Seine River, i.e. inversely to their suspended solid loads, the per capita loads of the Seine River are the greatest for Hg and Cd, the highest per capita Zn fluxes are observed in the Garonne River, while the lowest fluxes of all metals are in the Loire River. The top ranking of the Seine for Cd, although well-known mine tailings have also contaminated the Garonne for decades [79], is due to the numerous plating work-shops and industries in Paris in the 1960s until the 1980s. The particular pattern of Pb in the Rhone River suggests a shift in reference values. In this highly impounded basin whose solid discharge was considerably reduced during the last century, it is possible that a catchment with a high Pb background has become a higher contrib-utor to the Pb background in the last few decades.

These per capita figures for French rivers are compared to those of other industrial countries, at the peak contamination, i.e. the pre-regulation stage (Table 2C). They present a remarkable similarity.

8 Trajectories of the Environmental Issues Affecting River Basins and Society: Example of PCB Contamination in the Seine River Basin

The trajectory of the water quality situation (Fig. 1) includes the following compo-nents [74]: (1) the evolution of a state indicator measured during the targeted period; (2) the relevant pressure or driver indicator; (3) the first awareness of a deterioration of water quality, its social recognition, environmental monitoring and assessment and reporting; and (4) the technical and/or regulatory responses provided. Here, the trajectory is completed by a brief analysis of how scientific knowledge of this issue has evolved, in particular its controlling factors. For micropollutants, the state indicator might not always be derived from the regulatory survey (see Sect. 3) but from the sedimentary archives. Historical archives are used to reconstruct other components of trajectories: the scientific knowledge of the inhabitants' perception of the problem and the interactions between key actors, including controversies. This approach is therefore typically interdisciplinary. It has been tested in the Seine River basin for trace metals and requires a large amount of information to complete these components and the circulation of the contaminant within the anthroposphere, as for metals [36].

Figure 12 is an attempt to describe the trajectory of PCBs in the Seine River basin. The state indicator is the total PCB content in the dated core profile from [15], taken at the river outlet (65,000 km^2).

PCBs are one of the first industrial products to be regulated and then banned. Following the alert made by Rachel Carson (Silent Spring, 1962), PCBs and other organochlorines such as DDT and lindane were an early concern of some French scientists [80]. In 1972 these compounds were included on the blacklist of UNESCO's International Oceanographic Commission and, in turn, on the list of

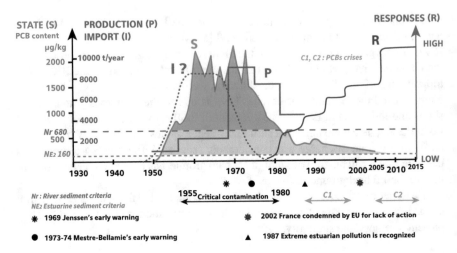

Fig. 12 Schematic trajectory of the PCBs dumped into the Seine River, based on the data recorded at the OUT1 coring site [15]

products to be carefully monitored in French estuaries and river basin outlets [51]. Marchand [81] stated that the Seine estuary sediments had the highest contamination in PCBs with regard to all other French study sites.

PCBs have been commonly used in France since 1950 in transformers and condensers. They were first imported; then in the late 1950s, they were manufactured in France by two chemical companies, outside the Seine River basin. In 1979, Colas [82] estimated that the total consumption of PCBs at that period was between 40,000 and 50,000 t, i.e. about 12,000–15,000 t for the Seine River basin, considering the proportion of industries in the Seine River basin compared to other river basins. The societal response to the PCB issues (Step K, Fig. 1) lags far behind the maximum contamination period, itself unnoticed until the first analysis of PCBs in the core.

The preliminary analysis of PCBs in Seine River basin fish revealed fish contamination [42]. The mussel watch established since 1979 in the estuary by Ifremer confirmed an extreme contamination by PCBs, when compared to other French and foreign estuaries [81]. However, this issue could not be handled at the level of a single river basin. At the national level, the banning of the further sale and use of PCB-containing devices was an answer to the first socio-political PCB crisis (C1, Fig. 12). The second national crisis (C2) in 2005 was triggered by a health regulation when PCB values exceeding the WHO criteria for edible fish were detected in bream fished in the Rhône River, near Lyon [83]. As a result, all fish from the Rhône basin were officially declared as non-edible, causing great concern on the overall quality in this basin. This event, in which fish were used as environmental indicators, triggered a national PCB plan (PCB 2008–2012) which forced the Seine River authorities to undertake specific assessments and accelerated the elimination of all former industrial devices still containing PCBs.

The current evolution of the contamination of the Seine River and the estuary is directly provided by the core analysis of the main PCB congeners ($n = 7$). The OUT$_1$

river outlet core [15] presents the following pattern (Fig. 12): (1) an emergence of contamination in 1950, as in the Rhône River [17], (2) a sharp increase until 1960, (3) then a 15-year plateau between 1,500 and 2,200 µg kg^{-1} until 1975 followed by (4) an exponential decrease to 200 µg kg^{-1} in 2005. The Lower Seine River was under critical contamination from 1955 to 1980 when applying the criteria for dredged sediments in rivers (Nr = 680 µg kg^{-1}) and from 1952 to 2005, when using those for estuaries (Ne$_2$ = 160 µg kg^{-1}) [84]. When the PCB content exceeds these levels, the sediment should be treated or appropriately stored [83]. This trend is confirmed by that observed in the estuarine Rouen core [26], which covers the period from 1968 to 2008: the maximum contamination is noted in 1972, as in the OUT$_1$core, with total PCBs at 4150 µg kg^{-1}, and then a regular 30-fold decrease to 140 µg kg^{-1}. The higher contents in the estuary can be attributed to direct sources and to finer sediments in this core (Al contents at 4% vs 3% in the Seine River SPM). Highly polluted estuarine sediments and agricultural soils in the Seine River floodplain are a long-term contamination heritage. When the second crisis was triggered in 2005, the PCB contamination was already reduced by a factor of 20–30 compared to the 1970s: at that time it was probably considerably above the current WHO criteria that could have resulted in health problems for regular fish and seafood consumers, but this issue was not detected. Sedimentary records unveil this past environmental history.

9 Conclusions

The emergence of new water quality issues is continuous as scientific knowledge progresses and analytical capacities are developed to detect new deleterious substances at reduced costs. Environmental regulations are evolving as well, but with a time-lapse of several years or decades between the start of use of a compound and its recognition as a potentially harmful agent to the environment or health. For the micropollutants studied in the PIREN-Seine programme, the medium time lapse is at least 30 years. Furthermore, the environmental release of many of these xenobiotics is not regulated, such as for antibiotics. It is common knowledge that resources are lacking to implement surveys dedicated to unregulated compounds. Finally, as in the case of PCBs, the regulatory survey which followed the water compartment was not adequate, missing the target media: biota and SPM. If the biota is now included as a target media in the WFD, SPM guidelines are still lacking.

Sedimentary archives make it possible to document the long-term state of the river quality for many issues. Water quality trajectories should then be complemented by substantial research on environmental history: (1) definition and recognition of the issue, with effects, causes, control factors, indicators, metrics, criteria and/or scales of water quality; (2) implementation of sustainable observatories for long-term monitoring and surveillance on proper media, at proper stations and frequencies, with appropriate analytical tools; (3) assessment of water quality and its declaration by relevant authorities; (4) identification of the origin(s) of the issue; (5) social consensus on the issue's importance, origins and solutions; and (6) funding and implementation of appropriate solutions.

None of the long-term contaminant trajectories in the Seine River – nor in any other French river – was available before the use of sediment archives, since the monitoring of these pollutants is very recent and/or has been inefficient, due to lack of appropriate analysed media and analytical difficulties. The studies presented here demonstrate the effectiveness of sedimentary archives, which give a comprehensive overview of the contamination in the Seine River basin. The results presented here are in agreement with previous studies dedicated to other European rivers and have highlighted for the first time the specificities of the Seine River basin compared to other large French and European basins. Sediment archives are still present for all European rivers, waiting for multidisciplinary teams, aware of the sedimentary, hydrological, chemical and historical issues, who will be able to elucidate their contamination history.

Acknowledgements This work is a contribution to the PIREN-Seine research programme (www.piren-seine.fr), which belongs to the Zone Atelier Seine part of the international Long-Term Socio-Ecological Research (LTSER) network.

References

1. Goldberg ED, Hodge V, Koide M et al (1978) A pollution history of Chesapeake Bay. Geochim Cosmochim Acta 42(9):1413–1425
2. Kemp A, Williams J, Thomas R et al (1978) Impact of man's activities on the chemical composition of the sediments of Lakes Superior and Huron. Water Air Soil Pollut 10 (4):381–402
3. Eisenreich SJ, Capel PD, Robbins JA et al (1989) Accumulation and diagenesis of chlorinated hydrocarbons in lacustrine sediments. Environ Sci Technol 23(9):1116–1126
4. Trefry JH, Metz S, Trocine RP et al (1985) A decline in lead transport by the Mississippi River. Science 230(4724):439–441
5. Santschi PH, Presley BJ, Wade TL et al (2001) Historical contamination of PAHs, PCBs, DDTs, and heavy metals in Mississippi river Delta, Galveston bay and Tampa bay sediment cores. Mar Environ Res 52(1):51–79
6. Horowitz AJ (2013) A review of selected inorganic surface water quality-monitoring practices: are we really measuring what we think, and if so, are we doing it right? Environ Sci Technol 47 (6):2471–2486
7. Valette-Silver NJ (1993) The use of sediment cores to reconstruct historical trends in contamination of estuarine and coastal sediments. Estuaries 16(3):577–588
8. Förstner UJN (1976) Lake sediments as indicators of heavy-metal pollution. Naturwissenschaften 63(10):465–470
9. Salomons W, De Groot A (1977) Pollution history of trace metals in sediments, as affected by the Rhine river. In: Krum-Bein W (ed) Environmental biogeochemistry, vol 1. Ann Arbor Science, Ann Arbor, pp 149–162
10. Middelkoop H (2000) Heavy-metal pollution of the river Rhine and Meuse floodplains in the Netherlands. Neth J Geosci 79(4):411–427

11. Grousset F, Jouanneau J, Castaing P et al (1999) A 70 year record of contamination from industrial activity along the Garonne River and its tributaries (SW France). Estuar Coast Shelf Sci 48(3):401–414

12. Audry S, Schäfer J, Blanc G, Jouanneau J-M (2004) Fifty-year sedimentary record of heavy pollution (Cd, Zn, Cu, Pb) in the Lot River reservoirs (France). Environ Pollut 132:413–426

13. Castelle S, Schäfer J, Blanc G et al (2007) 50-year record and solid state speciation of mercury in natural and contaminated reservoir sediment. Appl Geochem 22(7):1359–1370

14. Le Cloarec M-F, Bonte P, Lestel L et al (2011) Sedimentary record of metal contamination in the Seine River during the last century. Phys Chem Earth, Parts A/B/C 36(12):515–529

15. Lorgeoux C, Moilleron R, Gasperi J et al (2016) Temporal trends of persistent organic pollutants in dated sediment cores: chemical fingerprinting of the anthropogenic impacts in the Seine River basin, Paris. Sci Total Environ 541:1355–1363

16. Grosbois C, Meybeck M, Lestel L et al (2012) Severe and contrasted polymetallic contamination patterns (1900–2009) in the Loire River sediments (France). Sci Total Environ 435:290–305

17. Desmet M, Mourier B, Mahler BJ et al (2012) Spatial and temporal trends in PCBs in sediment along the lower Rhône River, France. Sci Total Environ 433:189–197

18. Ferrand E, Eyrolle F, Radakovitch O et al (2012) Historical levels of heavy metals and artificial radionuclides reconstructed from overbank sediment records in lower Rhône River (South-East France). Geochim Cosmochim Acta 82:163–182

19. Tamtam F, Le Bot B, Dinh T et al (2011) A 50-year record of quinolone and sulphonamide antimicrobial agents in Seine River sediments. J Soils Sediments 11(5):852–859

20. Thiebault T, Chassiot L, Fougère L et al (2017) Record of pharmaceutical products in river sediments: a powerful tool to assess the environmental impact of urban management? Anthropocene 18:47–56

21. Bajard M, Poulenard J, Sabatier P et al (2017) Long-term changes in alpine pedogenetic processes: effect of millennial agro-pastoralism activities (French-Italian Alps). Geoderma 306:217–236

22. Krishnaswamy S, Lal D, Martin J et al (1971) Geochronology of lake sediments. Earth Planet Sci Lett 11(1–5):407–414

23. Förstner U, Müller G (1973) Heavy metal accumulation in river sediments: a response to environmental pollution. Geoforum 4(2):53–61

24. Barjhoux I, Fechner LC, Lebrun JD et al (2018) Application of a multidisciplinary and integrative weight-of-evidence approach to a 1-year monitoring survey of the Seine River. Environ Sci Pollut Res 25:23404–23429

25. Meybeck M, Lestel L, Bonté P et al (2007) Historical perspective of heavy metals contamination (Cd, Cr, Cu, Hg, Pb, Zn) in the Seine River basin (France) following a DPSIR approach (1950–2005). Sci Total Environ 375(1):204–231. https://doi.org/10.1016/j.scitotenv.2006.12.017

26. Boust D, Lesueur P, Berthe T (2012) RHAPSODIS - reconstruction de l'historique des apports particulaires à la Seine par l'observation de leur intégratio, sédimentaire. Rapport GIP Seine Aval, Rouen

27. Meybeck M, Vörösmarty C (2005) External geophysics, climate and environment: fluvial filtering of land-to-ocean fluxes: from natural Holocene variations to Anthropocene. CR - Géosci 337(1):107–123. https://doi.org/10.1016/j.crte.2004.09.016

28. Vörösmarty CJ, Meybeck M, Fekete B et al (2003) Anthropogenic sediment retention: major global impact from registered river impoundments. Glob Planet Chang 39(1):169–190. https://doi.org/10.1016/S0921-8181(03)00023-7

29. Idlafkih Z, Meybeck M, Chiffoleau JF et al (1997) Comportement des metaux particulaires (Al, Fe, Mn, Cd, Cu, Hg, Pb et Zn) dans la Seine a Poses en periode de hautes eaux (1990–1995). Int Assoc Hydrol Sci Publ 243:45–58

30. Meybeck M, de Marsily G, Fustec É (1998) La Seine en son bassin: fonctionnement écologique d'un système fluvial anthropisé. Elsevier, Paris

31. Baccini P, Brunner PH (1991) Metabolism of the Anthroposphere, vol 53. Springer, Berlin
32. Lestel L, Meybeck M, Thévenot D (2007) Metal contamination budget at the river basin scale: a critical analysis based on the Seine River. Hydrol Earth Syst Sci Discuss 4:1795–1822
33. Lestel L (2012) Non-ferrous metals (Pb, Cu, Zn) needs and city development: the Paris example (1815–2009). Reg Environ Chang 12(2):311–323
34. Billen G, Garnier J, Le Noë J et al (2020) The Seine watershed water-agro-food system: long-term trajectories of C, N, P metabolism. In: Flipo N, Labadie P, Lestel L (eds) The Seine River basin, Handbook of environmental chemistry. Springer, Cham. https://doi.org/10.1007/698_2019_393
35. Meybeck M (2013) Heavy metal contamination in rivers across the globe: an indicator of complex interactions between societies and catchments. In: IAHS (ed) Understanding freshwater quality problems in a changing world, vol 361. IAHS-IAPSO-IASPEI Assembly, Gothenburg
36. Lestel L, Meybeck M, Thévenot D (2007) Metal contamination budget at the river basin scale: an original Flux-Flow Analysis (F2A) for the Seine River. Hydrol Earth Syst Sci 11 (6):1771–1781
37. Flipo N, Lestel L, Labadie P et al (2020) Trajectories of the Seine River basin. In: Flipo N, Labadie P, Lestel L (eds) The Seine River basin. Handbook of environmental chemistry. Springer, Cham. https://doi.org/10.1007/698_2019_437
38. Chiffoleau J, Claisse D, Cossa D et al (2001) La contamination métallique. Editions Ifremer, Paris
39. Horowitz AJ, Meybeck M, Idlafkih Z et al (1999) Variations in trace element geochemistry in the Seine River Basin based on floodplain deposits and bed sediments. Hydrol Process 13 (9):1329–1340
40. Meybeck M, Horowitz AJ, Grosbois C (2004) The geochemistry of Seine River Basin particulate matter: distribution of an integrated metal pollution index. Sci Total Environ 328 (1):219–236. https://doi.org/10.1016/j.scitotenv.2004.01.024
41. Grosbois C, Meybeck A, Horowitz A et al (2006) The spatial and temporal trends of Cd, Cu, Hg, Pb and Zn in Seine River floodplain deposits (1994-2000). Sci Total Environ 356 (1–3):22–37
42. Chevreuil M, Chesterikoff A, Letolle R (1987) PCB pollution behaviour in the river Seine. Water Res 21(4):427–434
43. Le Gall M, Ayrault S, Evrard O et al (2018) Investigating the metal contamination of sediment transported by the 2016 Seine River flood (Paris, France). Environ Pollut 240:125–139. https://doi.org/10.1016/j.envpol.2018.04.082
44. Barles S, Guillerme A (2014) Paris: a history of water, sewers, and urban development. A history of water, series III, vol 1. IB Tauris, London
45. Bouleau G, Marchal P, Meybeck M et al (2016) La construction politique d'un espace de commune mesure pour la qualité des eaux superficielles. L'exemple de la France (1964) et de l'Union Européenne (2000). Développement durable territoires
46. Dmitrieva T, Lestel L, Meybeck M et al (2018) Versailles facing the degradation of its water supply from the Seine River: governance, water quality expertise and decision making, 1852–1894. Water Hist 10(2–3):183–205
47. Seine-Normandie AdlE (1976) Les bassins de la seine et les cours d'eau, tome 2 : Besoins et utilisation de l'eau. vol Fascicule 8. Agence financière de Bassin Seine-Normandie. Mission déléguée de Bassin, Paris
48. Stanners D, Bourdeau P (1995) Europe's environment: the Dobris assessment. European Environment Agency, Copenhagen. Arsenic in Urine and Drinking Water We found the article by Calderon et al, "Excretion of Arsenic in Urine as a Function of Exposure to Arsenic in Drinking Water," vol 1
49. Chesterikoff A, Carru A, Garban B et al (1973) La pollution de la Basse-Seine par le mercure (du Pecq à Tancarville). La Tribune du CEBEDEAU 355(356):1–8

50. Lorenzi L (1975) Hydrogeochimie du mercure dans les eaux du bassin de Paris. Thèse de IIIème cycle, Paris 6, p 44
51. Martin J, Meybeck M, Salvadori F (1976) Pollution chimique des estuaires: etat actuel des connaissances; revue bibliographique arrêtée en Juin 1974. Rapports Scientifiques et Techniques CNEXO, vol 22
52. Robbe D (1981) Pollutions metalliques du milieu naturel: guide methodologique de leur etude a partir des sediments: rapport bibliographique. Rapport de Recherche LPC, vol 104. Ministère de l'urbanisme et du logement - Ministère des Transports
53. Pereira-Ramos L (1989) Exploitation critique des résultats d'analyses de métaux sur sédiments et bryophytes dans le bassin Seine-Normandie de 1979 à 1988. Agence de Bassin Seine Normandie, Institut d'Hydrologie et de climatologie
54. Oudin L, Maupas DJM, L'eau AD (2003) Système d'évaluation de la qualité de l'eau des cours d'eau (Seq-Eau), vol 40. MEDD Agences de l'eau
55. Miquel G, France (2003) Rapport sur la qualité de l'eau et de l'assainissement en France. Assemblée nationale, France
56. Miquel GJOPdEdCSeT (2001) Effet des métaux lourds sur l'environnement et la santé, rapport 261.365
57. Ollivon D, Garban B, Blanchard M et al (2002) Vertical distribution and fate of trace metals and persistent organic pollutants in sediments of the Seine and Marne rivers (France). Water Air Soil Pollut 134(1–4):57–79
58. Azimi S, Ludwig A, Thevenot DR et al (2003) Trace metal determination in total atmospheric deposition in rural and urban areas. Sci Total Environ 308(1–3):247
59. Avoine J, Boust D, Guillaud J-F (1986) Flux et comportement des contaminants dissous et particulaires dans l'estuaire de la Seine. Rapports et procès-verbaux des réunions, vol 186. Conseil international pour l'exploration de la mer
60. Vrel A (2012) Reconstitution de l'historique des apports en radionucléides et contaminants métalliques à l'estuaire fluvial de la Seine par l'analyse de leur enregistrement sédimentaire. Caen University
61. Vrel A, Boust D, Lesueur P et al (2013) Dating of sediment record at two contrasting sites of the Seine River using radioactivity data and hydrological time series. J Environ Radioact 126:20–31
62. Ayrault S, Priadi CR, Evrard O et al (2010) Silver and thallium historical trends in the Seine River basin. J Environ Monit 12(11):2177–2185
63. Ayrault S, Roy-Barman M, Le Cloarec M-F et al (2012) Lead contamination of the Seine River, France: geochemical implications of a historical perspective. Chemosphere 87(8):902–910. https://doi.org/10.1016/j.chemosphere.2012.01.043
64. Ayrault S, Lefèvre I, Bonté P et al (2009) Archives sédimentaires, témoignages de l'histoire du développement du bassin. Rapport PIREN-Seine
65. Estèbe A (1996) Impact de l'agglomération parisienne et de ses rejets de temps de pluie sur les concentrations en métaux des matières en suspension et des sédiments en Seine en période estivale. Université Paris-Est Créteil Val de Marne (UPEC)
66. Bussy A-L (1996) Mobilité des métaux dans un système fluvial urbain. Université Paris-Est Créteil Val de Marne (UPEC)
67. Bertrand O, Montarges-Pelletier E, Mansuy-Huault L et al (2013) A possible terrigenous origin for perylene based on a sedimentary record of a pond (Lorraine, France). Org Geochem 58:69–77
68. Joron J-L, Treuil M (2019) Cycles géochimiques: une histoire naturelle illustrée des éléments chimiques. http://iramis.cea.fr/ComScience/GeochimieCycles/GeochimieDonnees.php. Accessed 22 Jan 2019
69. Dauvin J-C (2006) Estuaires Nord-Atlantiques: problèmes et perspectives, September 2006. Groupement d'intérêt public Seine Aval
70. Meybeck M, Kummu M, Dürr H (2013) Global hydrobelts and hydroregions: improved reporting scale for water-related issues? Hydrol Earth Syst Sci 17(3):1093–1111

71. Froger C, Quantin C, Gasperi J et al (2019) Impact of urban pressure on the spatial and temporal dynamics of PAH fluxes in an urban tributary of the Seine River (France). Chemosphere 219:1002–1013

72. Meybeck M, Lestel L, Winklhöfer K et al (2017) Un exemple de trajectoire environnementale : la contamination métallique de la Seine, la Spree, la Senne et le Lambro (1950–2010). In: Lestel L, Carré C (eds) Les rivières urbaines et leur pollution. vol Collection Indisciplines. QUAE

73. Lehmann A, Rode MJWR (2001) Long-term behaviour and cross-correlation water quality analysis of the river Elbe, Germany. Wat Res 35(9):2153–2160

74. Meybeck M, Lestel L, Carré C et al (2018) Trajectories of river chemical quality issues over the Longue Durée: the Seine River (1900S–2010). Environ Sci Pollut Res 25(24):23468–23484

75. Lestel L, Eschbach D, Meybeck M et al (2020) The evolution of the Seine basin water bodies through historical maps. In: Flipo N, Labadie P, Lestel L (eds) The Seine River basin. Handbook of environmental chemistry. Springer, Cham. https://doi.org/10.1007/698_2019_396

76. Meybeck M (1982) Carbon, nitrogen, and phosphorus transport by world rivers. Am J Sci 282 (4):401–450

77. Winkels H, Kroonenberg S, Lychagin MY et al (1998) Geochronology of priority pollutants in sedimentation zones of the Volga and Danube delta in comparison with the Rhine delta. Appl Geochem 13(5):581–591

78. Gobeil C, Rondeau B, Beaudin L (2005) Contribution of municipal effluents to metal fluxes in the St. Lawrence River. Environ Sci Technol 39(2):456–464

79. Schäfer J, Blanc G, Lapaquellerie Y et al (2002) Ten-year observation of the Gironde tributary fluvial system: fluxes of suspended matter, particulate organic carbon and cadmium. Mar Chem 79(3–4):229–242

80. Mestres R, Belamie R, Aguesse P (1971) Rôle joué par les substances organochlorés dans la pollution des eaux douces, vol 31. Etude la région Centre, Trav Soc Pharm, Montpellier

81. Marchand M (1989) Les PCB dans l'environnement marin. Aspects géochimiques d'apports et de distribution. Cas du littoral français. Rev Sci Eau/J Water Sci 2(3):373–403

82. Colas L (1979) Détermination des valeurs limites applicables aux effluents de polychlorobiphényles et polychloroterphényles, compte tenu des meilleurs moyens techniques disponibles (trans: consommateurs Sdleedlpd). Communauté économique européenne, Bruxelles

83. Amiard J-C, Meunier T, Babut M (2016) PCB, environnement et santé. Tec & Doc. Lavoisier

84. Carpentier S, Moilleron R, Beltran C et al (2002) Quality of dredged material in the river Seine basin (France). II. Micropollutants. Sci Total Environ 299(1–3):57–72

Changes in Fish Communities of the Seine Basin over a Long-Term Perspective

Jérôme Belliard, Sarah Beslagic, and Evelyne Tales

Contents

1 Introduction 302
2 Various and Complementary Data to Address Time Trajectories 302
3 Non-native Species: An Accelerating Establishment 303
4 Migratory Fish: A Dramatic Decline Followed by a Still Limited Recovery 305
5 Changes in Local Communities Since the Nineteenth Century 309
6 The Seine River Axis During the Past Four Decades 312
7 Conclusion and Perspectives 317
References 319

Abstract Using both historical and current data, we retrace the long-term evolution of fish assemblages in the Seine River basin since 1880, from headwaters to upstream of the Seine River estuary. Successive phases are observed, related to anthropogenic impacts on habitat conditions and river water quality. Temporal trajectories were thus reconstructed on several reaches based on the change of the proportion of species' ecological traits, in order to detect the main drivers of alteration. Contrasted trends occur between large rivers and small streams of the basin. In this context, migratory fish declined, whereas the proportion of non-native species increased in the fish community of the Seine River.

Keywords Historical ecology, Migratory fish, Non-native species, Urban impact, Water pollution

---

The copyright year of the original version of this chapter was corrected from 2019 to 2020. A correction to this chapter can be found at https://doi.org/10.1007/698_2020_667

J. Belliard (✉), S. Beslagic, and E. Tales (✉)
Irstea UR HYCAR, Antony Cedex, France
e-mail: jerome.belliard@irstea.fr; evelyne.tales@irstea.fr

Nicolas Flipo, Pierre Labadie, and Laurence Lestel (eds.), *The Seine River Basin*, Hdb Env Chem (2021) 90: 301–322, https://doi.org/10.1007/698_2019_380,
© The Author(s) 2020, corrected publication 2020, Published online: 3 June 2020

1   Introduction

The interest in studying the long-term ecological functioning of rivers grew in the 1980s in an attempt to provide a better understanding of the current ecological status of water bodies [1, 2]. These historical approaches have been boosted since the 2000s with the implementation of the Water Framework Directive, which imposes the assessment of ecological status of water bodies relative to a former reference status. The question of defining ecological reference conditions was raised, and reconstructing trajectories of temporal changes of fish communities was a way of addressing this issue. European rivers and particularly the Seine River have been heavily modified by human activities. The period around 1850 is considered as a milestone with the beginning of channelisation for navigation purposes [3], although mills already greatly impacted streams as early as the Middle Ages, making historical data necessary to reconstructing long-term changes. Historical written documents are a valuable source of information, particularly for reconstructing long-term changes in fish communities, although their exploitation requires a critical assessment [4].

Given the many human impacts on rivers and their fish communities over the Seine basin, temporal trajectories may differ depending on the time period considered but also the geographical context (e.g. demographic impact, land use changes, stream size, etc.). In this chapter, we examined fish communities and their response to human impacts, over the past 150 years and more recent time periods, and at different spatial scales, from local river stretches to the whole Seine basin.

2   Various and Complementary Data to Address Time Trajectories

One of the main challenges in investigating long-term changes in fish communities is the availability of reliable data. The electrofishing approach, which is used widely nowadays to sample river fish communities, began to be applied in the basin in the 1960s. Unfortunately, the data collected at that time were rarely archived. Some of these data were nevertheless used to elaborate maps of the distribution and abundance of species in several areas of the Seine basin around the 1970s. From the 1980s onwards, electrofishing sampling methods began to be standardised, their application became more widespread and the data collected were better stored. Annual monitoring was implemented on the Seine River in the Paris region as from 1990 [5] and then extended to numerous rivers in the whole basin beginning in 1995 [6]. As a result, changes in fish communities over the past three or four decades can be analysed in detail.

To go back further in time, it is necessary to use other data sources such as historical written sources. These historical written sources are of multiple types (early scientific and naturalist literature, fishery laws, administrative surveys on fish stocks, catch records, river engineering projects, observations by fishermen and anglers, etc.), each type of source having its own advantages and limitations. For the Seine River basin, a database called CHIPS was developed to gather historical written sources on freshwater fish in order to improve knowledge about past species distribution and composition of communities [7]. While the oldest data stored in this database date back to the sixteenth century, most of the observations relate to the period from the nineteenth to the first half of the twentieth century and are therefore complementary to the more recent electrofishing data set.

Archaeological remains of fish, generally obtained from food waste or latrines in past habitats, provide another potentially useful source of data. They allow us to address older periods and provide additional information for the periods already covered by the written historical archives. In the Seine basin, numerous urban and rural sites have provided archaeological fish remains for medieval and early modern periods [8]. While archaeological remains provide insight into the past occurrence of species, their main limitation concerns the uncertainty about the proximity of fish supply sources. It can be suspected that some of the fish consumed locally were actually caught in distant waters. The abundance of marine fish remains, particularly on sites associated with the nobility and aristocracy, reinforces these doubts. Under these conditions, it can be assumed that a significant proportion of the archaeological remains, particularly when they concern rare species of high commercial value, were not obtained from local catches.

We combined these different sources of data (i.e. contemporary, historical and archaeological) to reconstruct the trajectories of fish communities in the Seine basin at different temporal and spatial scales.

3 Non-native Species: An Accelerating Establishment

Combining different sources of information (e.g. palaeontological, archaeological, written), Persat and Keith (1997) [9] estimated that, before human intervention, the native fish fauna of the Seine River basin was likely to include up to 23 strictly freshwater species, to which 11 diadromous species can be added. Through their fishing activity, their ability to transport fish and, more broadly, their action on their environment, humans are likely to have modified this native fauna very early. However, the first known introduction of an exotic species into the Seine River basin concerns the common carp (*Cyprinus carpio*). Its establishment, linked to the development of pond fish farming, has been documented since the thirteenth century [10, 11]. Since that time, the deliberate or accidental introductions of exotic species have continued to such an extent that they now account for nearly 45% of the species recorded in the basin (and even 55% of strictly freshwater species) (Fig. 1). The second half of the nineteenth century marked the first phase of an acceleration in the establishment of exotic species. It was partly the result of technical and scientific progress in fish breeding – in particular the control of artificial reproduction of fish – thereby facilitating the spread of species in the form of embryonated eggs, which are easier to transfer than the fish themselves. More generally, it was in line with a

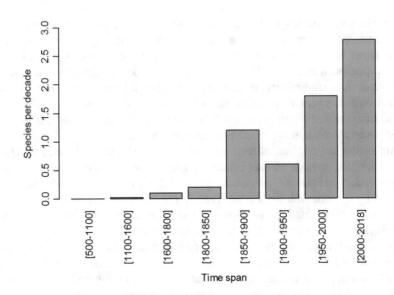

Fig. 1 Number of non-native species established per decade in the Seine River basin for successive periods. Species that were temporarily established and are no longer present in the basin are not taken into account

collective desire led by the public authorities (see the creation of the French zoological society of acclimatisation in 1854), to improve the production of natural systems through the establishment of species from other countries. This period was thus marked by numerous attempts to introduce species, particularly of North American origin, many of which resulted in viable establishments.

After a relative decline in the first half of the twentieth century, probably due to the two World Wars but also to the gradual awareness of the potentially detrimental consequences of alien species on ecosystems, the rate of establishment of non-native species has accelerated since 1950 (Fig. 1). Compared to the nineteenth century, the drivers of recent introductions are no longer part of a coordinated global approach. Some of them are still driven by zootechnical concerns or the development of recreational fishing; for example, in recent decades, the introductions of pikeperch (*Sander lucioperca*), grayling (*Thymallus thymallus*) or wels catfish (*Silurus glanis*) were clearly favoured by anglers. However, numerous introductions have recently taken place in an accidental manner in association with the acceleration of exchanges at the worldwide level promoting the spread of biological organisms [12]. In this respect, for the Seine basin and in line with what is happening for other European rivers, the establishment of navigation channels, which connect previously isolated river basins, acts as an accelerator of the arrival of exotic species (and not only fish) from Central and Eastern Europe [13]. This modality of colonisation seems to be involved in the case of the common nase (*Chondrostoma nasus*), which was observed as early as the nineteenth century and in the more recent arrival of species such as the ide (*Leuciscus idus*), the asp (*Leuciscus aspius*) and the round goby (*Neogobius melanostomus*).

Despite recent regulations to reduce or prohibit introduction of alien species, their proportion in the whole Seine basin is constantly increasing, and, since the beginning of the twenty-first century, the rate of implantation of new species has reached nearly three species per decade (Fig. 1).

Interestingly, there are discrepancies between written scientific sources and archaeological sources in terms of species introductions [8]. According to scientific and written sources, wels catfish and pikeperch were established in the Seine basin in the second half of the twentieth century, although attempts to introduce them were already mentioned as early as the nineteenth century. However, archaeological remains of these two species, dating from the late Middle Ages, have been discovered in the centre of the basin. This apparent contradiction may in fact reveal a possible limitation of archaeological sources, as mentioned above (uncertainty regarding the location of catches). However, these archaeological remains may also reflect an earlier presence than previously assumed, suggesting very old and repeated introduction attempts, before these two species really became established in the twentieth century.

4 Migratory Fish: A Dramatic Decline Followed by a Still Limited Recovery

Historically, 11 diadromous fish species occurred in the Seine River basin: eel (*Anguilla anguilla*), thinlip mullet (*Liza ramada*), flounder (*Platichthys flesus*), sea lamprey (*Petromyzon marinus*), river lamprey (*Lampetra fluviatilis*), European sturgeon (*Acipenser sturio*), Atlantic salmon (*Salmo salar*), sea trout (*Salmo trutta*), smelt (*Osmerus eperlanus*), allis shad (*Alosa alosa*) and twaite shad (*Alosa fallax*). For some of them, several authors have mentioned an early decline in Northwestern Europe as from the Middle Ages [14]. For example, in the medieval period, a sharp decline of salmon populations coincided with the widespread establishment of water mills, which prevented access to spawning grounds but also presumably encouraged overfishing [15].

However, the nineteenth century was a turning point for migratory fish in the Seine River basin, due to (1) the widespread development of waterways for commercial navigation and (2) the significant increase in populations and human activities, particularly around Paris. The building of weirs and locks to facilitate navigation began in the middle of the nineteenth century in the centre of the basin and gradually spread, over the following decades, on the Seine and its major tributaries, involving increasingly higher structures. This has led to a gradual increase in barriers to fish migration, which has been reinforced for some species by the simultaneous construction of dams in headwater catchments in order to regulate river flows. For example, the construction of the Settons Dam on the Cure River (19 m high), finished in 1858, definitively closed off access to the best upstream salmon spawning grounds [16].

On the lower Seine River, the Martot dam (currently destroyed) was constructed in 1864 just upstream of Rouen, followed by the Poses dam a few kilometres upstream. The latter has been operational since 1886, and, at that time, it was the largest navigation dam in the basin at 4 m high; it had a major impact by reducing the accessibility of a large part of the basin. This led, a few years later, to a collapse of salmon and allis shad stocks, whose spawning grounds were located further upstream, while species such as twaite shad and smelt, able to reproduce further downstream, were less adversely affected (Fig. 2). As a result, the most vulnerable species, i.e. European sturgeon, salmon and allis shad, became extinct in the Seine River basin at the beginning of the twentieth century.

The degradation of water quality has also contributed significantly to the decline of migratory fish. From the end of the nineteenth century (and probably well before), downstream of Paris, several kilometres of the Seine River were affected by a low oxygen concentration, due to wastewater spills directly into the river. With the growth of the Paris urban area, this situation deteriorated gradually until the early 1970s [17]. At that time, the entire lower Seine River stretch, downstream of Paris, was depleted with dissolved oxygen, with particularly harsh conditions in its estuarine course. In the estuary, a 125-km-long reach underwent severe hypoxia (<3 mg/L) for 200–300 days every year, sometimes reaching periods of total anoxia, particularly in summer and autumn [18, 19]. As a result, the populations of smelt and twaite shad, two fish species that are able to reproduce in the lower parts of the river and which had consequently been less affected by the establishment of dams, collapsed during the 1950s until their assumed extinction throughout the basin around 1965–1970.

Overall, the construction of dams on major rivers, in combination with the considerable increase in pollution downstream of the Seine since the mid-nineteenth century, has therefore caused a considerable decline of migratory fish communities, until the gradual extinction of seven species. At the end of the 1970s, only four diadromous species were still present in the Seine River basin, *L. fluviatilis*, *A. anguilla*, *Salmo trutta trutta* and *Platichthys flesus*, but their populations were very sparse at that time [20].

Since the mid-1970s, considerable efforts have been made to treat domestic and industrial sewage and to prevent their direct discharge into watercourses. This resulted in a gradual improvement in water quality along the Seine River, from the Paris urban area to the estuary, which allowed the return of diadromous species [17]. The first significant returns concerned species that reproduce in the downstream part of the river, such as twaite shad and smelt, whose catches have become regular since 1998 [21]. In the early 2000s, species migrating further upstream, such as salmon or allis shad, began to recolonise the basin, supported by the gradual installation of fish passes on navigation dams and by the improvement of water quality (in relation with more efficient wastewater treatment), but without any restocking operations [22, 23] (Fig. 3).

Since 2008, several tens of salmons and allis shads and several hundred sea lampreys have been observed each year, at the fish pass installed at Poses (furthest downstream dam, located about 150 km from the sea) [24]. For these three species,

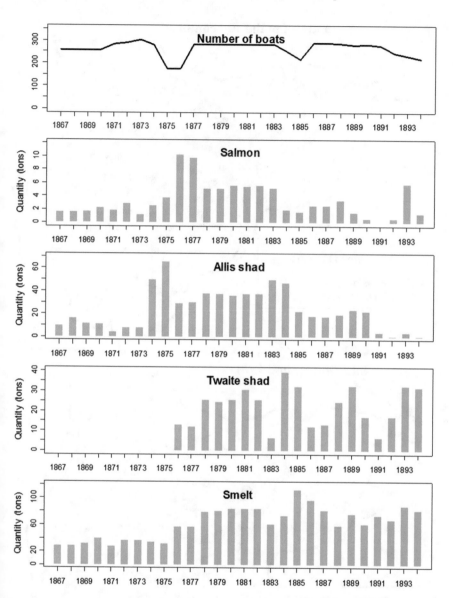

Fig. 2 Commercial landings of diadromous fish on the lower Seine River in Rouen between 1867 and 1894. The number of operating fishing boats is also indicated (from Refs. [21] and [25])

evidence of reproduction has recently been documented at different locations in the basin. Overall, among the 11 diadromous species historically present in the Seine basin, only the European sturgeon seems to be definitively extirpated. However, the species that have recently naturally recolonised the Seine River basin are still very vulnerable due to the small size of their populations.

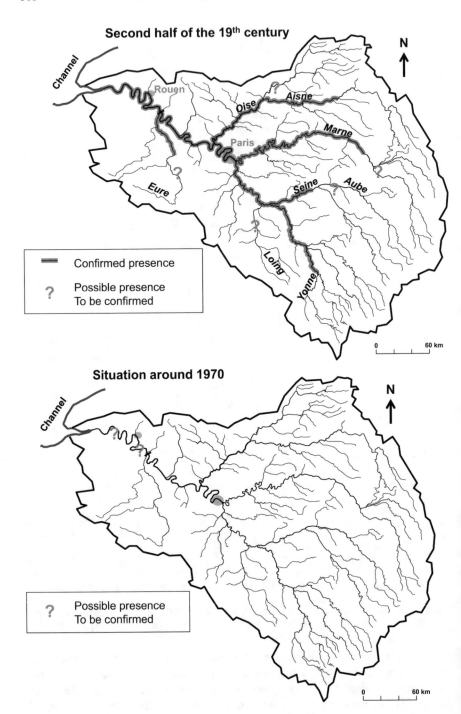

Fig. 3 Change in the distribution range of allis shad in the Seine River basin since the mid-nineteenth century

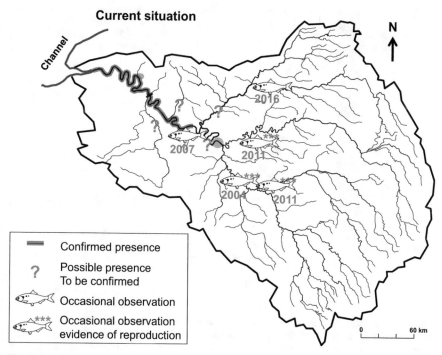

Fig. 3 (continued)

5 Changes in Local Communities Since the Nineteenth Century

Several studies have analysed changes in composition of fish communities in different localities of the Seine River basin since the late nineteenth and early twentieth century. Such studies relied on historical sources and more recent scientific sampling data [7, 26, 27], and, overall, they examined community changes in about 30 river stretches with varying environmental characteristics (i.e. from small rural rivers to the lower Seine River). Although the data and methods used in these studies were not always similar (e.g. analysis on taxonomic structure or consideration of functional traits, species presence/absence or integration of relative abundance data), a number of general trends can nevertheless be highlighted regarding the long-term trajectories of fish communities (Fig. 4).

These studies concluded that long-term changes vary greatly depending on the local natural and anthropogenic conditions. However, they also suggest that the most significant changes have been observed in both streams and larger rivers, while communities in mid-size rivers have remained much more stable.

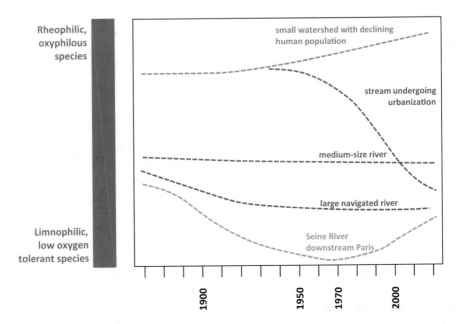

Fig. 4 Reconstruction of long-term trajectories of fish community composition (proportion of rheophilic/oxyphilic species vs. limnophilic/low-oxygen-tolerant species) for different watercourse types in the Seine basin. This schematic and simplified representation was established on the basis of historical data available for about 30 sites covering the Seine basin (see [7, 26, 27] for original studies)

Streams located in the periphery of the Seine River basin, in areas where the human population density has declined quite sharply since the nineteenth century, are characterised by a long-term increase in the proportion of rheophilic, lithophilic and oxyphilic species. This trend seems to be the result of the extirpation of some warm water, limnophilic, low-oxygen-tolerant species, initially present, rather than an actual increase in rheophilic/lithophilic fish. In such streams, the long-term reduction of human population density combined with the improvement of wastewater treatment during recent decades has certainly led to a reduction in organic pollution levels [28]. The impact of such a reduction of pollution on fish assemblages has already been described in other systems [29], and, overall, fish responses reported in these cases are consistent with what we have observed. In these areas of the Seine River basin, ponds were historically widespread [30]. They provided very suitable habitats for warm water limnophilic fish and certainly contributed to the establishment of these species in headwater streams. These aquatic habitats have declined considerably since the end of the eighteenth century, which may explain, at least in part, the long-term fish assemblage trends observed.

A stream located in the centre of the basin (i.e. the Viosne River, 40 km northwest of Paris) shows completely opposite long-term changes in fish assemblage, with a strong decline in the proportion of rheophilic, lithophilic and oxyphilic species. This trend is the result of both the collapse of some typical headwater species such as trout (*S. trutta*) and European bullhead (*Cottus perifretum*) and the establishment of eurytopic and warm water species. Compared with the streams mentioned above, located on the periphery of the Seine River basin, this site has undergone a completely different change in its environment, shifting in a few decades from a rural catchment to an urbanised area with a high human population density. This increase in human pressures has led to deterioration in water quality and transformations in the morphology and hydrology of the watercourse, which most likely triggered the major changes observed in the fish assemblage composition.

The large rivers currently used for commercial navigation and located upstream from the Paris conurbation (e.g. lower Yonne, Marne, Oise and Aisne rivers) show fairly homogenous long-term changes characterised by a generally modest increase in the proportion of limnophilic, eurytopic/phytophilic and warm water species. For these watercourses, the oldest historical sources used to describe fish communities date back to the late nineteenth century. At that time, the main infrastructures for navigation (dams and locks) were already in place, and the canalisation of river courses was already nearly completed. This situation could explain the small magnitude of long-term fish community changes detected in such rivers.

In contrast with the other navigated waterways, the lower Seine River, from Paris to the estuary, experienced much more pronounced and complex changes in fish communities. In this part of the Seine River, river sections for which historical data are available show very similar trajectories. But it is in the Seine section of Paris and immediately downstream that fish community changes can be reconstructed with the greatest precision, due to the greater richness of the historical archives (Fig. 5). Fish communities deteriorated continuously from the late nineteenth century to the 1970s or 1980s (depending on sectors and data availability). This deterioration was evidenced in particular by an increase in the proportion of eurytopic species at the expense of rheophilic species, an increase in the proportion of species tolerant to oxygen depletion and an increase in the proportion of omnivorous species at the expense of specialist species (invertivorous and piscivorous fish). This trend is consistent with the previously reported deterioration in water quality that affected the whole lower Seine River during this period. During the late 1960s and early 1970s, the level of pollution reached its maximum [17], and the Seine downstream of Paris experienced widespread and repeated periods of severe anoxia. As a consequence, only a small number of species, tolerant to low dissolved oxygen contents, remained [31]. This deterioration of fish assemblages was probably also accentuated by the development of the waterway during this period, which was marked in part by an increase in the river canalisation (removal of islands and side arms, deepening of the navigation channel) and the gradual increase in the height of dams and locks. The 1990s and subsequent decades have seen a shift in this trend, which is illustrated by a return of specialised and intolerant species (rheophilic, oxyphilic, invertivorous and piscivorous fish) (Fig. 5). As already reported for diadromous fish, this trend reversal, which affected the entire fish community, coincided with the improvement of water quality and in particular the decrease in the level of organic pollution resulting in a dissolved oxygen increase. However, the

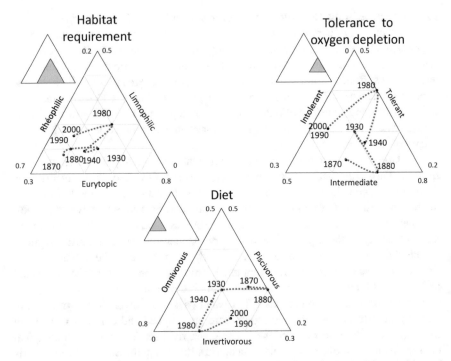

Fig. 5 Long-term changes in the proportion of ecological traits of species in the fish assemblage of the Seine River around Paris

fish community recovery process was not complete, probably because of the irreversible morphological alteration of the river (Fig. 5).

6   The Seine River Axis During the Past Four Decades

The existence of numerous electrofishing data since the early 1980s makes it possible to address in some detail the changes in fish communities along the Seine River axis over the past four decades, a pivotal period for a better understanding of the ongoing recovery processes. For this purpose, we focused our analysis on the Seine River course, from its confluence with the Aube River, about 200 km upstream from Paris, to the estuary. Variations of the French fish index (IPR) values and its metrics over time were examined.

The IPR (Indice Poissons Rivière, i.e. Fish River Index) [32] was developed to measure the state of river fish communities. It is composed of seven metrics, based on the number of species or fish abundance, reflecting the community's functional or structural features. A score is assigned to each metric measuring the deviation between the assessed situation and a reference with minimal human impact. The final IPR is the sum of the scores on the seven metrics. Its value is close to 0 when the community composition is close to the expected community under the reference condition. It increases as the human impact on the community grows.

For this analysis, the Seine course was partitioned into four sectors characterised by differences in the levels of anthropogenic pressures and how these pressures have changed over the past few decades.

The upstream sector corresponds to the Seine from the confluence of the Aube to its entrance into the Paris urban area. With the exception of its most upstream part, it was formerly canalised to allow commercial navigation. Water quality in this sector was already good about 40 years ago and has changed little since then (Fig. 6).

The next two sectors cover the Seine throughout the Paris urban area. The upstream urban sector (up to the sector immediately downstream of the centre of Paris, sector 2 in Fig. 6) was initially less polluted than the downstream urban sector, and its water quality improved earlier, returning to satisfactory levels in the early 1990s. The downstream urban sector received the main wastewater discharges from the metropolitan area, and, as a result, water quality reached an extreme level of degradation in the 1960s and the 1970s [17]. In addition to chronic pollution, this sector was frequently subjected to massive accidental pollution related to urban storm water discharges. By causing a sudden fall in dissolved oxygen, these events led to frequent catastrophic fish mortality and persisted until the first half of the 1990s [31, 33]. The establishment of new wastewater treatment plants and better management of sewer systems have gradually reduced organic pollution and consequently attenuated oxygen deficits.

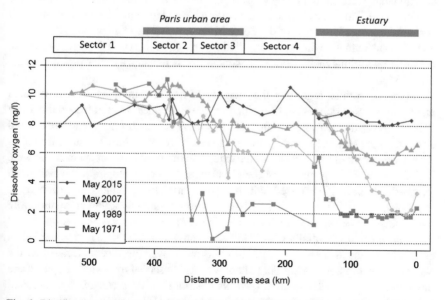

Fig. 6 Dissolved oxygen content measured in May along the Seine River from the confluence of the Aube to the estuary for the years 1971, 1989, 2007 and 2015

The last sector, downstream of the Paris urban area, was also strongly affected by upstream urban discharges and experienced a very comparable water quality trajectory. In the 1970s, the oxygen deficit was particularly severe as far as the estuary; only the area immediately downstream of navigation dams provided satisfactory oxygenation conditions in very local areas. The oxygen deficit then gradually decreased, reaching levels compatible with the survival of most fish species in the 2000s (Fig. 6).

Long-term variations in IPR scores across sectors are consistent with changes in water quality in the Seine River (Fig. 7, Table 1). Thus, the sectors that have experienced the greatest improvements in their water quality (sectors 4 and 3 and to a lesser extent sector 2) also showed the most substantial signs of recovery in their fish communities. In the upstream sector, IPR scores also show a significant decrease over time, but this improvement trend remains slight (only four out of seven metrics show improvement trends and one metric – density of omnivorous species – even shows the opposite trend) but without any comparison with the magnitude of the changes observed further downstream. The temporal trend curves (Fig. 7) suggest that in the upstream urban sector (sector 2), a considerable community improvement would have occurred in the early 1990s, while further downstream (sectors 3 and 4), the improvement would have been more pronounced later, in the early 2000s. This is consistent with the fact that organic pollution problems were solved earlier upstream of the urban area than farther downstream. However, this statement must be addressed with some caution given that the sites sampled may have changed over time, which may bias temporal trends.

Despite the general improvement in the fish community composition, fish tissues are still heavily contaminated by various pollutants such as mercury, PCBs and some pesticides [5]. This widespread contamination has led to regulations prohibiting fish consumption on a large part of the Seine River axis and its main tributaries. Despite a significant reduction in organic pollution and eutrophication in recent decades, pollution by various metal or organic contaminants remains a major problem in the basin [34–36].

Beyond questions regarding the integrity of fish communities, a trend towards increasing species richness has emerged in recent decades. First noted on the Seine River in the Paris metropolitan area [5], this trend is actually more widespread and also affects the sectors located upstream and downstream of the urban area. It implies the establishment of new species or the extension of populations of species that were initially much more restricted in their distributions. For example, the wels catfish (*S. glanis*), a species introduced quite recently, started to be sampled occasionally in the 2000s, mainly in the mid-Seine River. It is now commonly found along the entire river course. Other species such as bitterling (*Rhodeus amarus*) and pumpkinseed (*Lepomis gibbosus*), initially mainly located upstream of the urban part of the Seine, gradually extended their distribution downstream over the recent decades as the water quality improved. A similar pattern was also observed for some lithophilic and rheophilic species such as barbel (*Barbus fluviatilis*) or dace (*Leuciscus leuciscus*) [37]. In this respect, the case of the European bullhead (*C. perifretum*) is particularly remarkable: considered typical of the upstream zones, this oxyphilic species has

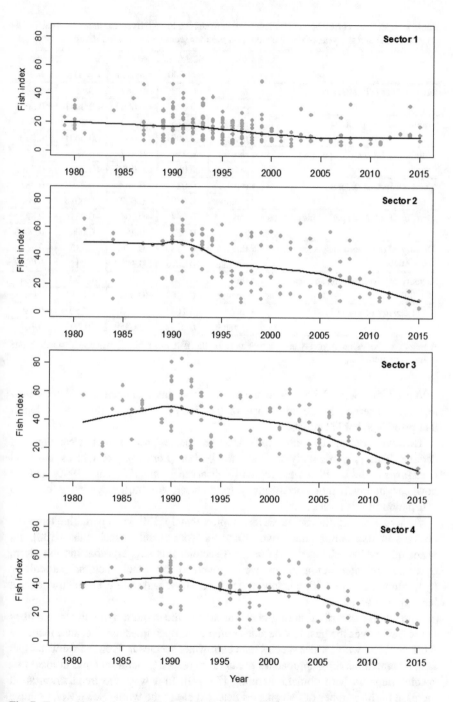

Fig. 7 Temporal variations in IPR scores in four successive sectors of the Seine River, from the confluence with the Aube River to the beginning of the estuary. A decrease in index values corresponds to an improvement in the state of the fish community. Smoothing curves were constructed using LOWESS tools (see Fig. 6 for location of the four sectors)

Table 1 Spearman correlation coefficient (rho) between the French river fish index (and each of its seven metrics) and the year of the evaluation on the four sectors of the Seine River

		Sector 1 $n = 200$	Sector 2 $n = 120$	Sector 3 $n = 125$	Sector 4 $n = 109$
French river fish index, IPR	rho	−0.412	−0.544	−0.563	−0.561
	p-value	<0.0001	<0.0001	<0.0001	<0.0001
Number of rheophilic species	rho	−0.315	−0.481	−0.550	−0.584
	p-value	<0.0001	<0.0001	<0.0001	<0.0001
Number of lithophilic species	rho	−0.450	−0.444	−0.531	−0.472
	p-value	<0.0001	<0.0001	<0.0001	<0.0001
Total number of species	rho	0.023	−0.362	−0.279	−0.383
	p-value	0.740	<0.0001	0.0015	<0.0001
Density of tolerant species individuals	rho	0.044	−0.104	−0.331	−0.323
	p-value	0.531	0.257	0.0001	0.0006
Density of omnivorous species individuals	rho	0.218	0.150	−0.112	−0.318
	p-value	0.0018	0.100	0.211	0.0007
Density of invertivorous species individuals	rho	−0.394	−0.403	−0.390	−0.464
	p-value	<0.0001	<0.0001	<0.0001	<0.0001
Total density of individuals	rho	−0.164	−0.203	−0.250	−0.135
	p-value	0.0196	0.0259	0.0047	0.161

A negative rho value suggests an improvement of the fish assemblage, whereas a positive one suggests degradation

colonised the Seine River since the recovery of high oxygen levels. It is now regularly observed in the urban part of the river, in rip-rap located along banks that provide suitable habitats.

However, during the same period, some species have shown clear signs of decline. This is particularly the case for burbot (*Lota lota*) and black bullhead (*Ameiurus melas*), which were regularly observed in the 1980s and 1990s and are now caught much more occasionally (particularly for burbot, which is close to extirpation in this river reach).

At two sites of the Seine River, approximately 150 km apart, the temporal changes in fish communities over 15 years were closely synchronised [38]. As water quality had changed little in these sectors, this suggests that the long-term dynamics of fish communities were governed by other factors, particularly hydroclimatic factors (flood, low water level, heat waves, etc.), which act on a large spatial scale.

The relative contribution of global warming to the evolution of fish communities in the Seine over the past few decades remains an open question. Because there are more cool and warm water species than cold water species in French freshwater fish fauna, some modelling approaches predict for rivers a general increase in local fish species richness with climate warming [39, 40]. In a way, the trend towards an increase in the number of fish species detected along the whole Seine River course over the past four decades could be part of this process. Studies focusing on recent changes in fish communities in French rivers have shown that climate change is already affecting their composition [40], particularly in large rivers where the proportion of warm water and southern species is increasing substantially [41].

Compared to the other large French rivers, the trend observed on the Seine River is more equivocal in so far as the species whose abundance has increased in recent decades include both warm water (e.g. bitterling, pumpkinseed and wels catfish) and cold water fish (e.g. European bullhead and dace) [38]. In the Seine River, the improvement in water quality has allowed the return of rheophilic and oxyphilic species, which are often also cold water species, undoubtedly leading to a blurring of the effect of global warming.

7   Conclusion and Perspectives

The results described above suggest that long-term changes in fish communities in the Seine River basin are driven by multiple factors operating at different spatial and temporal scales. Overall, the most significant trends concern (1) a constantly increasing establishment of new exotic species; (2) a dramatic decline in migratory fish until the extinction of several species during the twentieth century, followed by a significant but still limited recovery process over the last two decades; (3) and more locally, changes in species composition of communities that vary according to the context of anthropogenic pressures. In particular a severe deterioration in the state of fish populations occurred in the most human-impacted large rivers until the 1970s, followed by a partial recovery following the implementation of effective measures to control domestic and industrial pollutions.

In this respect, the changes observed in the Seine River basin are quite similar to those observed in other European river basins strongly marked by the density and intensity of human activities, such as the Rhine, the Thames, the Spree in Berlin or the Lambro in Milan [42–45]. All these rivers have indeed experienced an increase in exotic species, a massive decline in intolerant species, especially migratory species, compared to the preindustrial period and, in recent times, evidence of recovery with the development of pollution control policies.

The colonisation of rivers by exotic fish species is a global phenomenon, particularly acute in Europe, which continues to rise [46]. Considering that the spread of alien species is favoured by the globalisation of the circulation of people and commodities and by the artificialisation of ecosystems [47], it is very unlikely that the arrival of new exotic species on the Seine River basin will decline, at least for the next few decades. If the colonisation rate of alien species continues at its current level (and even more so if it keeps increasing), the number of non-native species noted in the Seine River basin could quickly exceed the number of native species.

In the 2000s, an established population of *Misgurnus anguillicaudatus* was identified in an ornamental pond in Paris (eradicated since then). This weather loach, native of East Asia, is frequently used as food or aquarium fish and has been recently detected in several European countries [48]. More recently, in 2016, a small population of fathead minnow *Pimephales promelas* was discovered in a stream near Paris (C. Houeix, pers. com.). This North American minnow was used as bait a few decades ago, and populations are already established in several rivers in Belgium [48]. These two examples illustrate the diversity of the origin of alien species likely to establish in the Seine basin and may be the first signs of larger-scale colonisation. The recent record, in 2015, of the round goby *N. melanostomus* in a gravel pit connected to the lower Seine and the subsequent spread of its distribution [49] deserve particular attention. This event is in line with a general process of extending distribution of several Ponto-Caspian gobiid species, facilitated by inland navigation and establishment of canals between rivers previously disconnected [48, 50]. Thus, three other species, *Proterorhinus semilunaris*, *Ponticola kessleri* and *Neogobius fluviatilis*, are present in Northeastern France and could colonise the Seine basin in the near future. Unlike the other non-native species already present in the Seine River, these species, in particular *N. melanostomus*, are highly invasive. For instance, on several sites in the Rhine basin, although recently colonised, gobiid species (mainly *N. melanostomus*) generally account for 60–90% of the fishes, and they seem to compete with several native species, particularly benthic species [49]. If the forthcoming colonisation of the Seine basin by these species is confirmed, in the coming decades, it could lead to a profound change in fish communities on most watercourses.

The return of migratory species in the Seine River basin over the past two decades is a remarkable event, especially since, unlike other major European rivers, this recolonisation has taken place naturally, from nearby populations, without any use of restocking operations [23]. This return is a direct response to improvements in water quality, but the establishment of fish passes on some navigation dams has also contributed to this recovery. In the future, it is planned to pursue the establishment of fish passes on the basin's main watercourses, which should reinforce the existing populations and extend their distribution area. The high potential of the basin for the recovery of migratory fish with the restoration of river continuity can be exemplified by the case of the Carandeau navigation dam located on the Aisne River (a sub-tributary of the Seine River), about 400 km from the sea. It was equipped with a fish pass in 2016, which in the subsequent months resulted in the passage of several salmon and shad individuals, when these species had no longer been observed on this axis since the nineteenth century.

Over the longer term, however, global warming could compromise the return of some migratory species. Modelling approaches suggest that the thermal conditions predicted by the end of the twenty-first century could jeopardise the sustainability of the smelt stock [20], while it has recovered a flourishing level in recent years in the estuary. The question of a negative impact of global warming also arises for salmon and sea trout, two cold water migratory species [51].

Beyond migratory species, climate change is expected to affect all freshwater fish species. At the scale of France, and more locally in the Seine basin, prospective approaches evaluating different warming scenarios predict a contraction of the distribution ranges of cold water species [39, 52, 53]. This contraction of distribution areas is reported to be particularly pronounced in the case of brown trout and

European bullhead, two of the most abundant species in the Seine River basin at present, but which could be extirpated from most rivers by the end of the century. In the Seine basin, the impact of global warming on cold water species seems even more severe given that the homogeneity of environmental conditions and the prevailing low elevations limit the potential refuge areas. In this context, chalk rivers, because their waters remain naturally cold in summer, could provide important refuge areas, especially since they are numerous in the basin. Conversely, many warm water species, such as barbel (*Barbus barbus*) and chub (*Squalius cephalus*), could benefit from warming of running waters to extend their distribution.

Over the whole basin, global warming is also expected to result in a decrease in river discharge, especially during summer, and a lowering of aquifer levels (locally quite high, >10 m), which may consequently multiply the frequency of drying periods in headwater streams [54]. This decrease in water resources in rivers and aquifers should logically induce a concentration of pollutants and therefore a degradation of the water quality available to biota. It is expected that these modified environmental conditions will have adverse impacts on river fish communities, especially on rheophilic and/or oxyphilic species.

The current global warming and its consequences on the hydrology, temperature and water quality of rivers will undoubtedly lead to deep changes in the composition of fish communities in the Seine River basin over the coming decades. While the broad outlines of these changes can already be sketched out, their details are still largely ignored, particularly because of the complexity of the interactions characterising river ecosystem functioning and the multiple uncertainties on the magnitude of future climate change and its implications for human activities.

Acknowledgements This work is a contribution to the PIREN Seine research program (www.piren-seine.fr), which belongs to the Zone Atelier Seine, a site of the international Long Term Socio Ecological Research (LTSER) network. This work was partly funded by the French Agency for Biodiversity (AFB).

References

1. Petts GE (1989) Historical analysis of fluvial hydrosystems. In: Petts GE (ed) Historical change of large alluvial rivers: Western Europe. Wiley, Chichester, pp 1–18
2. Magnuson JJ (1990) Long-term ecological research and the invisible present. Bioscience 40(7):495–501
3. Wolter C (2015) Historic catches, abundance, and decline of Atlantic salmon Salmo salar in the River Elbe. Aquat Sci 77(3):367–380
4. Haidvogl G, Lajus D, Pont D et al (2014) Typology of historical sources and the reconstruction of long-term historical changes of riverine fish: a case study of the Austrian Danube and northern Russian rivers. EcolFreshwat Fish 23(4):498–515
5. Azimi S, Rocher V (2016) Influence of the water quality improvement on fish population in the Seine River (Paris, France) over the 1990–2013 period. Sci Total Environ 542:955–964
6. Poulet N, Beaulaton L, Dembski S (2011) Time trends in fish populations in metropolitan France: insights from national monitoring data. J Fish Biol 79(6):1436–1452

7. Beslagic S, Marinval MC, Belliard J (2013) CHIPS: a database of historic fish distribution in the Seine River basin (France). Cybium 37(1–2):75–93
8. Beslagic S, Belliard J (2014) L'apport des sciences de l'environnement à la compréhension de l'histoire des milieux: l'exemple des peuplements de poissons du bassin de la Seine au regard des données archéologiques et historiques. In: Havre PudRed (ed) Journées archéologiques de Haute Normandie, Rouen, pp 191–198
9. Persat H, Keith P (1997) The geographic distribution of freshwater fishes in France: which are native and which are not? Bull Fr Pêche Piscic 344–345:15–32
10. Hoffmann RC (1996) Economic development and aquatic ecosystems in Medieval Europe. Am Hist Rev 101(3):631–669
11. Hoffmann RC (1995) Environmental change and the culture of common carp in medieval Europe. Guelph Ichthyol Rev 3:57–85
12. Leprieur F, Beauchard O, Blanchet S et al (2008) Fish invasions in the world's river systems: when natural processes are blurred by human activities. PLoS Biol 6(12):2940
13. Keller RP, Drake JM, Drew MB et al (2011) Linking environmental conditions and ship movements to estimate invasive species transport across the global shipping network. Divers Distrib 17(1):93–102
14. Hoffmann RC (2005) A brief history of aquatic resource use in medieval Europe. Helgoland Mare Res 59(1):22–30
15. Lenders HJR, Chamuleau TPM, Hendriks AJ et al (2016) Historical rise of waterpower initiated the collapse of salmon stocks. Sci Rep 6:29269
16. Moreau E (1898) Les poissons du département de l'Yonne. Bull Soc Sci Hist Nat Yonne (52):3–82
17. Meybeck M, Lestel L, Carre C et al (2018) Trajectories of river chemical quality issues over the Longue Durée the Seine River (1900s–2010). Environ Sci Pollut Res 25(24):23468–23484
18. Fisson C, Leboulenger F, Lecarpentier T et al (2014) L'estuaire de la Seine: état de santé et évolution. Fascicule Seine-Aval, vol 3.1. GIP Seine Aval, Rouen
19. Billen G, Garnier J, Ficht A et al (2001) Modeling the response of water quality in the Seine river estuary to human activity in its watershed over the last 50 years. Estuaries 24(6B):977–993
20. Rochard E, Pellegrini P, Marchal J et al (2009) Identification of diadromous fish species on which to focus river restoration: an example using an eco-anthropological approach (The Seine Basin, France). In: Haro A, Smith KL, Rulifson RA et al (eds) Challenges for diadromous fishes in a dynamic global environment. American fisheries society symposium, vol 69, p 691
21. Costil K, Dauvin J-C, Duhamel S et al (2002) Patrimoine biologique et chaines alimentaires. Fascicule Seine-Aval, vol 1.7. GIP Seine Aval, Rouen
22. Belliard J, Marchal J, Ditche JM et al (2009) Return of adult anadromous allis shad (*Alosa alosa* L.) in the River Seine, France: a sign of river recovery? River Res Appl 25(6):788–794
23. Perrier C, Evanno G, Belliard J et al (2010) Natural recolonization of the Seine River by Atlantic salmon (*Salmo salar*) of multiple origins. Can J Fish Aquat Sci 67(1):1–4
24. Comité de Gestion des poissons migrateurs du bassin Seine-Normandie (2016) Plan de gestion des poissons migrateurs du bassin Seine Normandie (2016–2021). Driee Ile de France
25. Euzenat G, Pénil C, Allardi J (1992) Migr'en Seine: stratégie pour le retour du saumon en Seine. Rapport Conseil Supérieur de la Pêche/SIAAP
26. Belliard J, Boet P, Allardi J (1995) Long-term evolution of the fish community of the Seine River. Bull Fr Pêche Piscic 337–339:83–91
27. Belliard J, Beslagic S, Delaigue O et al (2018) Reconstructing long-term trajectories of fish assemblages using historical data: the Seine River basin (France) during the last two centuries. Environ Sci Pollut Res 25(24):23430–23450
28. Billen G, Garnier J, Nemery J et al (2007) A long-term view of nutrient transfers through the Seine river continuum. Sci Total Environ 375(1–3):80–97
29. Eklov AG, Greenberg LA, Bronmark C et al (1998) Response of stream fish to improved water quality: a comparison between the 1960s and 1990s. Freshwat Biol 40(4):771–782

30. Passy P, Garnier J, Billen G et al (2012) Restoration of ponds in rural landscapes: modelling the effect on nitrate contamination of surface water (the Seine River Basin, France). Sci Total Environ 430:280–290

31. Boet P, Belliard J, Berrebi-dit-Thomas R et al (1999) Multiple human impacts by the City of Paris on fish communities in the Seine river basin, France. Hydrobiologia 410:59–68

32. Oberdorff T, Pont D, Hugueny B et al (2002) Development and validation of a fish-based index for the assessment of 'river health' in France. Freshwat Biol 47(9):1720–1734

33. Boët P, Duvoux B, Allardi J et al (1994) Incidence des orages estivaux sur le peuplement piscicole de la Seine à l'aval de l'agglomération parisienne (bief Andrésy-Méricourt). Houille Blanche (1–2):141–147

34. Chevreuil M, Blanchard M, Teil MJ et al (1998) Polychlorobiphenyl behaviour in the water/sediment system of the Seine river, France. Water Res 32(4):1204–1212

35. Meybeck M, Lestel L, Bonte P et al (2007) Historical perspective of heavy metals contamination (Cd, Cr, Cu, Hg, Pb, Zn) in the Seine River basin (France) following a DPSIR approach (1950–2005). Sci Total Environ 375(1–3):204–231

36. Teil MJ, Tlili K, Blanchard M et al (2014) Polychlorinated biphenyls, polybrominated diphenyl ethers, and phthalates in Roach from the Seine River basin (France): impact of densely urbanized areas. Arch Environ Contam Toxicol 66(1):41–57

37. Tales E, Belliard J, Rouillard J et al (2008) Les poissons de la Seine. Paris sous l'oeil des chercheurs. Belin, Paris, pp 59–73

38. Tales E (2008) Tendances d'évolution des peuplements de poissons de la Seine en réponse à la variabilité hydroclimatique. Hydroécol Appl 16:29–52

39. Buisson L, Grenouillet G (2009) Contrasted impacts of climate change on stream fish assemblages along an environmental gradient. Divers Distrib 15(4):613–626

40. Comte L, Buisson L, Daufresne M et al (2013) Climate-induced changes in the distribution of freshwater fish: observed and predicted trends. Freshwat Biol 58(4):625–639

41. Daufresne M, Boet P (2007) Climate change impacts on structure and diversity of fish communities in rivers. Glob Change Biol 13(12):2467–2478

42. Raat AJP (2001) Ecological rehabilitation of the Dutch part of the River Rhine with special attention to the fish. Regul Rivers Res Manage 17(2):131–144

43. De Leeuw J, Buijse A, Grift R et al (2005) Management and monitoring of the return of riverine fish species following rehabilitation of Dutch rivers. Arch Hydrobiol Large Rivers 15(1–4):391–411

44. Griffiths AM, Ellis JS, Clifton-Dey D et al (2011) Restoration versus recolonisation: the origin of Atlantic salmon (*Salmo solar L.*) currently in the River Thames. Biol Conserv 144(11):2733–2738

45. Tales E, Belliard J, Beslagic S et al (2017) Réponse des peuplements de poissons à l'urbanisation et aux altérations anthropiques à long terme des cours d'eau. In: Lestel L, Carré C (eds) Les rivières urbaines et leur pollution. QUAE Indisciplines, Paris, pp 242–252

46. Copp GH, Bianco PG, Bogutskaya NG et al (2005) To be, or not to be, a non-native freshwater fish? J Appl Ichthyol 21(4):242–262

47. Meyerson LA, Mooney HA (2007) Invasive alien species in an era of globalization. Front Ecol Environ 5(4):199–208

48. Kottelat M, Freyhof J (2007) Handbook of European freshwater fishes. Waterman, Berlin

49. Manné S (2017) Les gobies d'origine Ponto-Caspienne en France: détermination, biologie-écologie, répartition, expansion, impact écologique et éléments de gestion. Agence Française pour la Biodiversité

50. Manné S, Poulet N, Dembski S (2013) Colonisation of the Rhine basin by non-native gobiids: an update of the situation in France. Knowl Manag Aquat Ecosyst 411:2

51. Lassalle G, Rochard E (2009) Impact of twenty-first century climate change on diadromous fish spread over Europe, North Africa and the Middle East. Glob Change Biol 15(5):1072–1089

52. Buisson L, Grenouillet G, Casajus N et al (2010) Predicting the potential impacts of climate change on stream fish assemblages. In: Gido KB, Jackson DA (eds) Community ecology of

stream fishes: concepts, approaches, and techniques. American fisheries society symposium, vol 73, pp 327–346
53. Grenouillet G, Comte L (2014) Illuminating geographical patterns in species' range shifts. Glob Change Biol 20(10):3080–3091
54. Habets F, Boe J, Deque M et al (2013) Impact of climate change on the hydrogeology of two basins in northern France. Clim Chang 121(4):771–785

Bathing Activities and Microbiological River Water Quality in the Paris Area: A Long-Term Perspective

Jean-Marie Mouchel, Françoise S. Lucas, Laurent Moulin,
Sébastien Wurtzer, Agathe Euzen, Jean-Paul Haghe, Vincent Rocher,
Sam Azimi, and Pierre Servais

Contents

1   Introduction 324
2   Bathing in the Seine and Marne Rivers in Paris and Its Suburbs Has Recently Become
    a Health Issue 326
    2.1   Bathing in Paris: Washing and Cooling Off 326
    2.2   Bathing Despite Regulations and the Development of Industry and Harbors 327
    2.3   Bathing in Suburban Areas: From the Middle of the Nineteenth Century 327
    2.4   A Social Control More Than a Matter of Public Sanitation 329
3   The Concept of Fecal Indicator Bacteria 329

The copyright year of the original version of this chapter was corrected from 2019 to 2020.
A correction to this chapter can be found at https://doi.org/10.1007/698_2020_667

J.-M. Mouchel (✉)
Sorbonne Université, CNRS, EPHE, UMR Metis, Paris, France
e-mail: jean-marie.mouchel@sorbonne-universite.fr

F. S. Lucas
Faculté des Sciences et Technologie, Université Paris Est-Créteil, LEESU UMR-MA 102,
Créteil Cedex, France

L. Moulin and S. Wurtzer
R&D Department, Eau de Paris, Ivry sur Seine, France

A. Euzen
Université Paris Est, Ecole des Ponts ParisTech, CNRS, UMR LATTS, Marne-la-Vallée, France

J.-P. Haghe
Université Paris-1, UMR 8586 PRODIG, Paris, France

V. Rocher and S. Azimi
R&D Department, SIAAP, Colombes, France

P. Servais
Ecology of Aquatic Systems, Université Libre de Bruxelles, Brussels, Belgium

Nicolas Flipo, Pierre Labadie, and Laurence Lestel (eds.), *The Seine River Basin*,
Hdb Env Chem (2021) 90: 323–354, https://doi.org/10.1007/698_2019_397,
© The Author(s) 2020, corrected publication 2020, Published online: 3 June 2020

4   Long-Term Fluctuations of Microbiological Water Quality in the Seine and Marne
     Rivers 331
5   Sources of Fecal Contamination 333
6   The EU Bathing Water Directive 336
7   Present Level of Fecal Bacteria Indicators in the Seine and Marne Rivers
     in the Paris Area 339
     7.1   The Main Factors Explaining the Distribution of E. coli 339
     7.2   Longitudinal Distributions of E. coli in the Urban Sections of the Seine and Marne
           Rivers 340
8   Presence of Pathogens in the Seine and Marne Rivers in the Paris Area 343
     8.1   Viruses 343
     8.2   Parasites 345
9   Where to Now? 346
10   Conclusion 349
References 350

Abstract This chapter presents the historical aspects regarding swimming in rivers
in the Paris region since the seventeenth century, including the concept of fecal
contamination indicator bacteria (FIBs) developed at the very beginning of the
twentieth century, and historical contamination data covering more than one century
in the Paris agglomeration. The sources of microbiological contamination of river
waters are quantified, showing the importance of rain events. The present contamination levels are presented with reference to the European Directive for bathing
water quality. FIB levels show that the sufficient quality for bathing is not reached
yet in any of the monitored stations. A comprehensive data set regarding waterborne
pathogens (viruses, *Giardia*, *Cryptosporidium*) in the Seine and Marne rivers is
presented as a necessary complement to the regulatory FIB data to better evaluate
health risks. The last section concludes on the actions to be conducted to improve the
rivers' microbiological quality in the coming years.

Keywords Bathing, Fecal indicator bacteria, LTSER "Zone Atelier Seine",
Microbiology, Paris agglomeration, Pathogens, PIREN-Seine, Seine River

1   Introduction

Bathing activities have been observed for a very long time in the main rivers of the
Paris region (Fig. 1). They were progressively banned during the twentieth century
mainly because of the significant degradation of water quality, most particularly
microbiological quality of surface waters, which presents a major health challenge.
Contaminated water can contain various types of pathogenic microorganisms (bacteria, viruses, parasitic protozoa) likely to cause various types of infections (gastrointestinal, respiratory, ocular, cutaneous) in humans. The main source of these

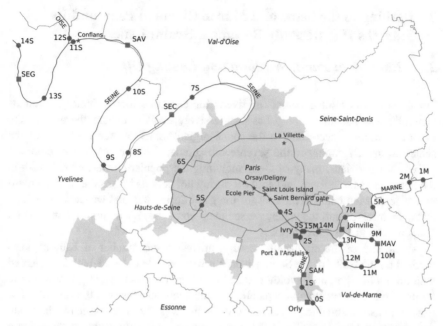

Fig. 1 Situation map. Dots indicate the stations where FIBs have been measured. Squares represent the inlet of drinking water production plants (Joinville, Orly, and Ivry) or the outlet of wastewater treatment plants (MAV, SAM, SEC, SAV, SEG). Present-day *département* boundaries are drawn with a dotted line; their names are in italics. The pink area represents the municipalities with a population density higher than 1,000 hab/km² in 1881. Nowadays, these municipalities would cover most of the map

microbial contaminants is the digestive tract of humans and warm-blooded animals or the aquatic environment itself. Microbial contaminants may reach the aquatic environment via wastewater treatment plant discharges, urban combined sewer overflows, sanitation system failures, agricultural and urban runoff, and direct dejection of breeding or wild animals.

In the Paris area, thanks to major improvements in wastewater treatment, the microbiological quality of water has improved significantly since the mid-1980s, and the desire to engage in swimming activities in rivers in the Paris agglomeration is spreading among the population. This objective is shared by the local authorities of Ile-de-France who wish to create or reopen bathing areas in urban rivers. Indeed, the Syndicat Marne Vive and the relevant actors of the urban Marne River have planned to open bathing areas in 2022. Similarly, the city of Paris supports a project to organize swimming events in open water in the Seine River, especially during the Olympic Games in 2024. The opportunity to bathe in urban river water motivates the development of renewed water policies in the urban area.

2 Bathing in the Seine and Marne Rivers in Paris and Its Suburbs Has Recently Become a Health Issue

2.1 Bathing in Paris: Washing and Cooling Off

Local populations have bathed in rivers for centuries for hygienic as well as symbolic reasons; this activity has progressively evolved toward the ages at the rate of modernity, according to the social organizations and the evolution of the social norms. Beginning in the seventeenth century, bathing was a very common activity for Parisians: many of them, including the popular classes, would wash in the Seine and Marne rivers and cool off during hot periods [1]. At the Saint Bernard gate where "Men bathe at the foot [of the gate] during the heat wave; we see them throw themselves into the water, we see them come out, it's an amusement," as La Bruyère says in *Les Caractères*.

At the end of the eighteenth century, an increasing number of bathing boats moored on the banks of the Seine River, allowing people to take a bath in a pierced bathtub out of sight, thus avoiding social or sexual promiscuity [1]. They were run by private entrepreneurs and separation of the social classes was the rule [2]. The least expensive, so-called "*quatre-sous*" baths, was the most basically laid out. According to Eugène Briffault [3], the show was not very raucous: "imagine a band of demons, blackened with coal and smoke, plunging into a bath that one moment becomes thick and muddy." In this description, the chronicler's gaze attributed the poor quality of the water to its contact with individuals of another social class and not to the water itself.

The most sophisticated bathing boats offered many services such as a private cabin, hot water, a restaurant, a massage parlor, and a hairdresser. Beginning in the 1820s, some of them developed swimming pools. All were supplied with water pumped directly in the Seine River with no filtration device. Comparing two swimming school boats, Briffault found that the water at Saint-Louis Island was "pure, clear and healthy because it has not yet crossed the Parisian cesspool" while the Orsay water was "dirty, cloudy, often foul and unhealthy because it has already rolled in the filth of the big city." But this situation was not detrimental to the reputation of the Deligny baths at Orsay, which were "opulent and majestic." The chronicler noted, however, that on days of heavy rains, the pools of Deligny became muddy and "that it is difficult to swim for a long time in these waters without having the chest soiled with a muddy deposit and ears plugged" [3]. This is a rare and early comment regarding bathing and the deterioration of water quality due to urban wet weather pollution.

2.2 Bathing Despite Regulations and the Development of Industry and Harbors

The occupation of the banks and the uses of the Seine River were regulated from the beginning of the eighteenth century. A law on June 5, 1711, thus "forbids anyone to bathe in the arm of the Seine River from the bridge of the Hotel Dieu hospital, to that called the Petit Pont, the fine was 20 pounds." The same prohibition was reiterated in 1724, 1737, and 1738 [4]. It was mainly for reasons of morality that these prohibitions were imposed following the excessive behavior of some bathers exposing themselves nude close to the Hôtel Dieu Hospital. Other reasons motivated the prohibitions to bathe in the Seine River. A police order dated June 28, 1726, stated: "It is forbidden for the water carriers to draw water during the summer from the Great Degrees to the Pont Neuf and from the Saint-Paul port to the Ecole pier (. . .) and all people to swim in this stretch of the river, punishable by a fine of four hundred pounds." Starting from this time until the first half of the nineteenth century, on its periphery the city tended to reject all the activities likely to encumber the banks and to hinder river traffic [5]. However, wild bathing in the Seine River remained the rule for the working classes, despite the regulations and the fines, until the 1930s, which saw the development of an alternative with the municipal baths and swimming pools.

The development of shipping and the installation of warehouses and factories on the banks during the nineteenth century did not seem to change the appetite for wild bathing. In 1822, a report from the canal manager reported the presence of 150 bathers along the Bassin de la Villette, who threatened and insulted the guards [5]. In 1877, the famous sanitary engineer Durand-Claye noted that "every year since time immemorial, the Bassin de la Villette has been invaded by the crowd of workers from the neighbourhood whose number sometimes amounts to 500, 600 or even more."

2.3 Bathing in Suburban Areas: From the Middle of the Nineteenth Century

From the mid-nineteenth century, to escape urban and industrial life, Parisians started to enjoy the banks of the Seine and Marne rivers in the suburbs of Paris to relax or party [6]. Trams and other suburban trains facilitated access to the capital's surrounding recreational areas for the working classes as well as upper-class families, who enjoyed a picnic on the grass, a walk, a fishing party, or boating activities. The pleasure of bathing that has always been enjoyed in both rivers largely developed in these improvised spaces, appropriated by bathers, between natural cove and sandbank. Later, the Ile-de-France municipalities gradually developed the banks to create beaches and whole complexes to accommodate not only bathers but also

Fig. 2 Bathing at Port-à-l'Anglais. Left: Engraving by AB Flamen, about 1,650, Bibliothèque Nationale de France. Right: Bathing at Port-à-l'Anglais in 1920, gallica.bnf.fr/Agence de presse Mondial Photo-Presse

boaters and walkers often coming from the capital for a "partie de campagne" on Sundays [1].

During the second half of the nineteenth century, Joinville-le-Pont, on the Marne River, became a major spot for water sports activities. The recognition of swimming as a sport owes much to the military. The teaching of swimming in the army had to be done in the river, because of a lack of other facilities. The Joinville Gymnastics School was created in 1852; swimming took place in the Marne River from a pontoon. For 80 years instructors at the military schools were trained here and then civilians. The first swimming and diving competitions at the national and international level were organized on the banks of the Marne River at Joinville. From the beginning of the twentieth century, various establishments were multiplying along the Marne River to meet the demand for an emerging sport: swimming schools, diving boards, and landscaped pools. During the summer, thousands of Parisians and suburbanites bathed in the water of the Marne River without a second thought. These practices were banned at the end of the summer of 1970 by a prefectural decree prohibiting bathing in the Marne River because of the poor quality of the water.

In Port-à-l'Anglais, on the banks of the Seine River, upstream from Paris, bathers had been swimming in white water since the seventeenth century, as illustrated by an engraving depicting bathers in the middle of a lush natural setting (Fig. 2). Later on, after the First World War, the shorelines were gradually developed, integrating beaches and bathing places to satisfy the recreational needs of local populations, which could use a new tramway constructed in 1920 to access this area. However, the idyllic landscape of the seventeenth century was replaced from 1850 to 1950 by a forest of smoking industrial chimneys (Fig. 2), while as the nineteenth century came to an end, only rare voices were raised about the poor water quality in this area because of the discharge of raw industrial and municipal effluents into the river [7].

Fig. 3 Bathers and onlookers at the Iena bridge in 1945. gallica.bnf.fr/ Agence de presse Mondial Photo-Presse

2.4 A Social Control More Than a Matter of Public Sanitation

The regulation of bathing in the Seine River can be divided chronologically into three periods [8]. The first corresponds to the eighteenth century: bathing is allowed in Paris but is confined to certain places chosen with the main intention of ensuring respect for morality and regulating conflicting uses of the river banks. The second period is a transitional period during which swimming became strictly forbidden in Paris in 1867, extended to the nearest suburbs in 1923, except for certain places authorized by the Police Prefecture. The last period begins in the middle of the twentieth century with repeated closures of bathing areas due to episodes of pollution. Bathing is completely prohibited in the Marne River in 1970 by prefectural decree. Finally in 2003, it was also forbidden in the Seine River within the limits of the Val-de-Marne department. Despite this legal and administrative regulation, the practice of wild bathing in the Seine and Marne rivers persisted until the end of the 1960s, a period that represents a shift in the public's perception of the health risk linked to bathing (Fig. 3).

3 The Concept of Fecal Indicator Bacteria

In the present EU bathing directive (2006/7/EC), the monitoring of bathing water quality is based on the concentrations of two bacterial types: *Escherichia coli* (EC) and intestinal enterococci (IE), which are used as fecal indicator bacteria (FIBs). The indicator concept was originally defined to evaluate the quality of drinking water and is based on several hypotheses. FIBs are residents of human and mammal intestines, they are shed in feces, and they are assumed not to grow in the soil and water environments. The presence of FIBs is assumed to be related to the presence of pathogens [9].

This fecal indicator paradigm developed at the end of the nineteenth century, because it was obvious that waterborne diseases were mostly gastroenteritis. Since it was impossible to monitor all enteric pathogens, microbiologists focused on FIBs as indirect predictors of their presence ([10–12]). In 1892, Schardinger [12] proposed using *Bacterium coli* which was isolated by Theodor Escherich in 1885 [13] and was renamed *Escherichia coli* in 1919 [14]. Selective media were later developed for the quantification of *E. coli* in drinking waters, surface waters, and wastewaters. Thermotolerant coliforms (also called fecal coliforms) that include *E. coli* were used for a while as indicators; however, their lack of specificity resulted in their being dropped in the revised Bathing Water Directive in 2006. Enterococci were integrated into the monitoring of FIBs when Slanetz et al. [15] proposed a selective medium for their detection; they were first named fecal streptococci [14].

Although the FIB concept proved to be useful for several decades to estimate potential health risk and contributed greatly to decreasing the occurrence of waterborne diseases in industrialized countries, it has many drawbacks. Some FIB strains survive and even grow in sand, sediment, soil, and waterbodies, which may serve as potential reservoirs [16–18]. Culture-based assays cannot distinguish whether the source of contamination is animal or human [19]. The correlation between coliforms and human pathogens has not been validated over a wide range of waterbodies with differences in water quality, geographic origin, climatic conditions, watershed morphology and hydrology, and local pollution inputs. In fact, the FIB/pathogen ratio in natural waters is mainly a function of the occurrence of pathogens in the population, different fates in water treatments, and different survival times in the environment. These ratios are therefore site-specific and seasonally dependent and show substantial spatial and temporal variations [20–22].

As a consequence, there is often no correlation between FIB and pathogen concentrations, which prevent efficient prediction of contamination [23–25]. In the Seine River in the Paris area, virus monitoring carried out in 2013–2014 and 2017–2018 showed no significant correlation between FIBs and a panel of various enteric viruses (Fig. 4). From January 2005 to August 2007, only a significant correlation could be found between IE and *Giardia* cysts in the Seine and Marne rivers [26]; other fecal bacteria were not correlated with the protozoans monitored.

To counter some of these drawbacks, new indicators were developed using molecular techniques. Microbial source tracking methods distinguish human and animal sources using genetic markers such as specific gut bacteria or comparison of bacterial communities [9, 27]. *Bacteroidetes* or *Firmicutes* species and fecal bacteriophages were identified as potential alternate indicators [19, 28]. Although somatic coliphage and F-specific RNA phage are sometimes used by stakeholders, these indicators are still not implemented in the EU Bathing Directive, and *E. coli* and IE remain the gold standard [29].

In conclusion, the presence of FIBs can predict the probable presence of viruses, *Giardia*, and *Cryptosporidium* in surface water affected by sewage inputs, but they cannot predict their concentration. This is in accordance with the original indicator concept in drinking water, which established FIBs as an index of fecal pollution and, therefore, the probability of the presence of pathogens and potential health risks.

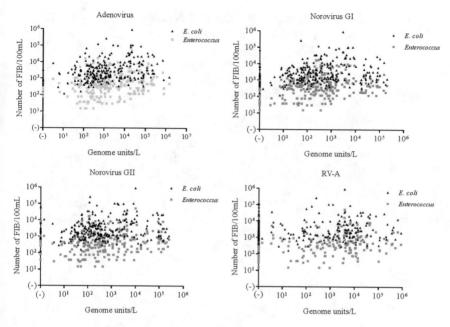

Fig. 4 Relationship between viral and FIB concentrations (*E. coli* and IE) observed in the Seine River from two campaigns (2013–2014 and 2017–2018) ($n = 257$). ($-$) means not detected. *AdV* adenovirus, *NVGI* norovirus group I, *NVGII* norovirus group II, *RV-A* rotavirus A [25]

It has often been interpreted and used in a different context than its original concept, resulting in a multitude of misapplications [30].

4 Long-Term Fluctuations of Microbiological Water Quality in the Seine and Marne Rivers

The microbiological quality of the Seine River in the Paris area has been studied regularly since 1911. All data were collected in historical lab books from the Ville de Paris laboratory (called CRECEP data) and from the drinking water treatment plant quality laboratory (called EdP data) and carefully double-checked. SIAAP measurements were also added to the timeline. *E. coli* measurements were used when available (mainly in more recent years), or alternatively thermotolerant coliform measurements (called *Bacterium coli*, *B. coli*, or *Th. coliforme* in the lab books). All fecal coliform concentrations were transformed into *E. coli* concentrations using a ratio of 0.77. This conversion factor was established by comparing the numbers of thermotolerant coliforms measured on plate counts and the numbers of *E. coli* enumerated using the most probable number microplate method (ISO 9308-3) in a set of samples collected in rivers in the Seine watershed [31]. This ratio is in accordance with the ratio observed (*E. coli*/thermotolerant coliforms = 0.8,

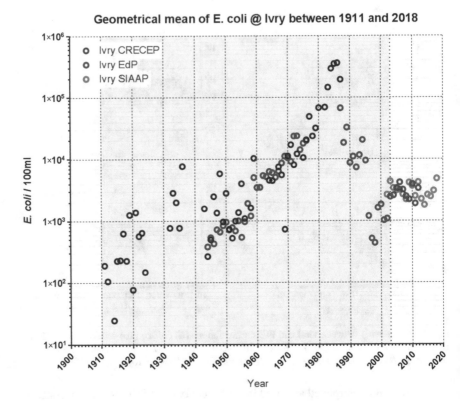

Fig. 5 Annual geometrical means of *E. coli* concentrations in the Seine River at Ivry between 1911 and 2018. Data were collected from the Ville de Paris laboratory (CRECEP in black open circles), the Ivry drinking water plant (EdP in blue open circles), and SIAAP (red open circles). The colored background indicates the period during which *E. coli* concentrations were estimated from the concentrations of thermotolerant coliforms converted to *E. coli* data using an *E. coli* to coliform ratio of 0.77

$n = 650$) in the historical data for the period during which both parameters were measured simultaneously. After verification of the data collected, *E. coli*'s annual geometrical means were calculated for each year to limit any influence of highly divergent results. The number of data taken into account per year varied from 4 to 254.

Figure 5 shows the evolution of the annual geometric mean of *E. coli* in the Seine River at Ivry just upstream from the confluence with the Marne River and from the entrance to Paris. Although caution is advised when comparing measurements made over a period of a century with methods that have evolved over time, it seems clear that there was deterioration in microbiological quality from the beginning of the century to the 1980s. For periods when two data sets are available, there is good concordance between the two annual geometric means calculated taking into account the fact that samples were not collected at the same frequency and on the same dates for both data sets; this gives us confidence in the data presented in Fig. 5.

During the 1950s to the 1980s, water quality deteriorated very rapidly, and the maximum annual geometric mean of concentrations (5×10^5 E. *coli*/100 mL) was reached in 1985. This accelerated quality degradation is related, on the one hand, to a significant increase in the urban population upstream from Paris (the urban population quadrupled between the 1950s and the 1980s [32]) and, on the other hand, to a slow implementation of domestic sewage treatment plants. After the mid-1980s, there was a very rapid and significant improvement of the microbiological quality at the Ivry sampling site. This is partly due to the commissioning of the Seine Amont WWTP located 4 km upstream from Ivry. The geometric means of E. *coli* concentrations continued to decrease until the beginning of the 2000s (Fig. 5). This improvement was first related to the widespread installation of new WWTPs and to subsequent improvement of their treatment processes. Since 2000, the microbiological quality has been quite stable, with annual geometric means around $3–5 \times 10^3$ E. *coli*/100 mL.

The evolution of the annual geometric mean of E. *coli* in the Seine River downstream from Paris at Conflans is presented in Fig. 6. The Conflans sampling station is located just upstream from the confluence of the Seine River with the Oise River. This station is situated downstream from the outfall of the effluents of the Seine Aval (SAV) WWTP, which today treats 70% of the wastewaters from the Paris agglomeration. During the entire 1911–1980 period, the E. *coli* concentrations were always significantly higher in Conflans compared to the concentrations measured in Ivry. The difference between the two stations was maximum in the 1950s (with annual geometric means sometimes two log units higher in Conflans than in Ivry). Unfortunately, data are not available for the period between 1980 and 2000, which probably displayed the highest contamination levels in Conflans. In recent years (since 2000), the E. *coli* concentrations have continued to decrease, with geometric means between 5×10^3 and 10^4 E. *coli*/100 mL. This can be attributed to the major improvements of the treatment processes in the SAV WWTP [33] and also to a reduction in the volume of effluents discharged by combined sewer overflows (CSOs) in the Seine River downstream from Paris (see the next section).

5 Sources of Fecal Contamination

When assessing the contamination of natural aquatic ecosystems by microorganisms of fecal origin, it is usual to distinguish point sources from nonpoint sources, also called diffuse sources. Microbial contaminants brought by surface runoff and soil leaching constitute the diffuse sources, which also include contamination linked to leakage from sanitation sewers and septic tanks and also direct defecation of certain animals (e.g., cows, waterfowl) in surface waters. The outfalls of raw urban wastewater, domestic WWTP effluents, industrial effluents, and urban runoff waters collected in separate sewer systems and CSOs constitute the point sources.

The inputs of fecal microorganisms via nonpoint sources are not easy to quantify. To assess the contribution of nonpoint sources to river fecal contamination,

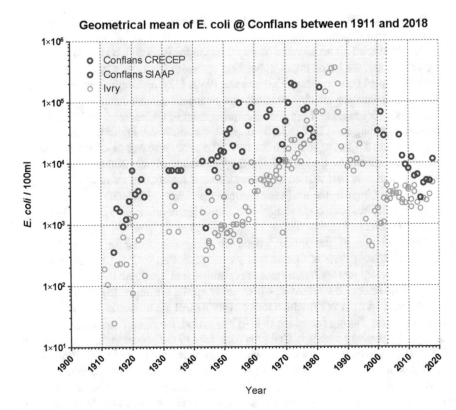

Fig. 6 Annual geometrical means of *E. coli* concentrations in the Seine River at Conflans (downstream from Paris) and Ivry (upstream from Paris) (light gray circles; see data from Fig. 5) between 1911 and 2018. Data at Conflans were collected from the Ville de Paris laboratory (CRECEP in blue open circles) and SIAAP (red open circles). The colored background indicates the period during which *E. coli* concentrations were estimated from the concentrations of thermotolerant coliforms and converted to *E. coli* data using an *E. coli* to coliform ratio of 0.77

concentrations of FIB in small streams upstream of any source of domestic contamination were measured, while these small streams were characterized on the basis of the land use of their watershed: forests and cultivated and grazed areas [34]. These data show that the concentrations measured depend greatly on the land use of the watershed [35, 36]. Small streams located in pastured areas presented contamination levels (median values of FIB concentrations: 250 *E. coli*/100 mL and 100 IE/ 100 mL) around one order of magnitude higher than the streams located in forest or crop areas (median values of FIB concentrations: 42 *E. coli*/100 mL and 11 IE/ 100 mL). This clearly demonstrates the importance of manure grazing livestock as a source of fecal contamination of surface waters in rural areas in the Seine River catchment.

Concentrations of microorganisms of fecal origin are very high in raw domestic wastewaters [37]. For example, Fig. 7 shows that median concentrations in raw

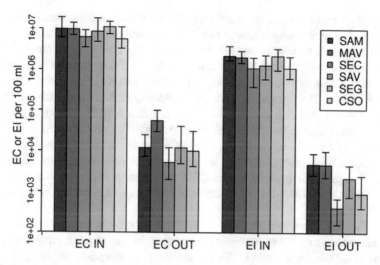

Fig. 7 *E. coli* (EC) and intestinal enterococci (IE), concentrations in raw (IN) and treated waters (OUT) of the five WWTPs located in the Paris area releasing their effluents in the Seine River (*SAM* Seine Amont, *SEC* Seine Centre, *SAV* Seine Aval, *SEG* Seine Grésillons, *MAV* Marne Aval) and in the two major combined sewer overflows (CSO, no treatment, raw water only) located at Clichy and La Briche. The gray blocks indicate the median values and the vertical black bars represent the range between the first and the third quartiles. Data from [33], data collected between 2012 and 2015, additional unpublished data for MAV IN were collected in 2010–2011

sewage entering Parisian WWTPs ranged between 6.2×10^6 and 1.1×10^7/100 mL for *E. coli* and between 1.2×10^6 and 2.2×10^6/100 mL for IE. WWTPs have not been specifically designed to remove fecal microorganisms but rather to remove suspended solids, organic matter (especially its biodegradable fraction), nitrogen, and phosphorus. They are quite efficient in removing FIBs, however. Studies conducted on the removal of fecal bacteria in WWTPs in the Seine watershed usually showed a 2-log-unit reduction of FIBs for plants using a primary settling treatment followed by an activated sludge process [32, 35, 38]. When a tertiary treatment is added to decrease nitrogen and phosphorus concentrations, an improvement in the removal of FIB is usually observed. In the Paris WWTPs, tertiary treatment is applied [33]: Fig. 7 shows log reductions varying between 2.9 and 3.1 log units for *E. coli* and between 2.7 and 3.4 for IE. Although FIBs are efficiently removed in the Paris WWTPs, their concentrations in treated effluents are still high (median values between 5.2×10^3 and 1.2×10^4 for *E. coli* and between 4.2×10^2 and 4.8×10^3 for IE). A way to decrease the concentrations of fecal microorganisms in the treated effluents is to polish the treatment using a disinfection treatment. In Europe, UV irradiation is the predominant option for wastewater disinfection because its main advantage is the absence of release of chemical products in the environment. Today, WWTPs equipped with a disinfection treatment unit are scarce in the Seine watershed.

The input to the aquatic natural environments of pathogenic microorganisms from domestic wastewater is not restricted to discharges of WWTP effluents. In cities equipped with a combined sewer network, the water flow in the sewers significantly increases during rain events due to runoff waters, and the transport capacity of the sewer system can be insufficient to allow all the water flow to reach the WWTP, or the treatment capacity of the WWTP can be insufficient to treat all the water flow. In such cases, CSOs can occur, resulting in the direct release of untreated waters into surface waters. The content of pathogenic microorganisms and FIBs in CSO is slightly lower compared to raw sewage entering the WWTPs during dry weather (Fig. 7). This is due to the dilution of domestic waters by runoff waters that are less contaminated by fecal microorganisms. Studies conducted after intense rain events in the Seine River downstream of one of the major CSO outfalls in Clichy have shown that the FIBs brought by the CSO can result in FIB concentrations in the Seine River exceeding the usual dry weather concentrations by two orders of magnitude [39].

In the last few years, the SIAAP (Syndicat Interdépartemental pour l'Assainissement de l'Agglomération Parisienne, in charge of the management of the sewer network and the WWTPs in the Paris agglomeration) has made a significant attempt to reduce the frequency and the volume of the CSO released into the Seine River (e.g., by building storage tunnels and detention basins). It resulted in a significant decrease of the annual volume of CSOs in the Paris area from 80~100 $\times$ 10^6 m^3 (volume discharged by the three main CSO outfalls located downstream from Paris at Clichy, La Briche, and La Frette) during the 2000–2002 period to around 15 $\times$ 10^6 m^3 in 2015 [40]. With knowledge of the annual CSO volume and the annual volume treated in the Paris WWTPs (800 $\times$ 10^6 m^3) and the FIB concentrations in CSOs and treated effluents (Fig. 7), the annual flux of FIBs brought to the Seine River by CSOs and WWTPs was calculated. For *E. coli*, an annual flux value of 1 $\times$ 10^{17} for the WWTPs and 8 $\times$ 10^{17} for CSOs were estimated for the very recent years. Even if these data are approximate values, they clearly demonstrate that CSOs are nowadays the major source of microbiological contamination of the Seine River in the Paris area.

6   The EU Bathing Water Directive

In the 1970s, the World Health Organization (WHO) stated the existence of a link between the microbiological quality of waters and the health of swimmers. Specific European regulations for the management of the water quality of bathing areas appeared in 1975 (76/160/EU). Based on WHO recommendations, the bathing directive provided guidelines for grading the water quality of bathing sites using FIB count data. Its main purpose was to safeguard public health against microbial pollution.

The directive was modified in 2006 (2006/7/EC) to consider recent advances in risk evaluation for bathers. In addition to previously existing epidemiological data [41, 42], a more recent epidemiological study was available for fresh water in Germany [43]. These epidemiological studies explore the most likely illnesses that can be attributed to contact with or ingestion of water while bathing [41]: gastrointestinal symptoms; eye infections; skin complaints; ear, nose, and throat infections; and respiratory illnesses. They are the major source of evidence, in marine and fresh water, that was used to define the criteria of the European Directive. In all studies a non-bathing population is used as a reference; surveys are conducted to characterize the bathing in great detail; the symptoms of swimmers and nonswimmers are queried about 1 week after the bathing event (up to 3 weeks for one study). Despite efforts to make the results as comparable as possible, differences may occur between studies. They may concern the criteria used to define the severity of the disease and to qualify bathing (one bath or one day at the beach, head immersed or not, duration of the bath, etc.). In a randomized study, Wiedenmann et al. [43] enrolled participants to bathe with a very precise protocol, which was not the case for other studies. Moreover, the data (number of diseases) can be given for sites with different levels of contamination or can be classified as a function of the recorded contamination level in case of a whole season survey at a single site; this is another source of discrepancy. Nevertheless, it is important to compare all these data to provide an integrated vision because epidemiological studies are rare, in particular for fresh water.

In Fig. 8, we gathered data from selected studies, conducted in fresh water (lake shores and rivers) in the USA and Europe, regarding the occurrence of highly probable gastroenteritis. *E. coli* or fecal coliform concentrations were selected to characterize water quality. Gastroenteritis is the most frequent disease reported among swimmers, although not the most dangerous. Figure 8 shows the variability that may exist between studies. However, above 200 *E. coli* per 100 mL (geometric mean), all studies show an excess risk of gastroenteritis, while one study showed an excess risk for concentrations as low as 10 *E. coli* per 100 mL.

The European Directive (2006/7/EC) defines a set of criteria based on *E. coli* and intestinal enterococci concentrations to characterize the excellent, good, sufficient, and poor status of bathing water (Table 1). More stringent criteria exist for marine water. Although bathing can be authorized in water with sufficient quality, and even a poor quality under specific circumstances, the directive recommends bringing as many bathing areas in Europe as possible to a good water quality status.

Water quality monitoring is required during at least four successive bathing seasons, and the data set to estimate water quality must include more than 16 samples. Bathing in an area with a poor status can be temporarily authorized, but it will be closed after 5 years of poor status. The directive also demands "bathing water profiles" to inform the public of the characteristics of the catchment, risks, and the occurrence of short-term events that might require the temporary closure of the bathing area and measures taken to limit risks (policies regarding spreading of manure or sewage sludge, sewer system management, wastewater disinfection, etc.). The bathing water profile should also inform on all additional potential risks (cyanobacteria or algal blooms, motor water sports, etc.).

For areas with a high frequency of dangerous short-term events, the directive allows disregarding no more than 15% of the samples with the highest concentrations before the estimation of the quality criteria. This imposes setting up efficient management measures including surveillance, early warning systems, and

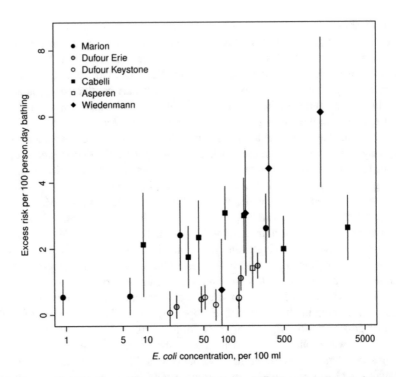

Fig. 8 Excess risk of highly credible gastrointestinal disease (HCGI), or a similar set of symptoms, from seven studies conducted from 1982 to 2010. The *x*-axis is the geometric mean of the concentration of *E. coli* or fecal coliforms in water. Excess risk was computed as the difference between the occurrence of HCGI in the bathing population and the non-bathing population. Data from [43–47]

Table 1 Quality criteria regarding *E. coli* and intestinal enterococci for inland water in the 2006/7/ EC directive

	Excellent	Good	Sufficient	Poor
E. coli	<500[a]	<1000[a]	<900[b]	>900[b]
IE	<200[a]	<400[a]	<330[b]	>330[b]

According to the European Directive, the percentiles are computed assuming a log-normal distribution of the data and must be computed as the percentiles of the log-normal distribution that fits the data. 90th percentile, antilog (μ + 1,282 σ); 95th percentile, antilog (μ + 1.65 σ)
[a]95th percentile
[b]90th percentile

monitoring, to prevent bathers' exposure by means of a warning or, where necessary, bathing prohibition.

But zero risk does not exist. The choice of water quality criteria results from a complex synthesis regarding the acceptability of risk and the existing epidemiological knowledge. Therefore, standards may differ in different countries. As stated by

AFSSE [48], the excess risk of gastroenteritis for the sufficient criteria based on *E. coli* in fresh water in Europe is in the range 1.5–2.2% per bath.

The 2006/7/EC directive should soon be reviewed. The WHO [29] made a series of recommendations to simplify quality criteria and strengthen monitoring. However, it is suggested that FIBs remain the basis for the water quality criteria since data are still lacking to develop criteria based on pathogens.

7 Present Level of Fecal Bacteria Indicators in the Seine and Marne Rivers in the Paris Area

Several monitoring networks provide information regarding the amount of FIB in the Marne and Seine rivers within Greater Paris: drinking water producers, municipalities, and institutions in charge of sanitation. Monitoring was recently intensified to prepare the bathing objective in 2022 in the Marne River and the organization of the Olympic Games in Paris in 2024. We selected data provided by the SIAAP, Eau de Paris, the Paris municipality, and Marne Vive to produce a comprehensive image of contamination. All data were obtained by the most probable number microplate methods for *E. coli* (ISO 9308-3) and intestinal enterococci (ISO 7899-1).

The distribution of *E. coli* and intestinal enterococci data in the Seine and Marne rivers shows that 85% of the samples have an EC/IE ratio higher than the ratio between the EC and IE criteria (900/330). Therefore, *E. coli* is a more critical parameter to define the bathing quality of both rivers, and the following sections will focus on *E. coli*.

7.1 The Main Factors Explaining the Distribution of E. coli

Long-term data are available at drinking water production intakes in the Seine (Orly station) and Marne (Joinville station) rivers upstream from Paris. Ten years of data covering the most recent period, and including about 1,000 data, were selected to identify the major factors impacting *E. coli* concentrations. At Orly, a regression between the log of *E. coli* concentrations and the log of the river discharge explained 33% of the log *E. coli* concentration variance (F test p-value $<10^{-8}$). Adding the local rain of the previous day improves the explained variance up to 45% (p-value $<10^{-15}$). The addition of rainfall of the second day before sampling did not significantly improve the regression. Highly similar results were obtained when summer data only were considered. In the Marne River at Joinville, the effect of river discharge is not significant, while the rainfall of the previous day explains 26% and 28% of the variance of log *E. coli* concentrations (p-values $<10^{-5}$ and 0.003).

The negative influence of rainfall events on water quality was expected since urban runoff is a known source of fecal contamination; in areas with separate sewers,

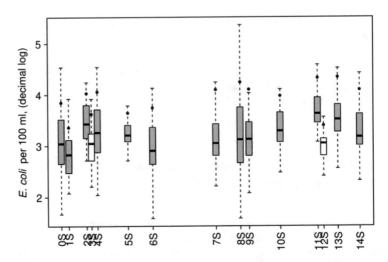

Fig. 9 *E. coli* concentrations during the 2014–2018 period; data were selected during the bathing season (June to September). At each station, except 0S, 4S, and 5S, mixed samples were produced from initial samples collected close to each bank and in the center of the river flow. The central bar shows the median, the box illustrates the 25% and 75% quantiles, and the whiskers illustrate the 99% interval, assuming a log-normal distribution. Dots outside the whiskers are extreme values. The black dot is the 90% percentile. The *x*-axis is representative of the distance (100 km between the most upstream and downstream stations). White boxes show the tributaries of the Seine River (Marne and Oise rivers). The number of data in each boxplot ranges from 28 to 75, except for Orly (station 0S, 335 data)

its importance is enhanced by former poor connection of housing outlets to the storm water network. An additional factor could be a decreased efficiency of WWTPs during the largest rain events [37]. The highly significant positive relationship between river discharge and the *E. coli* concentration in the Seine River likely results from a complex combination of factors. This effect is not systematically observed in all large rivers [49]. Dilution of point sources may provoke a negative correlation. The positive correlation of *E. coli* concentrations can be explained by an increased runoff and erosion in the whole river basin, notably from pastures with cattle breeding. Another very important factor is the reduced residence time of water during high-flow periods, which also reduces the extent of the degradation processes that strongly influence on *E. coli* concentrations.

7.2   Longitudinal Distributions of E. coli in the Urban Sections of the Seine and Marne Rivers

A detailed longitudinal profile along the Seine River can be drawn from the available data set. Data obtained for samples collected from June to September were selected to determine water quality during the summer bathing season (Fig. 9). Most data were collected bimonthly, more frequently at some stations. The 2014–2018 period was selected to limit the effect of the strong temporal trends shown in Sect. 4.

The longitudinal trends result from a combination of factors whose effect cannot all easily be separated. Inputs from wastewater discharges or tributaries can increase or dilute the *E. coli* concentrations, while the natural elimination processes (grazing, settling, etc.) progressively decrease the concentration. The impact of point discharges or tributaries can be compared to concentrations at the nearest downstream station, but the combined impact of diffuse sources and elimination processes between recognized inputs cannot be disentangled.

Estimates of the impact of WWTPs were computed from the median concentrations discussed in Sect. 5, from the nominal daily volumes treated by WWTPs and river discharges at the low stage (monthly average with a 2-year return period, data from the French national hydrological database). The expected concentration increases after dilution in the river were 1,080, 130, 1,810 and 1,390 *E. coli* per 100 mL for the SAM, SEC, SAV, and MAV WWTPs, respectively.

The strong increase of median concentration and flux at Ivry is twice as high as the discharge of the SAM WWTP only. This suggests the existence of additional sources. During summer 2017, contamination levels were analyzed in the Seine and Marne rivers during well-established dry weather periods [50]. Samples were taken every kilometer, near each bank, and in the center of the flow. This revealed a substantial increase of *E. coli* concentrations upstream from the Ivry station (2S), but the source was situated on the left bank, whereas the SAM WWTP is on the right bank. A combined sewer overflow, which also drains non-contaminated groundwater during dry weather, was identified as the wastewater discharging outlet. It was concluded that a malfunctioning of the sewer network in the municipalities connected to the network was probably the cause of the presence of wastewater in the CSO during a dry weather period, but the problem could not be precisely identified.

Downstream from Paris, a progressive increase of concentrations is observed, from Suresnes (6S) to Conflans (11S). More than 50% of the concentration jump at Conflans-Seine can be attributed to the SAV WWTP, while the expected impact of the SEC WWTP is negligible. Like upstream from Paris, it is likely that much of the observed increase downstream from Paris is not only due to WWTP discharges but to malfunctions of the sewer network during dry weather, with a possible contribution of well-known major CSOs in this area and a possible delayed transport of FIBs. Dilution by the Oise River would explain most of the decrease at Poissy (13S), but the downstream decrease at Triel (14S) is mainly due to natural decay processes affecting *E. coli* in fresh water. The input of the SEG WWTP is a very minor contribution to the concentration at the Triel station.

The situation is different inside Paris and just downstream of it: *E. coli* concentrations tend to decrease from Tolbiac (4S) to Suresnes (6S). It is likely that better control of the sewer network has been achieved in this area, where the network is fully accessible, which may have facilitated the control of dry weather discharges.

The longitudinal profile in the Marne River (Fig. 10) is quite regular with a decrease of concentrations starting at station 5 M, which can be explained by natural

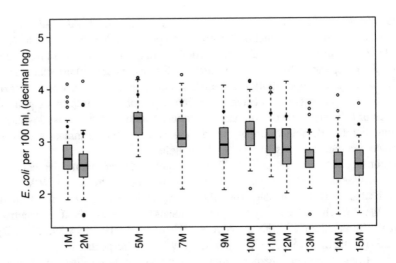

Fig. 10 Boxplot of *E. coli* concentrations in the Marne River during summers 2015 and 2017. The number of data in each box is within the 39–69 range. The distance between the most upstream and downstream stations is 28 km. (see also the legend of Fig. 9)

decay processes, since the observed slope is fully coherent with the decay rates obtained from laboratory experiments. The small increase at station 10 M down-stream of the MAV WWTP can be totally explained by the estimated inputs by the outlet of the treatment plant. However, between stations 2 and 5 M, a very strong increase could be mostly explained by uncontrolled discharges during dry weather, as demonstrated by the local authority (Syndicat Marne Vive), which estimated the dry weather flux by direct measurements in a number of outlets of the sewer network in this area. As in the Seine River upstream of Ivry, erroneous connections of black water to small urban streams or storm drainage networks seem to be the major cause of the water degradation. An estimate of $2-4 \times 10^{10}$ *E. coli* per inhabitant per day can be obtained from data on the average concentration of *E. coli* in raw wastewater in the Paris suburban area and the volumes treated in the WWTPs. The concentration increase observed between stations 2 and 5 M could be explained by the untreated wastewater flux of a few more than 2,000 inhabitants, which is negligible compared to the several million inhabitants living in the urbanized part of the Marne watershed but clearly deserves action to improve the wastewater collection system in the area concerned.

The data set analyzed above shows that the water quality inside the Paris agglomeration does not yet respect the EU Bathing Water Directive. Not to introduce too much complexity, the analysis is mainly based on so-called median fluxes computed from the median concentration and from representative low water flow in rivers. It shows that inputs of *E. coli* by WWTP discharges are important components of the overall contamination but that they would generally not explain the observed median flux increase. Uncontrolled dry weather fluxes from sewers are a major additional source. Additionally, as shown by the analysis of long-term time series at Orly and Joinville, the strong influence of rain events on the highest concentrations is clearly demonstrated.

8 Presence of Pathogens in the Seine and Marne Rivers in the Paris Area

Many waterborne diseases are not due to bacteria but to pathogenic viruses or parasites [51], as illustrated in Table 2 for the most recent period (data collected by the Centers for Disease Control and Prevention in the USA [52]). Such data sets suffer from substantial potential bias. Although foodborne outbreaks are easily identified, infection cases associated with waterborne outbreaks are brought to light with great difficulty due to different parameters: incubation time before symptom appearance, distance between contamination site and patient residence site, and also the lack of specialized medical consultation and systematic reporting when symptoms appear. Although waterborne outbreaks are probably underestimated, these statistics show the variety of potential etiological agents and require extensive information regarding additional classes of pathogens.

8.1 Viruses

Human enteric viruses are the leading cause of non-bacterial acute gastroenteritis worldwide [53]. Transmission occurs through the oral-fecal wayroute, either by contact contamination or by consumption of contaminated food or drinking water. Although most of the time the ingestion events lead to asymptomatic infections, the enteric viruses can also cause various symptoms such as intestinal and respiratory disorders, hepatitis, and conjunctivitis. These viruses can be associated with higher morbidity and mortality for at-risk populations such as young children, the elderly, and immunocompromised patients [54]. Human enteric viruses are characterized by their ability to infect their host through the gastrointestinal pathway. Some of them can use this gateway in the body for migrating to other organs (e.g., liver, nervous system, heart) and can be responsible for specific disorders. These viruses are then

Table 2 Percentage of waterborne diseases due to recreational activities in non-treated waterbodies reported to the CDC for the 1994–2014 period (4,640 identified cases, 1,322 unidentified)

Etiologic factors	Enteric bacteria	Enteric parasites	Enteric viruses	Other
Number of patients as a percentage of identified cases	28.4%	10.5%	38.4%	22.7%

Enteric bacteria were mainly *Shigella* (73%), enteric parasites were mainly *Cryptosporidium* (64%), and enteric viruses were mainly *Norovirus* (84%). "Other" includes waterborne parasites (*Schistosomes*, 49%), and the non-enteric bacteria *Leptospira* (42%)

shed in feces in high quantity (up to 10^{11} particles per gram of stool) for a few days up to several weeks [55]. Consequently, domestic wastewaters are often highly contaminated with human enteric viruses [56, 57]. These viruses are relatively resistant to various water treatments and environmental conditions and persist in the environmental waters for long periods of time. Moreover, the viruses can infect a host at a low infectious dose, from 10 to 1,000 viral particles, independently of individual immunity and genetic susceptibility to infection [58].

Estimating the risk associated with the presence of enteric viruses in the water is not a simple task. While the microbiological quality of drinking water and recreational waters is mainly based on the presence of FIBs, as described in Sect. 3, these bacteria are generally less able to persist in water and resist water treatments at a lower level than human enteric viruses. Consequently, these bacterial indicators could not be directly used to assess the viral risk [59, 60].

The estimation of risk for human health, linked to viruses in the environment, is based on the detection of infectious viruses that infect permissive cells. However, most viruses are not easily cultivable. The detection based on virus genome amplification bypasses the cultivation limitation, but genome detection does not systematically imply health risk since partially intact genomes from noninfectious particles could be amplified. The evaluation of virus capsid integrity offers complementary information. A study conducted with samples from the Seine River in the Paris area showed that few differences were observed between the detection of total genomes and encapsidated genomes [61]. These results suggest that most viruses detected in natural waters in the Paris agglomeration have intact capsids and are therefore potentially infectious. This study also suggests that virus detection by molecular methods such as real-time PCR provides a true image of the viral danger present in surface waters.

Human enteric viruses were extensively monitored using an integrity assay in the Seine River (Paris area) for several years. The results presented in Fig. 11 showed significant contamination with the viruses the most frequently implicated in gastroenteritis in humans. In the Seine River, the median value was 10,470 genome units (GU)/L for total adenovirus and 1,026 GU/L for type F adenovirus. The type F adenovirus (serotypes 40 and 41) accounted for about 10% of total adenoviruses. The main virus responsible for gastroenteritis in humans (norovirus) is titrated at 216 GU/L and 146 UG/L for human norovirus GI and GII, respectively. The median concentration for type A rotavirus was 4,082 GU/L. Somatic coliphages were detected by culture with a median value of 1,350 PFU/L and F-specific RNA coliphages titrated at 460 GU/L.

Similar results were obtained for the Marne River (data not shown). The median values were 39,800 GU/L and 2,200 GU/L for total and type F adenoviruses, respectively; 1,550 UG/L and 976 GU/L for human noroviruses GI and GII, respectively; and 7,026 GU/L for type A rotavirus. Viral indicators were titrated at 1700 GU/L and 15,262 GU/L for somatic and F-specific RNA coliphages, respectively. In both rivers, human enteroviruses (including poliovirus) were infrequently detected.

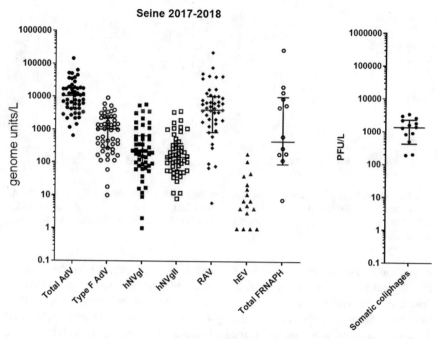

Fig. 11 Concentration of infectious human enteric viruses in the Seine River in 2017–2018. Adenovirus (AdV), type F adenovirus, human norovirus groups I and II (hNVgI and hNVgII), rotavirus (RAV), human enteroviruses (hEV), F-specific RNA (FRNAPH), and somatic coliphages. Horizontal bars show the median and interquartiles

Wastewater treatment is a crucial issue to reduce the risk of contamination, although, by design, WWTPs are generally not built to reduce viral contamination [62]. The results of intensive monitoring of viral contamination in the Seine River over 1 year [25] showed that treated wastewater from WWTPs was the main sources of pathogenic viruses. The contribution of CSOs, which could not be sampled, remains unknown. The equilibrated balance of virus flows in both sections studied suggests a very limited loss of viruses during their transit in the river (Fig. 12).

8.2 Parasites

Other well-known pathogens in water are the parasites, especially *Cryptosporidium* and *Giardia*. These two taxa are intestinal protozoa causing mild to severe infections in humans [63]. *Cryptosporidium* oocysts and *Giardia* cysts are often excreted in large quantities with the feces of infected animals and humans. Runoff from contaminated soils and WWTP discharges are responsible for surface water contamination [64]. Oocysts and cysts remain infective for months in environmental waters and are highly resistant to oxidative disinfectants [65]. Although the minimum quantity

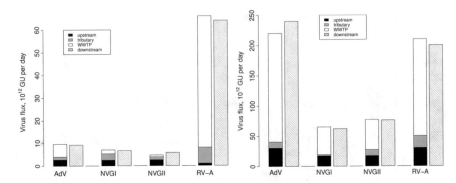

Fig. 12 Average infectious virus fluxes (GU per day) over 1 year (May 2013 to May 2014); 25 samples were collected and analyzed at each station. Left, sector A (1S–6S), right, sector B (8S–13S). Data from [25]. AdV, adenovirus; NVGI and NVGII, norovirus, group I and II; RV-A, rotavirus A

of ingested cysts necessary to infect humans is often debated in the medical literature, it is well known that humans can be infected by a low level of cysts in drinking water and during recreational activities such as swimming. Since 2001, *Cryptosporidium* oocysts and *Giardia* cysts are monitored in water supplies and drinking waters using a normalized method with a concentration of 20 L of surface water followed by IMS concentration and microscope counting after immune staining of the concentrate (NF T90-455, AFNOR, 2014).

In the Paris area, a large monitoring project helped to describe the contamination of surface water by these parasites. Results from 2005 to 2008 [26, 64] showed that if the level of parasites is relatively low (about 1–4 *Cryptosporidium* oocysts/10 L and 30–50 *Giardia* cysts/10 L), specific events could induce significant contamination (Fig. 13). Moreover, the contamination of surface water is not totally related to WWTPs and probably more to agricultural practices upstream in the Seine catchment. Monitoring of *Cryptosporidium* and *Giardia* in the Seine and Marne rivers between 2010 and 2018 shows that the concentration of such microorganisms has been relatively stable since 2005 and follows the same pattern (see Fig. 13) with substantial variation of parasites, unrelated to the increase in fecal indicators. The small increase of the mean values between the two periods is not statistically significant.

9   Where to Now?

As shown in Sect. 7, most of the stations in the Marne and Seine rivers do not presently attain the *E. coli* 90th percentile required for bathing. To illustrate the necessary effort, a simplified evaluation of two complementary strategies will be presented below. The first strategy is the reduction of the base dry weather input of *E. coli* into the river, from WWTPs and diffuse leakages off the sewer system.

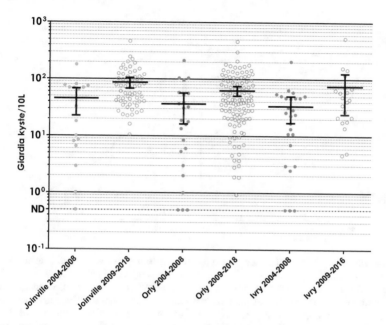

Fig. 13 *Giardia* concentration (in cysts/10 L) for the Marne and Seine rivers at Orly and Ivry for two sampling campaigns (2005–2008 and 2009–2018, except for Ivry 2009–2016). Mean and 95% confidence intervals are shown in black

The second strategy is to improve the management of urban runoff by increasing infiltration or storage to reduce the impact of storm inputs into the river.

Simulation results are presented in Fig. 14 for station 7 M in the Marne River. This station was chosen because of its symbolic value, because Joinville was one of major bathing places in the Marne River at the beginning of the century, and because its contamination level is representative of that of the whole area: 60% of all stations studied in the Seine and Marne rivers have *E. coli* 90th percentiles lower than the 7 M station.

We hypothesize that the concentrations lower than the median value are mostly representative of dry weather conditions and that the upper concentrations are additionally influenced by wet weather inputs. The reduction of the dry weather flux is simulated by a reduction factor globally simulating the efficiency of disinfection in WWTPs and of the necessary elimination of black water connections to the urban runoff network. The improved urban runoff management is simulated by additional storage or infiltration capacity expressed as a fraction of the maximum wet weather additional impact.

Two situations were tested, with and without the existence of a reliable bathing management plan making it possible to remove the 15% most degrading events to compute the statistics, following the European Directive. These results show that for a rather contaminated station, the required contamination level for bathing can be reached for a reduction of 80% of the dry weather inputs and increased storage or

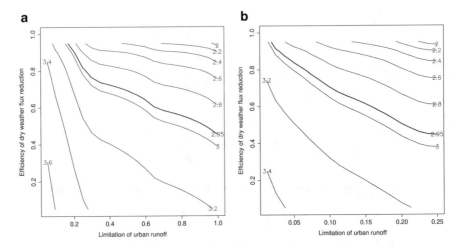

Fig. 14 Illustration of the efficiency of two policy lines to improve water quality at station 7 M. Isocurves are the 90th percentiles of *E. coli* concentrations on a decimal log scale. The sufficient criterion (900 *E. coli* per 100 mL) means a log value of 2.95 (bold line). Left, simulation (**a**), all data; right, simulation (**b**) 15% of the most degrading events removed. The *x*-axis represents a reduction of the runoff volume expressed as a fraction of the most degrading recorded event, 1 means the elimination of all rain events. The *y*-axis represents the reduction of dry weather inputs (all included) a fraction of the present situation, 1 means 100% reduction. Note the modified scale on the *x*-axis on the figure on the right

infiltration capacity equivalent to 7–8% of the volume of the most contaminating event to reduce runoff (Fig. 14, simulation (b)). Such a dry weather input reduction seems feasible regarding the potential improvement of treatments in WWTPs (see below) but requires careful improvements of bad connections in sewer networks for stations, such as station 7 M, where this additional source of dry weather contamination is much significant. With a policy regarding runoff focusing on the disconnection of black water inputs to urban runoff, efficacy as high as 60% would be required to reach the same objective (scenario not shown on Fig. 14).

To reduce the base dry weather input to the rivers, the addition of disinfection to the WWTP is an option being examined. Two techniques have been tested at full scale and have succeeded in very significantly inactivating discharged bacteria. It is now mostly a matter of political will requiring a decision on whether the construction and maintenance costs are worth the expected improvement in the river's water quality.

A full-scale UV treatment was installed in one of the WWTPs with irradiation rates about 50 mJ/cm². This reduced *E. coli* and IE by 1.7 and 1.3 log units, respectively [33]. However, the removal efficiency of the process varied significantly over time in relation to the suspended solids (from 4 to 180 mg/L in this study) contained in the treated wastewater; it was greatly improved (2.6 and 2.1 log units for *E. coli* and IE, respectively) for situations below 10 mg/L.

The addition of chemical disinfectants is another option to consider. More recently, full-size experiments were conducted in one of the Paris WWTPs by adding performic acid. Performic acid degrades very rapidly, producing free radicals that inactivate microorganisms. Reduction rates up to 2.5 and 2.7 log units have been obtained for *E. coli* and IE, respectively, with a concentration at the outlet much lower than 100 bacteria per 100 mL for each of the FIBs (unpublished data). Endocrine disruption tests on amphibian and fish larvae (up to 4 days exposure) did not show any significant additional impact of the chemically oxidized water. These encouraging results, for chemical and UV disinfection, must be confirmed by an analysis of their potential for degradation of viruses and other pathogens given that many of them require a more stringent treatment for the same efficacy.

A second source of fecal contamination to rivers is poor connection to sewers. This most important issue will both affect base dry weather contamination fluxes and contribute to the contamination brought by rain events. Solving this problem requires a very detailed, time-consuming, and complex project to check all connections, starting from the most impacting catchments and organizing the proper reconnection of identified the buildings identified, which is under the responsibility of private individuals. It is certainly a highly delicate challenge for the recovery of a sufficient microbial water quality in the Marne and Seine rivers. Meanwhile, a complementary effort is needed to reduce urban runoff by infiltration or storage of rain water.

10 Conclusion

The water quality of the rivers that cross urban spaces is a concern for all [66]. Affected in their quality of life and their need for "nature," Parisians and the Ile-de-France region's inhabitants wish to reclaim the river to satisfy their needs and participate in new leisure and sports activities. The idea of a "swimmable city" was born in 2015 [67] with considerable interest on the part of local actors along the Marne River. In Paris and in its region, it became a strong political issue for the mayor of Paris and in the perspective of the 2024 Olympic Games. The expected activities also involve mobilization of all users of the Seine and Marne rivers (e.g., boatmen, industrialists, sports associations) to better preserve the quality of water and avoid conflicts of use.

This approach is part of a geopolitical context (competition between major cities regarding the attractiveness of the living environment and the development of the sustainable city) and local politics (institutional power related to the reorganization of the area, economic development, etc.). Meeting the demand of the population in their desire to win back the Seine and Marne rivers without compromising their health is one of the mobilizing issues of all the actors involved, from scientists to managers, as well as users and decision-makers, all responsible for the quality of the Seine and Marne rivers and their usage.

Acknowledgments The authors thank Paris STEA and the Syndicat Marne Vive who provided a substantial amount of the FIB data used in this study. This research was conducted within the framework of the PIREN-Seine program, a component of the LTSER "Zone Atelier Seine."

References

1. Duhau I (2011) Les baignades en rivière en Ile-de-France. 22èmes journées scientifiques de l'environnement (Créteil, 2011)
2. Lombardo D (2011) Se baigner ensemble. Les corps au quotidien et les bains publics parisiens avant 1850 selon Daumier. Histoire Urbaine 31:47–68
3. Briffault E (1844) Paris dans l'eau, illustrated by Bertall. Initial publication by J. Hetzel. Re-edited, Bibliothèque Nationale de France, Collection XIX
4. Pardailhé-Galabrun A (1983) Les déplacements des Parisiens dans la ville aux XVIIème et XVIIIème siècles. Un essai de problématique. Histoire, Économie et Société 2:205–253
5. Backouche I (2010) Mesurer le changement urbain à la périphérie parisienne. Histoire & mesure XXV-1 I On line. http://histoiremesure.revues.org/3929
6. Csergo J (2004) Parties de campagne. Loisirs périurbains et représentations de la banlieue parisienne, fin XVIIIe-XIXe siècles. Sociétés Représentations 17:15–50
7. Colin L (1885) Paris, sa topographie, son hygiène, ses maladies. Masson, Paris
8. Teysseire R, Corbière R, Arnac A (2013) Histoire de la baignade en Seine: les modalités de sa réglementation, Mémoire. AgroParisTech, Paris
9. Harwood VJ, Staley C, Badgley BD et al (2014) Microbial source tracking markers for detection of fecal contamination in environmental waters: relationships between pathogens and human health outcomes. FEMS Microbiol Rev 38:1–40
10. Horrocks WH (1901) An introduction to the bacteriological examination of water. Kessinger Publishing, London, p 346
11. Prescott SC, Winslow CEA (1904) Elements of bacteriology with special reference to sanitary water analysis. Kessinger Publishing, London, p 272
12. Schardinger F (1982) Uber das Vorkommen Gahrung erregender Spaltpilze im Trinkwasser und ihre Bedeutung fur die hygienische Beurtheilung desselben. Wien Klin Wochenschr 5:403–405
13. Escherich T (1885) Die Darmbakterien des Neugeborenen und Säuglings. Fortschr Med 3 (47–54):515–522
14. Edberg SC, Rice EW, Karlin RJ et al (2000) *Escherichia coli*: the best biological drinking water indicator for public health protection. Symp Ser Soc Appl Microbiol 29:106S–116S
15. Slanetz LW, Bent DF, Bartley CH (1955) Use of the membrane filter technique to enumerate enterococci. Public Health Rep 70:67
16. Hazen TC, Toranzos GA (1990) Tropical source water. In: McFeters GA (ed) Drinking water microbiology. Springer, New York, pp 32–52
17. Jamieson RC, Gordon RJ, Sharples KE et al (2002) Movement and persistence of fecal bacteria in agricultural soils and subsurface drainage water: a review. Can Biosyst Eng 44:1.1–1.9
18. Whitman RL, Nevers MB (2003) Foreshore sand as a source of *Escherichia coli* in nearshore water of a Lake Michigan Beach. Appl Environ Microbiol 69:5555–5562
19. Griffin DW, Lipp EK, McLaughlin MR et al (2001) Marine recreation and public health microbiology: quest for the ideal indicator. Bioscience 51:817–825
20. Hurst CJ, Crawford RL, Garland JL et al (2007) Manual of environmental microbiology. ASM Press, Washington, DC, p 1293
21. Dechesne M, Soyeux E (2007) Assessment of source water pathogen contamination. J Water Health 5:39–50

22. Wilkes G, Edge T, Gannon V et al (2009) Seasonal relationships among indicator bacteria, pathogenic bacteria, Cryptosporidium oocysts, Giardia cysts, and hydrological indices for surface waters within an agricultural landscape. Water Res 43:2209–2223
23. Lund V (1996) Evaluation of E. coli as an indicator for the presence of Campylobacter jejuni and Yersinia enterocolitica in chlorinated and untreated oligotrophic lake water. Water Res 30:1528–1534
24. Bonadonna L, Briancesco R, Ottaviani M et al (2002) Occurrence of Cryptosporidium oocysts in sewage effluents and correlation with microbial, chemical and physical water variables. Environ Monit Assess 75:241–252
25. Prévost B, Lucas FS, Goncalves A et al (2015) Large scale survey of enteric viruses in river and waste water underlines the health status of the local population. Environ Int 79:42–50
26. Mons C, Dumètre A, Gosselin S et al (2009) Monitoring of Cryptosporidium and Giardia river contamination in Paris area. Water Res 43(1):211–217
27. Unno T, Staley C, Brown CM et al (2018) Fecal pollution: new trends and challenges in microbial source tracking using next-generation sequencing. Environ Microbiol 20 (9):3132–3140
28. Barrios ME, Blanco Fernández MD, Cammarata RV et al (2018) Viral tools for detection of fecal contamination and microbial source tracking in wastewater from food industries and domestic sewage. J Virol Methods 262:79–88
29. WHO (2018) WHO recommendations on scientific, analytical and epidemiological developments relevant to the parameters for bathing water quality in the Bathing Water Directive (2006/7/EC). World Health Organization, Geneva, p 96
30. Payment P, Locas A (2011) Pathogens in water: value and limits of correlation with microbial indicators. Ground Water 49:4–11
31. Garcia-Armisen T, Prats J, Servais P (2007) Comparison of culturable fecal coliforms and Escherichia coli enumeration in freshwaters. Can J Microbiol 53(6):798–801
32. Servais P, Billen G, Goncalvez A et al (2007) Modelling microbiological water quality in the Seine river drainage network: past, present and future situations. Hydrol Earth Syst Sci 11:1581–1592
33. Rocher V, Azimi S, Bernier J et al (2017) Qualité bactériologique des eaux en agglomération parisienne. Des eaux usées aux eaux de Seine. Edition Johanet, Paris, p 94
34. George I, Anzil A, Servais P (2004) Quantification of fecal coliforms inputs to aquatic systems through soil leaching. Water Res 38:611–618
35. Servais P, Garcia-Armisen T, George I et al (2007) Fecal bacteria in the rivers of the Seine drainage network: source, fate and modeling. Sci Total Environ 375:152–167
36. Garcia-Armisen T, Servais P (2007) Respective contributions of point and non-point sources of E. coli and enterococci in a large urbanized watershed (the Seine River, France). J Environ Manag 82(4):512–518
37. Lucas F, Thérial C, Goncalves A et al (2014) Variation of raw wastewater microbiological quality in dry and wet weather conditions. Environ Sci Pollut R 21:5318–5328
38. George I, Crop P, Servais P (2002) Fecal coliforms removal in wastewater treatment plants studied by plate counts and enzymatic methods. Water Res 36:2607–2617
39. Passerat J, Ouattara K, Mouchel JM et al (2011) Impact of an intense combined sewer overflow event on the microbiological water quality of the Seine River. Water Res 45:893–903
40. Rocher V, Azimi S, Paffoni C et al (2018) Evolution de la qualité de la Seine en lien avec les progrès de l'assainissement de 1970 à 2015. Edition Johanet, Paris, p 76
41. Prüss A (1998) Review of epidemiological studies on health effects from exposure to recreational water. Int J Epidemiol 27:1–9
42. Pena L, Zmirou D, Le Tertre A et al (2001) Critères microbiologiques de qualité des eaux de baignade. Institut de Veille Sanitaire, Saint Maurice, p 39

43. Wiedenmann A, Krüger P, Dietz K et al (2006) A randomized controlled trial assessing infectious disease risks from bathing in fresh recreational waters in relation to the concentration of *Escherichia coli*, intestinal enterococci, *Clostridium perfringens*, and somatic coliphages. Environ Health Perspect 114:228–236
44. Marion JW, Lee J, Lemeshow S et al (2010) Association of gastrointestinal illness and recreational water exposure at an inland U.S. beach. Water Res 44:4796–4804. https://doi.org/10.1016/j.watres.2010.07.065
45. Dufour AP (1984) Health effects criteria for fresh recreational waters. US Environmental Protection Agency, Cincinnati, p 42
46. Cabelli VJ, Dufour AP, McCabe LJ et al (1982) Swimming-associated gastroenteritis and water quality. Am J Epidemiol 115:606–616
47. van Asperen IA, Medema G, Borgdorff MW et al (1998) Risk of gastroenteritis among triathletes in relation to faecal pollution of fresh waters. Int J Epidemiol 27:309–315
48. Duboudin C (2004) Qualité microbiologique des eaux de baignade. Analyse statistique des niveaux de risque et des seuils proposés par le projet de révision de la directive 76/160/CEE. Agence Française de Sécurité Sanitaire et Environnementale, Paris, p 114
49. Vermeulen LC, Hofstra N (2014) Influence of climate variables on the concentration of *Escherichia coli* in the Rhine, Meuse, and Drentse Aa during 1985-2010. Reg Environ Chang 14:307–319. https://doi.org/10.1007/s10113-013-0492-9
50. Mouchel JM, Colina-Moreno I, Kasmi N (2018) Évaluation des teneurs en bactéries indicatrices fécales en Seine dans l'agglomération parisienne par temps sec. Report for the City of Paris and the Water Agency Seine-Normandy, p 79
51. Craun GF, Calderon RL, Craun MF (2003) Waterborne outbreaks in the United States, 1971–2000. In: Pontius FW (ed) Drinking water regulation and health. Wiley, Hoboken, pp 45–69
52. https://www.cdc.gov/healthywater/surveillance/rec-water-surveillance-reports.html and references herein
53. Enserink R, Lugner A, Suijkerbuijk A et al (2014) Gastrointestinal and respiratory illness in children that do and do not attend child day care centers: a cost-of-illness study. PLoS One 9(8): e104940
54. Gerba CP, Rose JB, Haas CN (1996) Sensitive populations: who is at the greatest risk? Int J Food Microbiol 30(1–2):113–123
55. Blacklow N, Greenberg H (1991) Medical progress – Viral gastroenteritis. N Engl J Med 325:252–264
56. Cantalupo PG, Calgua B, Zhao G et al (2011) Raw sewage harbors diverse viral populations. mBio 2(5):e00180–e00111
57. Lodder WJ, Rutjes SA, Takumi K et al (2013) Aichi virus in sewage and surface water, the Netherlands. Emerg Infect Dis 19(8):1222–1230
58. Kirby AE, Teunis PF, Moe CL (2011) Two human challenge studies confirm high infectivity of Norwalk virus. J Infect Dis 211(1):166–167
59. Contreras-Coll N, Lucena F, Mooijman K et al (2002) Occurrence and levels of indicator bacteriophages in bathing waters throughout Europe. Water Res 36(20):4963–4974
60. Tree JA, Adams MR, Lees DN (2003) Chlorination of indicator bacteria and viruses in primary sewage effluent. Appl Environ Microbiol 69(4):2038–2043
61. Prevost B, Goulet M, Lucas FS et al (2016) Viral persistence in surface and drinking water: suitability of PCR pre-treatment with intercalating dyes. Water Res 91:68–76
62. Kitajima M, Haramoto E, Phanuwan C et al (2012) Molecular detection and genotyping of human noroviruses in influent and effluent water at a wastewater treatment plant in Japan. J Appl Microbiol 112(3):605–613
63. Fayer R (2004) Cryptosporidium: a water-borne zoonotic parasite. Vet Parasitol 126 (1–2):37–56

64. Moulin L, Richard F, Stefania S et al (2010) Contribution of treated wastewater to the microbiological quality of Seine River in Paris. Water Res 44:5222–5231
65. Betancourt WQ, Rose JB (2004) Drinking water treatment processes for removal of Cryptosporidium and Giardia. Vet Parasitol 126:219–234
66. Euzen A, Haghe JP (2012) What kind of water is good enough to drink? The evolution of perceptions about drinking water in Paris from the seventeenth to twentieth century. Water History 4(3):231–244
67. Haghe JP, Euzen A (2018) Une nouvelle catégorisation politique des eaux: la baignade en eau libre. L'exemple de Paris. IS River, 4 juin 2018, Lyon

Contaminants of Emerging Concern in the Seine River Basin: Overview of Recent Research

Pierre Labadie, Soline Alligant, Thierry Berthe, Hélène Budzinski, Aurélie Bigot-Clivot, France Collard, Rachid Dris, Johnny Gasperi, Elodie Guigon, Fabienne Petit, Vincent Rocher, Bruno Tassin, Romain Tramoy, and Robin Treilles

Contents

1   Introduction 356
2   Macro- and Microplastics 357
    2.1   Context 357
    2.2   Objectives 358
    2.3   Macroplastics in the Seine River 358
    2.4   Microplastic Sources and Fluxes in Greater Paris and the Seine River 360
3   Poly- and Perfluoroalkyl Substances (PFASs) 362
    3.1   Context 362

The copyright year of the original version of this chapter was corrected from 2019 to 2020. A correction to this chapter can be found at https://doi.org/10.1007/698_2020_667

P. Labadie (✉) and H. Budzinski
UMR 5805 EPOC, CNRS, Université de Bordeaux, Talence, France
e-mail: pierre.labadie@u-bordeaux.fr

S. Alligant, F. Collard, R. Dris, J. Gasperi, B. Tassin, R. Tramoy, and R. Treilles
LEESU (UMR MA 102, Université Paris-Est, AgroParisTech), Université Paris-Est Créteil, Créteil Cedex, France

T. Berthe and F. Petit
Université de Rouen, UMR CNRS 6143 M2C, FED SCALE 4116, UFR des Sciences, Mont Saint-Aignan Cedex, France

Sorbonne Université, CNRS, EPHE, PSL University, UMR 7619 METIS, Paris, France

A. Bigot-Clivot
Université de Reims Champagne Ardenne, Unité Stress Environnementaux et BIOsurveillance des milieux aquatiques, UMR-I 02 (SEBIO), Reims, France

E. Guigon
Sorbonne Université, CNRS, EPHE, PSL University, UMR 7619 METIS, Paris, France

V. Rocher
SIAAP, Direction Innovation et Environnement, Colombes, France

Nicolas Flipo, Pierre Labadie, and Laurence Lestel (eds.), *The Seine River Basin*, Hdb Env Chem (2021) 90: 355–380, https://doi.org/10.1007/698_2019_381,
© The Author(s) 2020, corrected publication 2020, Published online: 3 June 2020

 3.2 Dynamics of PFASs in the Seine River 363
 3.3 Transfer of PFASs to Biota in the Seine River Basin 365
4 Pathogenic Protozoa 367
 4.1 Context 367
 4.2 Occurrence of Pathogenic Parasites in the Seine River 368
5 Antibiotics and Bacterial Antibiotic Resistance 370
 5.1 Context 370
 5.2 Sources of Antibiotics in the Seine Watershed 370
 5.3 Antibiotic Contamination in the Seine River 371
 5.4 The Resistome and Antimicrobial Resistance 372
6 Conclusions and Perspectives 374
References 375

Abstract For over 30 years, the sources and the transfer dynamics of micropollutants have been investigated in the PIREN-Seine programme. Recent works included a wide range of chemicals and biological contaminants of emerging concern (i.e. contaminants whose occurrence, fate and impact are scarcely documented). This chapter presents a brief overview of research recently conducted on contaminants as diverse as macro- and microplastics, poly- and perfluoroalkyl substances (PFASs), pathogenic protozoa, antibiotics and the associated antibiotic resistance. The multiscalar study of plastics and PFASs at a large spatial scale is rare; the results produced in recent years on the Seine River catchment have provided an original contribution to the investigation of the dynamics of these contaminants in urban environments. The results also highlighted that pathogenic protozoa are ubiquitous in the Seine River basin and that the contamination of bivalves such as *Dreissena polymorpha* could reflect the ambient biological contamination of watercourses. The widespread occurrence of antibiotics in the Seine River was demonstrated, and it was shown that the resistome of biofilms in highly urbanised rivers constitutes a microenvironment where genetic support for antibiotic resistance (clinical integrons) and resistance genes for trace metals are concentrated.

Keywords Antibiotic resistance, Antibiotics, Bioaccumulation, Contaminants of emerging concern, Macro- and microplastics, Pathogenic protozoa, PIREN-Seine, Poly- and perfluoroalkyl substances, Sediment, Water, Zone Atelier Seine

1 Introduction

The Seine River basin is under severe anthropogenic pressure for a number of reasons, including the emission of micropollutants by industrial, agricultural and urban activities in combination with relatively low river water flow per capita (cf. [1]). Thus, the sources and the transfer dynamics of both organic micropollutants and trace metals have been investigated over several decades within the PIREN-Seine programme. As regards organics, previous studies addressed this issue at different temporal and spatial scales, focusing, for instance, on polycyclic aromatic hydrocarbons (PAHs) (see [2]) or legacy organohalogens [3, 4]. However, due to evolving chemical regulations, improved analytical capabilities and progress in ecotoxicological assessment, recent studies have included a wide range of chemicals and biological contaminants of emerging concern. A small selection of studies conducted since the 2010s is presented hereafter to illustrate this work, with a brief overview of recent studies on macro- and microplastics, fluoroalkyl substances, pathogenic protozoa as well as antibiotics and antibiotic resistance.

2 Macro- and Microplastics

2.1 Context

Water pollution by plastics and microplastics is often described as an emerging concern. However, in the Seine River basin, boatmen have been complaining since the 1990s about the presence of plastic bags blocking the cooling circuits of barges. Prior to 1994, SIAAP (Greater Paris Sanitation Authority) had installed floating booms to trap macro-waste, including plastics floating on the water surface. However, this highly visible pollution did not attract the attention of public authorities or aquatic environment managers (e.g. it was not mentioned in the 2000 European Water Framework Directive) or even environmental protection organisations. However, within the EU Marine Strategy Framework Directive released in 2008, litter was listed as one of the descriptors of good ecological status. Following observations in the ocean environment (e.g. "great garbage patch" in 1997), this issue, both abroad and in France, has become a social issue and a research subject for scientists.

Plastics observed in freshwater reach several orders of magnitude in size and a wide spectrum of shapes. Microplastics were brought to light in 2004 [5] and were defined as plastic particles whose longest dimension is less than 5 mm [6]. As a consequence, macroplastics have their longest dimension greater than 5 mm. Macroplastics are either manufactured plastic items (primary plastics) or fragments (secondary plastics), and microplastics are mostly secondary plastics categorised depending on their dimensions: fibres (length $\gg$ diameter) and fragments (characteristic length $\gg$ thickness) composed of different irregular shapes and spheres [6]. More generally, anthropogenic particles (APs) cover a very broad category of particles produced directly or indirectly by human activities. In the case of the Seine River, APs are mostly small pieces, fragments or fibres and, regardless of their size, originating from plastics, dyed particles or textile fibres. Special attention is being paid to plastic fibres whose production increases by approximately 4% per year (60,000 tons in 2016, 20% of the world plastic production [7]). Although fibres are often not included in the key figures concerning plastic materials, they are used in several industrial sectors including the textile industry. Plastic synthetic fibres represent the main fraction of the world fibre production, which also includes other artificial fibres made from natural raw material (e.g. rayon made from cellulose) and natural fibres (e.g. cotton and wool). Fibres are present in the environment and are produced by the wear and tear of larger items.

2.2 Objectives

Since the beginning of the 2010s, research on plastic litter in continental environments has developed, even if it remains limited compared to work on the marine environment. This research has revealed the widespread occurrence of plastic debris in environmental compartments including the atmosphere [8] and food: salt [9], possibly honey [10], mussels [11, 12] and also bottled water [13, 14]. These findings have raised concerns about the human health effects of plastics [15] but also receiving systems [16] and aquatic ecosystems [17, 18].

In 2013, the Leesu research unit launched a research project on microplastics in urban hydrosystems, in partnership with the PIREN-Seine programme and the Observatoire des Polluants Urbains en Ile-de-France (OPUR), which has expanded to include the issue of macro-pollution by plastics. This project had several objectives, including (1) the estimation of macroplastic mass fluxes from Paris to the mouth of the Seine and their dynamics and (2) the identification of the sources and fluxes of microplastics at the scale of the Parisian Metropolis, from the atmosphere to the Seine River and the assessment of its plastic contamination from the Paris conurbation to its mouth.

2.3 Macroplastics in the Seine River

In contrast to microplastics, macroplastic pollution is highly visible. There is a need to investigate the occurrence and dynamics of macroplastics, in order to implement efficient multiscale reduction policies in the Marine Strategy Framework Directive [19]. Various methods have been proposed recently [20, 21], based on waste production and different leakage rates into the environment; these approaches have the major interest to allow, relatively easily, for the estimation of estimating macroplastic flux to the ocean at the global scale.

The evaluation of these methods on data collected in watersheds is obviously the next step and is currently underway in the Seine River basin. A first rough attempt estimated the yearly macroplastic mass fluxes between 1,000 and 10,000 metric tons [22]. More recently, two conceptual modelling approaches, based on the data available in the Seine catchment, have been implemented [23]. The first one (CM1) was based on the extrapolation of the retention efficiency of a network of floating booms installed by the Greater Paris Sanitation Authority since 1994, which removes 27 tons of plastics yearly [24], accounting for 0.9–6.3% of the total mass of debris. A significant proportion of these macroplastics consists of food wrappers and containers and plastic cutlery, most likely associated with recreational activities. The second approach (CM2), based on the Jambeck and Lebreton methodologies [20, 21], has calculated the amount of mismanaged plastic waste (MMPW) at the basin scale, of which 15% and 40% are assumed to be transferred to the English Channel. MMPW is based on (1) the population (GIS data), (2) the economic level of

Table 1 Annual plastic mass flux to the English Channel from the Seine catchment areas

	Dris [22]	CM1	CM2	Lebreton et al. [21]
Annual flux estimate (metric ton year^{-1})	1,000–10,000	1,800–5,400	1,800–5,900	~20

the territory considered (World Bank data), (3) per capita plastic waste production (ADEME and ORDIF reports) and (4) an estimated 2% of littering [20].

The corresponding estimates (Table 1) were compared to the Lebreton theoretical estimate [21]. Despite their simplicity, these methods yielded similar results, especially the two conceptual models, based on completely different conceptual representations, i.e. between 1,800 and 5,900 metric tons a year.

At the same time, non-governmental organisations that harvest plastic litter on river banks in the estuary only collect up to 88–128 metric tons per year [23]. The discrepancy between these values raises new questions: does the fraction stranded on river banks really account for such a small fraction of the total flux transported by the river? Is the harvested fraction only a very small fraction of the stranded fraction? Are the two conceptual models and the Lebreton approach totally erroneous?

To provide preliminary answers to these fundamental questions, these annual fluxes were first converted to per capita fluxes. Over 1 year, plastic leakage into the Seine River reached 0.01–0.4 kg of plastic per capita, which is far less than estimations for the Nhieu Loc–Thi Nghe River, a tributary of the Saigon River in Vietnam, with a median load equal to 1.6 kg per capita. It is also lower than the average annual input of plastic for the coastal population worldwide, which reaches 0.7–2 kg per capita, with the highest values observed in South-East Asia. These values only confirm the fact that plastic leakage in Western countries is small compared to that reported for other parts of the world, Asia especially [20]; they do not, however, provide any relevant information to solve the questions mentioned above.

Other approaches are presently under investigation, which aim to understand more precisely the dynamics and the trajectories of macroplastics in the Seine River and especially in the Seine estuary. Such approaches might help in the implementation of new plastic harvesting strategies as well as new stringent regulations regarding plastic litter to drastically reduce ocean plastic pollution. Pathways and routes relevant to plastic debris remain partially unknown; in particular, the role of floods, runoff, combined sewer overflows (CSO) and bypass of wastewater treatment plants (WWTP) in plastic leakage must be investigated.

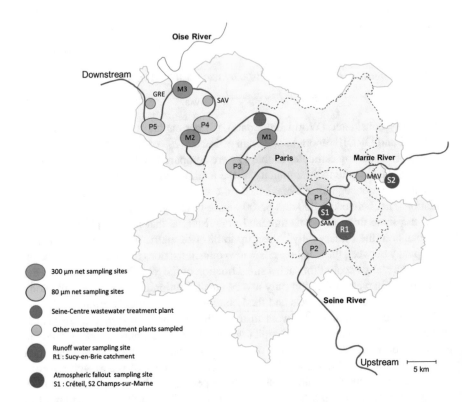

Fig. 1 Map of Paris megacity and location of the various sampling sites

2.4 Microplastic Sources and Fluxes in Greater Paris and the Seine River

Between 2014 and 2016, an investigation of sources and fluxes of microplastics was carried out on the Paris megacity. The following sources were investigated: atmospheric fallout, runoff water, grey water, wastewater outfall and CSO. Moreover, from April 2014 to December 2015, monthly monitoring was carried out at four sampling stations (P2–P5) on the Seine River from upstream to downstream of Paris plus one station on the Marne River (P1) [11] (Fig. 1).

This survey aimed at (1) estimating the various annual fluxes of microplastics in a large urban area and (2) linking the urban fluxes to the microplastic concentration in the Seine River and estimating the annual flux transported by this river.

Various sampling techniques (using nets or bulk water samples) were used. All analytical details are provided elsewhere [22]. The results are summarised in Fig. 2, which presents a first attempt at a mass balance of microplastics at the Greater Paris scale. Concentrations of fibres and fragments (in items L^{-1}) are provided, as well as plastic fluxes (metric tons year^{-1}).

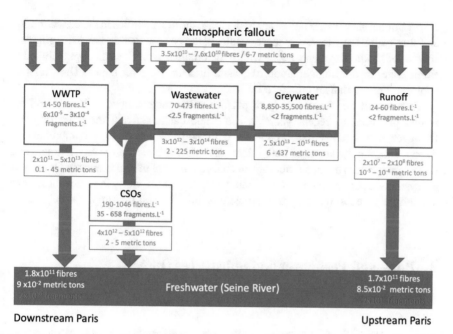

Fig. 2 Mass balance of microplastics at the Greater Paris scale (fibres and fragments)

Unexpected results were obtained. First of all, the number of plastic fibres (30% of the total number of fibres) exceeded the number of fragments by several orders of magnitude. The atmospheric fibre fallout represents a significant flux of plastic, around 10 metric tons per year at the Greater Paris scale. Fibres are also mainly present in grey and wastewater, and the input to WWTPs is estimated at several hundred tons per year. WWTPs contribute significantly to the reduction of plastic fluxes from the urban hydrosystem, and only 10% of the incoming flux is released into the receiving system (i.e. the Seine River). During wet weather periods, CSOs discharge huge fluxes of plastic fibre, which, based on a yearly average, are greater than the fluxes associated with treated wastewater discharge. In separate sewer system sectors, runoff exhibits concentration similar to those observed in WWTP effluents, but the corresponding flux is smaller by several orders of magnitude.

Microplastic fragment concentrations are small compared to fibre concentrations in the various compartments sampled, except in CSOs where the highest fragment concentration is observed: their concentration reaches 50% of the plastic fibre concentration. Thus, concerning the microplastic concentration and flux in the Seine River, two main conclusions can be drawn:

- There is no significative difference between the upstream and downstream concentration or flux.
- The flux observed in the Seine River is much smaller than the sum of urban incoming fluxes for fibres and much larger for fragments.

Additional surveys are necessary to decrease the uncertainties in the concentration and flux estimates. However, these results clearly show that additional significant sources of plastic fragments occur within the urban area, such as the inputs linked to storm water or the fragmentation of plastic litter on the river banks. Their relative contributions are undetermined so far.

This type of multiscalar study of plastics at the catchment scale is rare, and the results produced in recent years in the Seine catchment are therefore original. However, additional studies are clearly required to achieve a more comprehensive overview of the dynamics of both micro- and macroplastics in the environment and further insight into the ecotoxicological consequences of their presence in freshwater ecosystems. From a water quality management point of view, the relevant figures and numbers necessary to engage efficient actions are still missing.

3   Poly- and Perfluoroalkyl Substances (PFASs)

3.1   Context

Poly- and Perfluoroalkyl Substances (PFASs) constitute a vast family of molecules bearing a fluorinated aliphatic chain (C_nF_{2n+1}) [25]. The industrial synthesis of these compounds began around 1950, and world production exceeded three million tons in 2000. The numerous applications of PFASs include additives in the synthesis of fluoropolymers, water and oil repellents for textiles, firefighting foams, lubricants, coatings and food packaging [25]. Less than 20 years ago, perfluorooctane sulfonate (PFOS) was found to be globally distributed in wildlife [26] and humans [27], while concerns were raised about its adverse effects [28] before it was officially classified as a Persistent Organic Pollutant in 2009 [29]. Since then, a large number of studies have addressed the issues of PFAS sources and environmental fate [30]. Besides airports and military bases, industrial sites such as fluorochemical facilities, metal plating industries, textile mills and power plants, urban areas are also considered as key sources of PFASs to hydrosystems [31, 32], due to either point source contamination (e.g. wastewater discharge) or diffuse contamination (e.g. urban runoff). However, the dynamics of these chemicals in urban rivers still remain poorly understood. In this context, the aims of the studies conducted within the PIREN-Seine programme since 2010 were twofold: (1) investigate the occurrence and the spatio-temporal dynamics of PFASs in the Seine River under contrasted hydrological conditions and (2) investigate the transfer of these chemicals to biota in urban rivers.

3.2 Dynamics of PFASs in the Seine River

A longitudinal upstream–downstream concentration gradient was previously observed for various contaminants (e.g. PAHs, organohalogens, phthalates) in relation to the impact of the Greater Paris conurbation [33]. The dynamics of selected PFASs in the Seine River were investigated using a dual strategy: (1) the time trends and mass flows were determined for a single study site during a flood cycle, and (2) both the seasonal and spatial fluctuations at the water year scale were studied at the regional scale.

Changes in river flow may have a large impact on the concentration of chemical point sources, and coordinating water quality monitoring with the analysis of hydrological conditions is essential to understand the fate and transport of trace organics in surface waters [34]. Thus, water samples were first collected weekly over a 4-month period in 2011 (January–May), in the centre of Paris (Quai d'Austerlitz), right at the heart of the conurbation. This sampling site was deemed representative of the impact on water bodies of urban inputs. Selected PFASs, including C_4–C_{14} carboxylates (PFCAs) and C_4–C_{10} sulfonates (PFSAs), were analysed in both the particulate and dissolved phases of water samples using liquid chromatography coupled with tandem mass spectrometry (LC-MS/MS) [32].

Over the period considered, the river flow rate ranged between 150 and 640 m^3 s^{-1}, and several spate events were recorded. Suspended solid (SS) levels were generally low, in good agreement with previous reports [4]; these levels exhibited large variations (3–60 mg L^{-1}) and were strongly correlated with the river flow rate [32]. Among the targeted PFASs, perfluorooctane sulfonate (PFOS) and perfluorohexane sulfonate (PFHxS) were the dominant compounds. $\sum$PFASs varied between 30 and 90 ng L^{-1} (average, 55 ng L^{-1}), and PFASs were mainly found in the dissolved phase due to the relatively low SS levels. These PFAS concentrations were in good agreement with those determined in the Seine downstream of Paris during low-flow periods [35] and were close to the average value determined for 122 watercourses at the European level (59 ng L^{-1}) [36], which suggests that the Seine River in Paris is not a hotspot of contamination of PFASs by European standards. However, the median $\sum$PFASs for this site was approximately three times higher than that determined at a nationwide level in France (i.e. 7.8 ng L^{-1}) [31]. In addition, PFOS levels (10–40 ng L^{-1}) were lower than the maximum allowable concentration (EQS-MAC, 35 µg L^{-1}) but consistently above the environmental quality standard expressed as an annual average value (EQS-MA) of 0.65 ng L^{-1} in Directive 2013/39/EU.

$\sum$PFASs and the river flow rate were negatively correlated, which suggests a dilution of the contributions from point sources when flow increases. However, the selected linear model explained only 25% of the observed variation of concentration. Other contributions via the sewerage network or urban runoff are therefore likely to influence the levels observed in the Seine River. In addition, the PFHpA to PFOA concentration ratio, indicative of the contribution of direct (i.e. non-atmospheric) inputs [37], was also positively correlated with flow rate. This suggests an increase of indirect inputs such as atmospheric deposition and subsequent urban runoff during high-flow events, which also correspond to periods of heavier rainfall. The cumulative PFAS mass flow was estimated based on the relationship between PFAS concentrations and the river flow rate. The weekly PFAS mass flow ranged between 4 and 15 kg, and the total flow over the 15-week study period was estimated at about 140 kg (i.e. approximately 500 kg year^{-1}). Such figures are in reasonable agreement with previous estimates for the Seine River, but nearly 200 times lower than for the Po River (Italy) [35].

To achieve further insight into the dynamics of PFASs in this system, PFAS concentrations were monitored at the regional scale over a longer time period [38]. For sample collection, three sites previously used as pilot sites for nearly two decades were selected (Marnay, Bougival and Triel, from upstream to downstream; see [1] for site location). A wider range of chemicals was analysed: 11 PFCAs, 5 PFSAs, FOSA and its N-alkylated derivatives (MeFOSA, EtFOSA) and 1 fluorotelomer sulfonate (6:2 FTSA). Four 1-month campaigns were undertaken over a 1-year period (September 2011 to December 2012), and the results based on grab samples (2–4 per campaign) confirmed the ubiquitous character of PFASs, since they were detected in all samples.

In the water column, total PFAS levels ranged from 2 to 90 ng L^{-1}. A significant upstream–downstream gradient was observed, associated with the increase in anthropogenic pressure on the fluvial ecosystem (Fig. 3). The levels observed at Triel were on average ten times higher than those observed at Marnay, with intermediate levels being observed at Bougival. In accordance with previous obser- vations, PFOS, PFHxS and PFOA were the dominant compounds at Marnay and Bougival, but a non-negligible contribution of the shorter-chain (C$_5$–C$_7$) carboxylic acids was also reported. A notable feature was the sharp increase in the relative abundance of 6:2 FTSA at Triel. This compound has been used as an alternative to PFOS (metal plating), but it may also result from the degradation of more complex fluorotelomer-based compounds used in firefighting foams or food packaging [25]. The high levels of 6:2-FTSA observed at Triel, on average nine times higher than those observed at Bougival, could also result from the influence of an industrial source located in the nearby Oise basin. The hypothesis of the existence of distinct

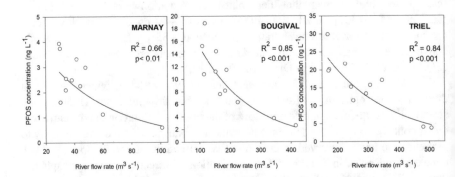

Fig. 3 Correlation between PFOS levels and river flow rate in the Seine River

sources for this compound downstream of the basin is reinforced by the lack of correlation between its levels and those of major PFASs (e.g. PFOS).

During this monitoring study, the river flow rate was in the range 30–100 m^3 s^{-1} at Marnay and 170–500 m^3 s^{-1} at Triel. At each site, PFAS levels appeared constant under stable hydrological conditions over short monitoring periods (e.g. 4 weeks), while the molecular pattern remained unchanged for a given site. Seasonal variations were examined at the water year scale, and the highest PFAS concentrations appeared to be associated with low-flow conditions. At the downstream sites (e.g. under the influence of the Paris conurbation), all major PFASs displayed negative correlations with flow rate, strongly suggesting that point sources were predominant. However, at the farthest upstream site, a few compounds only were negatively correlated with flow rate (e.g. PFOS; see Fig. 3), and no significant trend was found between ΣPFASs and flow rate. Thus, the upstream site (Marnay), located in a rural area between the urban centres of Troyes and Paris, appeared to be rather under the combined influence of point and diffuse PFAS sources.

The average PFAS daily flow in the Seine River increased by a factor of about 80 between Marnay (upstream) and Triel (downstream), thereby providing evidence of the major contribution of the Paris conurbation. The order of magnitude of the annual total PFAS mass flow estimate was 10 and 800 kg year^{-1} at Marnay and Triel, respectively. These figures are in good agreement with those previously determined in the centre of Paris (around 500 kg year^{-1}), especially considering the methodological differences between the two studies (e.g. sampling frequency).

3.3 Transfer of PFASs to Biota in the Seine River Basin

Considering the current mechanistic understanding of PFAS bioaccumulation, experimental studies are still needed to characterise the transfer of these chemicals within food webs. In the PIREN-Seine programme, the bioaccumulation of PFASs was investigated through several field studies that addressed this issue at different levels of biological organisation. In particular, two model organisms were selected, namely, (1) common chub, *Leuciscus cephalus*, a fish species widely used for water quality monitoring, and (2) periphytic biofilm. Note that only the results obtained with the latter are shown below.

Periphytic biofilms are mainly composed of both heterotrophic and autotrophic microbial cells embedded in an exopolymer matrix comprising polysaccharides and proteins [39]. At the interface of the water column and solid substrates such as bed sediment, they may play a central role in controlling contaminant bioavailability and transfer to consumers such as invertebrates or fish that graze on it.

Biofilm samples (3–4 per site, $n = 11$) were collected on artificial support (low-density polyethylene) colonised in situ. High detection frequencies (80–100%) were observed for C_8–C_{12} PFCAs, PFHxS and PFOS. ΣPFASs were in the range 4.3–33 ng g^{-1} dw, and concentrations in biofilm samples largely exceeded those of sediment samples [38]. The linear isomer of PFOS dominated

Table 2 Log BCF (mean $\pm$ SD) calculated for compounds detected in both water (dissolved phase) and biofilm [38]

Log BCF	PFOA	PFNA	PFDA	PFHxS	PFUnDA	PFOS	6:2 FTSA
Marnay	2.9 $\pm$ 0.1	4.0	3.6 $\pm$ 0.6	2.0 $\pm$ 0.2	–	3.8 $\pm$ 0.1	3.6
Bougival	2.3 $\pm$ 0.1	3.5 $\pm$ 0.4	3.0 $\pm$ 0.6	1.4 $\pm$ 0.2	–	3.1 $\pm$ 0.2	1.5
Triel	2.3 $\pm$ 0.1	3.6 $\pm$ 0.6	3.2 $\pm$ 0.1	1.6 $\pm$ 0.1	3.9	3.2 $\pm$ 0.1	1.4 $\pm$ 0.2

the molecular pattern (>50% of $\sum$PFASs), in agreement with profiles commonly reported for aquatic wildlife [40]. It can be speculated that PFASs were not only absorbed at the biofilm cell surface, but that they may also have been incorporated within the extracellular polymeric matrix or undergo intracellular accumulation [41]. These results imply that periphyton may constitute a key entry point for PFASs at the base of riverine food webs and a major source of PFASs for grazers.

The bioconcentration factor (BCF) is a useful metric to assess the bioaccumulation potential of a chemical. BCFs were calculated for PFASs detected in both dissolved phase and biofilm samples. The log BCF values ranged between 1.0 and 4.1 (Table 2), which is consistent with observations for other trace organics, e.g. pesticides [42].

Moderate $BCF_{biofilm}$ was reported for PFNA, PFDA and linear PFOS, larger than those of shorter-chain compounds such as PFHxS (Table 2). This highlights the importance of perfluoroalkyl chain length and functional group on PFAS bioaccumulation potential; the influence of such structural features on PFAS bioaccumulation in fish, as well as on the sediment–water partitioning, was extensively investigated in a tributary of the Seine River and is discussed in detail elsewhere [43]. The upstream–downstream gradient of PFAS levels in biofilm was comparable to the contamination gradient observed for water samples, i.e. displaying maximum values downstream of Paris. However, BCFs were significantly higher at the upstream sites than at the downstream sites (Table 2). Principal component analysis revealed the dependence of BCF on the dissolved phase concentration, which would be consistent with the conceptual model developed by Liu et al. [44] (i.e. adsorption-like process or PFAS–protein interaction). In addition, negative correlations were also observed between BCF and major cations (except for Ca^{2+}, unaffected by the longitudinal gradient), as observed for another model organism, the planktonic crustacean *Daphnia magna* [45]. However, this does not provide evidence of a causal relationship, and this result may be coincidental due to the collinearity between PFAS levels and major cations. Finally, the organic C/N ratio (i.e. proxy of bacteria/algae relative abundance) [46] was the only descriptor of biofilm characteristics positively correlated with BCF. Altogether, these results suggest that biofilm community characteristics may also be a determinant of PFAS bioaccumulation in periphyton.

4 Pathogenic Protozoa

4.1 Context

The increase of anthropogenic pressures on ecosystems has led to the increased frequency of pollution episodes by biological agents. Among these pollutions, the faecal contamination of aquatic environments affects numerous regions of the world, with proven risks to human health [47].

Three protozoan parasites are clearly identified as public health priorities: *Cryptosporidium* spp., *Giardia duodenalis* and *Toxoplasma gondii*. *Cryptosporidium* and *Giardia* are responsible for cryptosporidiosis and giardiasis, respectively. They can cause significant morbidity in immunocompetent patients, and *Cryptosporidium* can lead to death in immunocompromised patients [48]. *T. gondii* is responsible for toxoplasmosis, and 30% of the entire human population is chronically infected. An infection during pregnancy may lead to serious malformations of the foetus. In humans, the main vector of these biological agents is water contaminated by human or animal faeces, subsequently used for drinking or irrigating crops [49]. Their parasitic stages of transmission, i.e. oocysts and cysts, are very robust under environmental conditions, and they are ubiquitous in aquatic habitats. *Cryptosporidium* spp. and *G. duodenalis* are the protozoan parasites most often involved in water-related epidemics (i.e. due to the ingestion of drinking water or the accidental ingestion of contaminated water during recreational activities).

The assessment of the microbiological water quality is based on the monitoring of the occurrence of two bacterial indicators of faecal contamination – *Escherichia coli* and *Enterococcus* – according to the World Health Organization and European regulations (2006/7/EC) [50]. However, they can be quickly removed from the environment and are more sensitive than protozoa to environmental stresses (e.g. temperature variations, pollutants) and disinfection treatments [51]. Consequently, the abundance of these bacterial indicators does not reflect, or very little, the overall sanitary quality [52, 53]. Indeed, previous studies conducted on the Seine River demonstrated the lack of correlation between *Cryptosporidium* and *Giardia* concentrations and bacterial indicators in wastewater and river water [54, 55]. The authors suggest that viral and bacterial indicators are not appropriate to predict parasite loads in surface waters.

To investigate the occurrence of *Cryptosporidium* and *Giardia* in filtered water samples, the AFNOR NF T 90-455 standard (July 2001) proposes a detection technique based on both immunocapture on beads and immunofluorescence revelation. This detection is therefore highly specific, but it has not been applied yet to the detection of *T. gondii*. Other limitations have been identified: this method requires large volumes of water and high concentrations of parasites; it is expensive, and it does not allow for the rapid routine detection of parasites. It is therefore urgent to improve analytical tools for the detection of these biological contaminants for the purpose of monitoring water masses, thereby improving the assessment and management of health risks.

In this context, numerous studies have highlighted the value of bivalves for aquatic environment monitoring. These organisms are sedentary, have a high filtration rate and are characterised by their ability to accumulate environmental contaminants. For example, the use of bivalves revealed the contamination with pathogens, while direct measurements on water samples were negative [56]. In addition, oocysts of *Cryptosporidium parvum* have been detected in mussels (*Mytilus galloprovincialis*) and cockles (*Cerastoderma edule*) from a shellfish-producing region in Spain. The authors counted up to 5×10^3 oocysts in the tissues of bivalves [52]. The bioaccumulation of protozoa in bivalves is fairly well documented in marine environments; in contrast, few studies have been conducted on continental aquatic environments, despite their direct connections to pollutant sources (e.g. discharge of effluents from water treatment plants, direct discharge of livestock effluents, runoff or leaching from contaminated soil). The need for a better understanding of the protozoa ecology in freshwater ecosystems is increasingly spotlighted. Therefore, particular interest was focused on the freshwater bivalve *Dreissena polymorpha* (zebra mussel). Laboratory exposures have shown that *D. polymorpha* was capable of (1) bioaccumulating cysts of *G. duodenalis* and oocysts of *C. parvum* and *T. gondii* and (2) retaining *T. gondii* oocysts in its tissues in amounts close to those found in tanks after 14 days of exposure [57, 58].

4.2   Occurrence of Pathogenic Parasites in the Seine River

In a previous study, Mons et al. [54] assessed the protozoan contamination in the Seine River at sampling points located near the entry of drinking water plants (Ivry and Orly) or farther downstream in Paris (Tolbiac, Alma, Garigliano) and its periphery (Suresnes and Clichy). *Cryptosporidium* and *Giardia* were detected in filtered water in 45% (67/149 samples) and 94% (140/149 samples) of samples, respectively. *Giardia* was found more frequently and in larger quantities than *Cryptosporidium*. Thus, downstream of Paris, maximum concentrations reached 245 *Cryptosporidium* oocysts $10 \, L^{-1}$ and 512 *Giardia* cysts $10 \, L^{-1}$. These authors suggested that protozoan contamination in the Seine River was not linked to urban runoff but to land application of cattle manure and heavy rainfalls, which contribute to protozoan runoff from contaminated soils. Thus, *Cryptosporidium* and *Giardia* probably originate from rural areas, not from the Paris conurbation itself.

In this context, a field study was conducted within the PIREN-Seine programme, to further investigate the occurrence and sources of pathogenic protozoa in the Seine River using active biomonitoring with *D. polymorpha*. This field survey was carried out on three sites along the Seine River, following an upstream–downstream gradient: Marnay-sur-Seine (rural site), Bougival and Triel-sur-Seine (urban sites) (see [1]). Zebra mussels were collected in April 2016 at the Lac du Der-Chantecoq (N 48°36'10.0728" E 4°44'57.408"). Mussels measuring 2 ± 0.2 cm were acclimated in the dark in mineral water at 12°C for 2 weeks with two water changes a week to ensure that they were protozoan-free. Bivalves were caged in May 2016 for

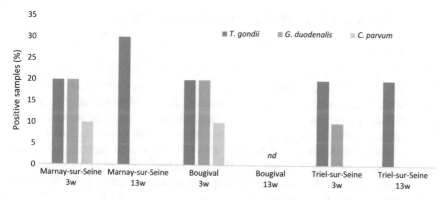

Fig. 4 Percentage of zebra mussels positive for *T. gondii*, *G. duodenalis* or *C. parvum* after 3 and 13 weeks of caging at Marnay, Bougival and Triel ($n = 10$). *nd* not detected

3 and 13 weeks, and three protozoa were quantified in the tissues of bivalves using molecular biology techniques. After 3 weeks of exposure, biological contamination by the three protozoa was observed at Marnay-sur-Seine and Bougival; *T. gondii* and *G. duodenalis* were detected in bivalves caged in Triel (Fig. 4). Thus, these results suggested that a 3-week caging period was sufficient to demonstrate the water contamination by protozoa. After 13 weeks, no protozoa were detected in zebra mussels caged at Bougival, and only *T. gondii* was detected in tissues of mussels caged at Marnay and Triel.

Bougival and Triel-sur-Seine are urban sites, and the biological contamination could be related to the high population density in this part of the Seine River basin. Prevalence rates of cryptosporidiosis in humans range from 1% to 20%; giardiasis is endemic in humans, and the prevalence of *Giardia* ranged from 1 to 5% [59]. Concerning *T. gondii*, felids are the definitive hosts, and toxoplasmosis is present in every country, with human seropositivity rates ranging from less than 10% to over 90% [48].

The biological contamination in Marnay-sur-Seine can be related to substantial agricultural and farming activities in this area. Livestock, particularly cattle, are an important source of *C. parvum*. In a Canadian farm animal analysis, the presence of *Cryptosporidium* was detected in faeces samples of cattle (20%), sheep (24%), pigs (11%) and horses (17%) [60]. Infected calves can excrete up to 10^7 oocysts per gramme of faeces [61].

These different studies highlighted the fact that protozoan parasites are ubiquitous in the Seine River and that bivalves, as sedentary organisms, could reflect ambient biological contaminations of watercourses. More specifically, *D. polymorpha* could be used as a new bioindicator in sanitary biomonitoring of freshwater bodies.

5 Antibiotics and Bacterial Antibiotic Resistance

5.1 *Context*

Since the first synthesis of antibiotics in 1940s, numerous molecules have been discovered, and nowadays there are about 10,000 antibiotics on the market [62]. Although these molecules have reduced mortality from infectious diseases and thus increased life expectancy, the use of these molecules has also induced environmental contamination. The first detection of pharmaceuticals in surface waters occurred in 1976 with the detection of clofibric acid and salicylic acid in a lake in Nevada. Since then, almost all categories of pharmaceutical substances have been found in surface waters [63].

The actual impact of antibiotic discharge on ecosystem functioning is still unknown. For instance, the concentrations of antibiotics observed in water (on the order of ng L^{-1}) are too low to affect the growth of fish such as Japanese medaka (EC50 100 mg L^{-1} for sulfonamides) or algae (in the range of 0.1–1 mg L^{-1}). At the microbial community level, concentrations measured in situ are below the minimum inhibitory concentration required to exert selection pressure on environmental microorganisms (on the order of mg L^{-1}); however, subinhibitory concentrations may promote mutagenesis or modify gene expression and may significantly influence bacterial physiology [64–67].

Since 2003, several studies of the PIREN-Seine programme have targeted pharmaceutical residues, particularly antibiotics. The main objectives of these studies are to determine the pathways of contamination in the natural environment, the environmental behaviour of these substances and the potential risk to ecosystems.

5.2 *Sources of Antibiotics in the Seine Watershed*

In rural areas, no contamination by antibiotics is observed in forest streams, but the contamination appears when streams flow through agricultural or breeding areas. Tamtam et al. [68] measured a concentration of 20 ng L^{-1} of enrofloxacin in a small river (the Blaise). This compound is exclusively used in veterinary medicine, and therefore the finding shows the contribution of this use to river contamination.

Antibiotic inputs to rivers in agricultural/rural areas may also come from their use in fish farming [69]: in such farms, antibiotics may be mixed with fish food and dispersed directly into the breeding ponds. Thus, the presence of fish farms may generate the discharge of antibiotics into rivers. During antibiotic treatment in fish farms, flumequine was quantified as high as 7 µg L^{-1} in the effluent of the treatment pond, and 2 days after the end of the treatment, flumequine concentrations were below the limits of quantification. However, 20 days later, this molecule was still measurable in the sediments of the river downstream of the fish farm discharge [68].

Antibiotics in rivers also have an urban origin, associated with urban or hospital wastewater treatment plants (WWTPs). Hospital effluents may contain numerous compounds, with individual concentrations ranging from a few 100 ng L^{-1} up to 47 µg L^{-1} (norfloxacin), and specific antibiotics such as vancomycin that are used exclusively in hospital facilities [70]. These concentrations are very high compared to those observed in the domestic effluents of residential areas. The mean hospital effluent concentration was 90-fold higher than that of the domestic effluent. However, since the volumes of hospital effluents are about six times lower than those of domestic effluents, the mass flow of antibiotics from domestic wastewater was approximately 1.5 times higher at the inlets of WWTPs [71].

WWTPs therefore play an important role in the life cycle of pharmaceutical products. Since these molecules are not completely eliminated by WWTPs, urban effluent outfalls are considered as point sources of antibiotics into the environment [63]. In WWTPs located in a small catchment in the Seine River basin, the antibiotics more frequently detected in influents were sulfamethoxazole, norfloxacin, ofloxacin and trimethoprim [71]. These compounds have different behaviours in WWTPs: fluoroquinolones (norfloxacin and ofloxacin) are mainly eliminated by adsorption onto sludge. Sulfamethoxazole and trimethoprim are, respectively, poorly or not adsorbed on particles, and their elimination through sorption is less efficient [72]. Thus, WWTPs only partly remove antibiotics, and the discharge of treated water into rivers can lead to increased concentrations downstream of the discharge, depending on river flow. Dinh et al. [70] observed the occurrence of fluoroquinolones and sulfonamides (sulfamethoxazole) in water downstream of WWTP discharge outlet. Fluoroquinolones are gradually adsorbed into the sediments, and only sulfonamides are detected far away from the discharge point. Therefore, fluoroquinolones accounted for up to 90% of antibiotics in the sediment [71].

5.3   Antibiotic Contamination in the Seine River

The main antibiotics quantified in the Seine River are sulfonamides, fluoroquinolones, macrolides and diaminopyrimidines [68]. These compounds are used in human as well as veterinary medicine. For example, sulfamethoxazole is the main sulfonamide, with concentrations ranging from 6 to 544 ng L^{-1} throughout the year. Their concentrations of sulfonamide increase from upstream to downstream with a maximum observed downstream of the main WWTP discharge outfalls of the Paris conurbation (Poissy) (Fig. 5). Besides sulfamethoxazole, fluoroquinolones are the main family of antibiotics measured in the Seine. Norfloxacin shows the same pattern as sulfamethoxazole, and ofloxacin is only detected at the farthest downstream site.

Trimethoprim (diaminopyrimidines) is often quantified in the Seine with a background level around 10 ng L^{-1} and is mainly present downstream of WWTP discharges because of its low elimination by WWTPs [72]. The decrease of these

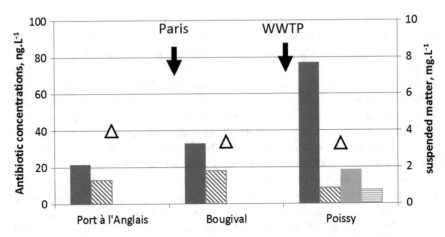

Fig. 5 Antibiotic concentrations (ng L^{-1}) in September 2009 in the Seine River: upstream of Paris (Port à l'anglais), downstream of Paris (Bougival) and downstream of the main WWTP discharge outfalls of the Paris conurbation (Poissy)

antibiotic concentrations during high-flow events seems to confirm that the intake is mainly related to medical uses and the origin of contamination is point sources. Overall, antibiotic concentrations measured in the Seine River are similar to those measured in Europe [73].

5.4 The Resistome and Antimicrobial Resistance

The contamination of water with antibiotics, which results from their prescription in human or veterinary medicine, is accompanied by a contamination of waterbodies by antibiotic-resistant bacteria. The occurrence of such resistant strains may be explained by the selection pressure exerted on the gut microbiota of humans and animals receiving antibiotic therapy. In the Seine River, Servais and Passerat demonstrated the presence of antibiotic-resistant faecal bacteria, the abundance of which reflects the level of anthropisation of the watershed [74].

In this context, particular attention is focused on biofilms or periphyton, which are microenvironments likely to concentrate chemical and microbiological contaminants [75–77]. In the aquatic environment, biofilms are ecological niches, where microbial communities experience chronic multiexposure to chemical contaminants (organic or metallic) including antibiotics. To this exposure is added a continuous supply of antibiotic-resistant bacteria of human or animal origin and therefore of genetic support involved in the dissemination of antibiotic resistance, such as clinical integrons [78]. Today, clinical integrons, considered as xenogeneic contaminants, are believed to be bioindicators of the risk of dissemination of antimicrobial resistance in the environment [79, 80]. Moreover, within these biofilms, the presence of metallic contaminants has been shown to be favourable for the spread of antimicrobial resistance [81, 82].

Within the PIREN-Seine programme, the resistome (genes conferring antibiotic and/or trace metal resistance) of microbial communities in biofilms has been studied as an indicator of vulnerability or environmental resilience to chemical or microbiological contaminants. In the Seine River, observations in situ and in the laboratory have shown that the acquisition of a trace metal tolerance of microbial biofilm communities depends on several factors. At the cellular level, the increase of genes encoding resistance to heavy metals, such as silver (*silA* gene) or cadmium/zinc/cobalt (*czcA* gene), suggests a selection of resistant bacteria in response to chronic exposure to toxic thresholds in Ag^+, Zn^{2+}, Co^{2+} or Cd^{2+}. At the microbial community level, the resistance to antibiotics may be related to a change in microbial diversity, with an increase in the abundance of bacterial genera able to grow in contaminated environments, such as *Burkholderiales*, *Cytophagales* and *Sphingobacteriales* [83]. Within these phyla, there are autochthonous bacterial genera but also bacteria that could be opportunistic pathogens such *Burkholderia*. Moreover, there is a permanent presence of antibiotic resistance genetic supports (class 1 clinical integron), whose abundance increases with the degree of anthropisation of the watershed, the maximum values being observed downstream from Paris and the discharge of the main treatment plant in the Paris region (Triel), regardless of the season (Fig. 6).

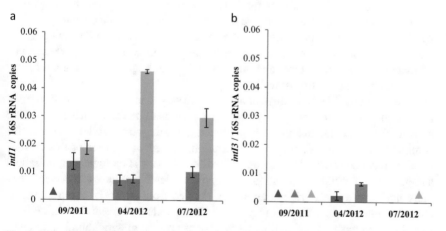

Fig. 6 Seasonal variation of relative abundance of clinical integrons in total DNA extracted from periphyton (artificial disposals) along the upstream–downstream transect, (■) upstream of Paris (Marnay), (■) downstream of Paris (Bougival), (■) downstream of Paris and impacted by the treated effluent of the Seine Aval WWTP Triel). (**a**) Class 1 integron (intI1/16S rRNA copy number) and (**b**) class 3 integron (intI3/16S rRNA copy number). Filled triangle: Detectable but not quantifiable. INSERM UMR 102 Limoges

These results show that the resistome of biofilms in highly urbanised rivers, such as the Seine River, constitutes a microenvironment where genetic support for antibiotic resistance (i.e. clinical integrons) is concentrated. It would be advantageous to determine whether antibiotic concentrations within these biofilms are consistent with increased mutagenesis and genetic transposition events. Biofilms would then constitute micro-niches or "hotspots" that are favourable for the transfer of genes and thus for the dissemination of the genes involved in antimicrobial resistance, within indigenous communities.

6 Conclusions and Perspectives

Research undertaken within the PIREN-Seine programme over 30 years has considerably improved the understanding of the sources and dynamics of a wide range of chemical and biological contaminants in the Seine River basin. A few examples of recent research were briefly exposed in this chapter.

In the near future, research prospects should include the investigation of additional contaminants of emerging concern or recently regulated chemicals; upon prioritisation, the list of newly targeted chemicals could encompass, for instance, chlorinated paraffins, novel flame retardants, biocides, antimony, manufactured nanoparticles and nanoplastics, etc. Further studies should better investigate the fate (i.e. transfer processes and fluxes) of micropollutants in relation to the hydrodynamic conditions, especially during extreme events related to climate change, such as floods or low flow/drought. To this end, it is anticipated that modelling approaches (1) would greatly improve the quantitation of gross fluxes transported by the Seine River, (2) could prove useful to better assess future contamination trends based on contaminant emission and hydrological scenarios and (3) would help estimate the exposure of biota and humans while enabling the investigation of key factors controlling this exposure.

Whenever possible, a more systemic approach should be implemented at the river basin scale, including various environmental compartments, to achieve a more holistic view of contaminant fate. In particular, the atmosphere–soil–river–estuarine continuum should be taken into consideration. The global impact of this chronic multi-contamination should be assessed at different levels of biological organisation, and human and social science should also be considered to address such issues in a more holistic way. An original approach, derived from territorial ecology, would consist in interpreting chemical fluxes at the basin scale by considering the connections between material production, trade or consumption of agricultural and household goods, emissions, stocks constituted in environmental compartments and transfers between compartments.

New methodological approaches should also be implemented. Suspect or nontarget screening using high-resolution mass spectrometry would greatly help characterise the chemical fingerprints of diverse environmental compartments and their temporal and spatial variability, thereby contributing to a better understanding of the human and biota exposome. Using passive sampling, in combination with both target and nontarget methods, would allow for the acquisition of time-averaged, low-frequency data to build up long-term data sets (plurennial or even decadal scale). This approach provides data that are complementary to high-frequency sampling and that are needed to help estimate the efficiency of regulations on the occurrence and dynamics of chemicals of interest. Biota could also be used to this end, provided well-known sentinel organisms are used (e.g. gammarids or freshwater mussels).

Finally, future research should also address emerging issues regarding biological contaminants, especially those that are not or are poorly related to the abundance of faecal bacterial indicator. The study of pathogenic protozoa transfer in freshwater ecosystems, or the emerging pathogens such *Leptospira*, is fundamental to improve the microbial risk assessment of surface waters. In addition, protozoa may lead to the modulation of physiological responses in sentinel organisms (e.g. bivalves), thereby potentially leading to erroneous interpretations in environmental monitoring studies. Thus, the influence of confounding factors such as the infection by protozoa represents a major issue as regards the use of biomarkers for environmental quality assessment (for further details regarding this issue, see [84]). Furthermore, the transfer of genetic element encoding resistance to antibiotics (1) from environmental microbial communities to strains that are pathogenic for humans or (2) from genes of clinical origin to environmental pathogenic bacteria that are opportunistic for humans (e.g. *Pseudomonas*, *Aeromonas* or *Burkholderia*) is identified as a major risk to public health related to the environment. The assessment of such a transfer of environmental resistance to humans (retro-transfer) is crucial to evaluate; it is, however, challenging because it involves rare events occurring on a time scale that remains difficult to determine.

Acknowledgments The authors would like to acknowledge the support of EPHE and R2DS Ile-de-France (i.e. Paris regional research network on sustainable development), which both provided a PhD grant. The authors would also like to thank the Aquitaine Region and the European Union (CPER A2E project) for their financial support, as well as the French National Research Agency (ANR) for its funding through IdEx Bordeaux (ANR-10-IDEX-03-02, PhD grant), the Investments for the Future Program (Cluster of Excellence COTE, ANR-10-LABX-45) and the SEQUADAPT project (headed by L. Fechner). The authors wish to thank Marie Cécile Ploy (UMR INSERM 1092) for the molecular quantification of clinical integron. This work was conducted in the framework of the PIREN-Seine research programme (www.piren-seine.fr), a component of the Zone Atelier Seine within the International Long-Term Socio Ecological Research (LTSER) network.

References

1. Flipo N, Lestel L, Labadie P et al (2020) Trajectories of the Seine River basin. In: Flipo N, Labadie P, Lestel L (eds) The Seine River basin. Handbook of environmental chemistry. Springer, Cham. https://doi.org/10.1007/698_2019_437

2. Gateuille D, Gaspery J, Briand C et al (2020) Mass balance of PAHs at the scale of the Seine River basin. In: Flipo N, Labadie P, Lestel L (eds) The Seine River basin. Handbook of environmental chemistry. Springer, Cham. https://doi.org/10.1007/698_2019_382

3. Chevreuil M, Carru AM, Chesterikoff A et al (1995) Contamination of fish from different areas of the river Seine (France) by organic (PCB and pesticides) and metallic (Cd, Cr, Cu, Fe, Mn, Pb and Zn) micropollutants. Sci Total Environ 162:31–42

4. Teil M, Blanchard M, Chesterikoff A, Chevreuil M (1998) Transport mechanisms and fate of polychlorinated biphenyls in the Seine River (France). Sci Total Environ 218:103–112

5. Thompson RC, Olsen Y, Mitchell RP et al (2004) Lost at sea: where is all the plastic? Science 304:838

6. Arthur C, Baker J, Bamford H (2008) Proceedings of the international research workshop on the occurrence, effects and fate of microplastic marine debris. Proceedings of the international research workshop on the occurrence, effects and fate of microplastic marine. University of Washington Tacoma campus in Tacoma, Washington

7. SPI – About Plastics – SPI Resin Identification Code – Guide to Correct Use. http://www.plasticsindustry.org/AboutPlastics/content.cfm?ItemNumber=823&navItemNumber=2144. Accessed 12 Apr 2016

8. Dris R, Gasperi J, Saad M et al (2016) Synthetic fibers in atmospheric fallout: a source of microplastics in the environment? Mar Pollut Bull 104:290–293. https://doi.org/10.1016/j.marpolbul.2016.01.006

9. Karami A, Golieskardi A, Choo CK et al (2017) The presence of microplastics in commercial salts from different countries. Sci Rep 7:46173. https://doi.org/10.1038/srep46173

10. Liebezeit G, Liebezeit E (2013) Non-pollen particulates in honey and sugar. Food Addit Contam Part A 30:2136–2140. https://doi.org/10.1080/19440049.2013.843025

11. Browne MA, Dissanayake A, Galloway TS et al (2008) Ingested microscopic plastic translocates to the circulatory system of the mussel, Mytilus edulis (L.). Environ Sci Technol 42:5026–5031

12. Li J, Qu X, Su L et al (2016) Microplastics in mussels along the coastal waters of China. Environ Pollut 214:177–184. https://doi.org/10.1016/j.envpol.2016.04.012

13. Oßmann BE, Sarau G, Holtmannspötter H et al (2018) Small-sized microplastics and pigmented particles in bottled mineral water. Water Res 141:307–316. https://doi.org/10.1016/j.watres.2018.05.027

14. Schymanski D, Goldbeck C, Humpf H-U, Fürst P (2018) Analysis of microplastics in water by micro-Raman spectroscopy: release of plastic particles from different packaging into mineral water. Water Res 129:154–162. https://doi.org/10.1016/j.watres.2017.11.011

15. Wright SL, Kelly FJ (2017) Plastic and human health: a micro issue? Environ Sci Technol 51:6634–6647. https://doi.org/10.1021/acs.est.7b00423

16. Dris R, Gasperi J, Rocher V, Tassin B (2018) Synthetic and non-synthetic anthropogenic fibers in a river under the impact of Paris megacity: sampling methodological aspects and flux estimations. Sci Total Environ 618:157–164. https://doi.org/10.1016/j.scitotenv.2017.11.009

17. Scherer C, Weber A, Lambert S, Wagner M (2018) Interactions of microplastics with freshwater biota. Freshwater microplastics. Springer, Cham, pp 153–180

18. Sanchez W, Bender C, Porcher J-M (2014) Wild gudgeons (Gobio gobio) from French rivers are contaminated by microplastics: preliminary study and first evidence. Environ Res 128:98–100. https://doi.org/10.1016/j.envres.2013.11.004

19. European Union (2008) Directive 2008/56/EC of the european parliament and of the council of 17 June 2008 establishing a framework for community action in the field of marine environmental policy (Marine Strategy Framework Directive)

20. Jambeck JR, Geyer R, Wilcox C et al (2015) Plastic waste inputs from land into the ocean. Science 347:768–771. https://doi.org/10.1126/science.1260352
21. Lebreton LCM, van der Zwet J, Damsteeg J-W et al (2017) River plastic emissions to the world's oceans. Nat Commun 8:15611. https://doi.org/10.1038/ncomms15611
22. Dris R (2016) First assessment of sources and fate of macro- and micro- plastics in urban hydrosystems: case of Paris megacity. Université Paris-Est, Champs-sur-Marne
23. Tramoy R, Gasperi J, Dris R et al (2019) Assessment of the plastic inputs from the Seine Basin to the sea using statistical and field approaches. Front Mar Sci 6:151. https://doi.org/10.3389/fmars.2019.00151
24. Gasperi J, Dris R, Bonin T et al (2014) Assessment of floating plastic debris in surface water along the Seine River. Environ Pollut 195:163–166. https://doi.org/10.1016/j.envpol.2014.09.001
25. Buck RC, Franklin J, Berger U et al (2011) Perfluoroalkyl and polyfluoroalkyl substances in the environment: terminology, classification, and origins. Integr Environ Assess Manag 7:513–541. https://doi.org/10.1002/ieam.258
26. Giesy JP, Kannan K (2001) Global distribution of perfluorooctane sulfonate in wildlife. Environ Sci Technol 35:1339–1342. https://doi.org/10.1021/es001834k
27. Kannan K, Corsolini S, Falandysz J et al (2004) Perfluorooctanesulfonate and related fluorochemicals in human blood from several countries. Environ Sci Technol 38:4489–4495. https://doi.org/10.1021/es0493446
28. Renner R (2001) Growing concern over perfluorinated chemicals. Environ Sci Technol 35:154A–160A. https://doi.org/10.1021/es012317k
29. United Nations Environmental Programme (2009) Recommendations of the persistent organic pollutants review committee of stockholm convention to amend annexes A, B or C of the convention. http://chm.pops.int/TheConvention/ThePOPs/TheNewPOPs/tabid/2511/Default.aspx
30. Ahrens L (2011) Polyfluoroalkyl compounds in the aquatic environment: a review of their occurrence and fate. J Environ Monit 13:20–31. https://doi.org/10.1039/c0em00373e
31. Munoz G, Giraudel JL, Botta F et al (2015) Spatial distribution and partitioning behavior of selected poly- and perfluoroalkyl substances in freshwater ecosystems: a French nationwide survey. Sci Total Environ 517:48–56
32. Labadie P, Chevreuil M (2011) Biogeochemical dynamics of perfluorinated alkyl acids and sulfonates in the River Seine (Paris, France) under contrasting hydrological conditions. Environ Pollut 159:3634–3639
33. Teil MJ, Tlili K, Blanchard M et al (2014) Polychlorinated biphenyls, polybrominated diphenyl ethers, and phthalates in roach from the Seine River basin (France): impact of densely urbanized areas. Arch Environ Contam Toxicol 66:41–57
34. Johnson AC (2010) Natural variations in flow are critical in determining concentrations of point source contaminants in rivers: an estrogen example. Environ Sci Technol 44:7865–7870
35. McLachlan MS, Holström KE, Reth M, Berger U (2007) Riverine discharge of perfluorinated carboxylates from the European continent. Environ Sci Technol 41:7260–7265
36. Loos R, Gawlik BM, Locoro G et al (2009) EU-wide survey of polar organic persistent pollutants in European river waters. Environ Pollut 157:561–568
37. Simcik MF, Dorweiler KJ (2005) Ratio of perfluorochemical concentrations as a tracer of atmospheric deposition to surface waters. Environ Sci Technol 39:8678–8683. https://doi.org/10.1021/es0511218
38. Munoz G, Fechner LC, Geneste E et al (2018) Spatio-temporal dynamics of per and polyfluoroalkyl substances (PFASs) and transfer to periphytic biofilm in an urban river: case-study on the River Seine. Environ Sci Pollut Res 25:23574–23582
39. Sabater S, Guasch H, Ricart M et al (2007) Monitoring the effect of chemicals on biological communities. The biofilm as an interface. Anal Bioanal Chem 387:1425–1434

40. Houde M, De Silva AO, Muir DCG, Letcher RJ (2011) Monitoring of perfluorinated compounds in aquatic biota: an updated review. Environ Sci Technol 45:7962–7973. https://doi.org/10.1021/es104326w
41. Gerbersdorf SU, Cimatoribus C, Class H et al (2015) Anthropogenic trace compounds (ATCs) in aquatic habitats – research needs on sources, fate, detection and toxicity to ensure timely elimination strategies and risk management. Environ Int 79:85–105
42. Ruhí A, Acuña V, Barceló D et al (2016) Bioaccumulation and trophic magnification of pharmaceuticals and endocrine disruptors in a Mediterranean river food web. Sci Total Environ 540:250–259
43. Labadie P, Chevreuil M (2011) Partitioning behaviour of perfluorinated alkyl contaminants between water, sediment and fish in the Orge River (nearby Paris, France). Environ Pollut 159:391–397
44. Liu C, Gin KYH, Chang VWC et al (2011) Novel perspectives on the bioaccumulation of PFCs – the concentration dependency. Environ Sci Technol 45:9758–9764. https://doi.org/10.1021/es202078n
45. Xia X, Rabearisoa AH, Dai Z et al (2015) Inhibition effect of Na+ and Ca2+ on the bioaccumulation of perfluoroalkyl substances by Daphnia magna in the presence of protein. Environ Toxicol Chem 34:429–436
46. McMahon RF, Hunter RD, Russell-Hunter W (1974) Variation in aufwuchs at six freshwater habitats in terms of carbon biomass and of carbon: nitrogen ratio. Hydrobiologia 45:391–404
47. WHO (World Health Organization) (2002) Emerging issues in water and infectious disease. World Health Organization, London
48. Pappas G, Roussos N, Falagas ME (2009) Toxoplasmosis snapshots: global status of toxoplasma gondii seroprevalence and implications for pregnancy and congenital toxoplasmosis. Int J Parasitol 39:1385–1394. https://doi.org/10.1016/j.ijpara.2009.04.003
49. Castro-Hermida JA, Garcia-Presedo I, González-Warleta M, Mezo M (2010) Cryptosporidium and Giardia detection in water bodies of Galicia, Spain. Water Res 44:5887–5896. https://doi.org/10.1016/j.watres.2010.07.010
50. Figueras MJ, Borrego JJ (2010) New perspectives in monitoring drinking water microbial quality. Int J Environ Res Public Health 7:4179–4202. https://doi.org/10.3390/ijerph7124179
51. Chauret C, Armstrong N, Fisher J et al (1995) Correlating Cryptosporidium and Giardia with microbial indicators. Am Water Works Assoc 87:76–84
52. Gomez-Bautista M, Ortega-Mora LM, Tabares E et al (2000) Detection of infectious Cryptosporidium parvum oocysts in mussels (Mytilus galloprovincialis) and cockles (Cerastoderma edule). Appl Environ Microbiol 66:1866–1870
53. Pachepsky YA, Blaustein RA, Whelan G, Shelton DR (2014) Comparing temperature effects on Escherichia coli, Salmonella, and Enterococcus survival in surface waters. Lett Appl Microbiol 59:278–283. https://doi.org/10.1111/lam.12272
54. Mons C, Dumètre A, Gosselin S et al (2009) Monitoring of Cryptosporidium and Giardia river contamination in Paris area. Water Res 43:211–217. https://doi.org/10.1016/j.watres.2008.10.024
55. Moulin L, Richard F, Stefania S et al (2010) Contribution of treated wastewater to the microbiological quality of Seine River in Paris. Water Res 44:5222–5231. https://doi.org/10.1016/j.watres.2010.06.037
56. Ayres PA, Burton HW, Cullum ML (1978) Sewage pollution and shellfish. In: Lovelock DM, Davies R (eds) Techniques for the study of mixed populations. Society for applied bacteriology technical series. Academic Press, London, pp 51–62
57. Palos Ladeiro M, Aubert D, Villena I et al (2014) Bioaccumulation of human waterborne protozoa by zebra mussel (Dreissena polymorpha): interest for water biomonitoring. Water Res 48:148–155. https://doi.org/10.1016/j.watres.2013.09.017

58. Palos Ladeiro M, Bigot-Clivot A, Aubert D et al (2015) Assessment of toxoplasma gondii levels in zebra mussel (Dreissena polymorpha) by real-time PCR: an organotropism study. Environ Sci Pollut Res 22:13693–13701. https://doi.org/10.1007/s11356-015-4296-y
59. Pond K, Rueedi J, Pedley S (2004) Pathogens in drinking water sources. In: Microbiological risk assessment: a scientific basis for managing drinking water safety from source to tap. Robens Centre for Public and Environmental Health, University of Surrey
60. Olson ME, Thorlakson CL, Deselliers L et al (1997) Giardia and Cryptosporidium in Canadian farm animals. Vet Parasitol 68:375–381. https://doi.org/10.1016/S0304-4017(96)01072-2
61. Smith HV, Rose JB (1990) Waterborne cryptosporidiosis. Parasitol Today 6:8–12. https://doi.org/10.1016/0169-4758(90)90378-H
62. Ventola CL (2015) The antibiotic resistance crisis: part 1: causes and threats. P T 40:277–283
63. Kümmerer K (2003) Significance of antibiotics in the environment. J Antimicrob Chemother 52:5–7. https://doi.org/10.1093/jac/dkg293
64. Davies J, Spiegelman GB, Yim G (2006) The world of subinhibitory antibiotic concentrations. Curr Opin Microbiol 9:445–453. https://doi.org/10.1016/j.mib.2006.08.006
65. Kohanski MA, DePristo MA, Collins JJ (2010) Sublethal antibiotic treatment leads to multidrug resistance via radical-induced mutagenesis. Mol Cell 37:311–320. https://doi.org/10.1016/j.molcel.2010.01.003
66. Baharoglu Z, Mazel D (2011) Vibrio cholerae triggers SOS and mutagenesis in response to a wide range of antibiotics: a route towards multiresistance. Antimicrob Agents Chemother 55:2438–2441. https://doi.org/10.1128/AAC.01549-10
67. Gullberg E, Cao S, Berg OG et al (2011) Selection of resistant Bacteria at very low antibiotic concentrations. PLoS Pathog 7:e1002158. https://doi.org/10.1371/journal.ppat.1002158
68. Tamtam F, Mercier F, Le Bot B et al (2008) Occurrence and fate of antibiotics in the Seine River in various hydrological conditions. Sci Total Environ 393:84–95. https://doi.org/10.1016/j.scitotenv.2007.12.009
69. Thurman EM, Dietze JE, Scribner EA (2002) Occurrence of antibiotics in water from fish hatcheries. U.S. Department of the Interior/U.S. Geological Survey, USGS Fact Sheet 120-02
70. Dinh QT, Moreau-Guigon E, Labadie P et al (2017) Fate of antibiotics from hospital and domestic sources in a sewage network. Sci Total Environ 575:758–766. https://doi.org/10.1016/j.scitotenv.2016.09.118
71. Dinh QT, Moreau-Guigon E, Labadie P et al (2017) Occurrence of antibiotics in rural catchments. Chemosphere 168:483–490. https://doi.org/10.1016/j.chemosphere.2016.10.106
72. Rosal R, Rodríguez A, Perdigón-Melón JA et al (2010) Occurrence of emerging pollutants in urban wastewater and their removal through biological treatment followed by ozonation. Water Res 44:578–588. https://doi.org/10.1016/j.watres.2009.07.004
73. Gros M, Petrović M, Barceló D (2009) Tracing pharmaceutical residues of different therapeutic classes in environmental waters by using liquid chromatography/Quadrupole-linear ion trap mass spectrometry and automated library searching. Anal Chem 81:898–912. https://doi.org/10.1021/ac801358e
74. Servais P, Passerat J (2009) Antimicrobial resistance of fecal bacteria in waters of the Seine river watershed (France). Sci Total Environ 408:365–372. https://doi.org/10.1016/j.scitotenv.2009.09.042
75. Kovac Virsek M, Hubad B, Lapanje A (2013) Mercury induced community tolerance in microbial biofilms is related to pollution gradients in a long-term polluted river. Aquat Toxicol 144–145:208–217. https://doi.org/10.1016/j.aquatox.2013.09.023
76. Aubertheau E, Stalder T, Mondamert L et al (2017) Impact of wastewater treatment plant discharge on the contamination of river biofilms by pharmaceuticals and antibiotic resistance. Sci Total Environ 579:1387–1398. https://doi.org/10.1016/j.scitotenv.2016.11.136

77. Marti E, Jofre J, Balcazar JL (2013) Prevalence of antibiotic resistance genes and bacterial community composition in a river influenced by a wastewater treatment plant. PLoS One 8: e78906. https://doi.org/10.1371/journal.pone.0078906
78. Koczura R, Mokracka J, Taraszewska A, Łopacinska N (2016) Abundance of Class 1 Integron-Integrase and sulfonamide resistance genes in river water and sediment is affected by anthropogenic pressure and environmental factors. Microb Ecol 72:909–916. https://doi.org/10.1007/s00248-016-0843-4
79. Gillings MR, Gaze WH, Pruden A et al (2015) Using the class 1 integron-integrase gene as a proxy for anthropogenic pollution. ISME J 9:1269–1279
80. Borruso L, Harms K, Johnsen PJ et al (2016) Distribution of class 1 integrons in a highly impacted catchment. Sci Total Environ 566–567:1588–1594. https://doi.org/10.1016/j.scitotenv.2016.06.054
81. Seiler C, Berendonk TU (2012) Heavy metal driven co-selection of antibiotic resistance in soil and water bodies impacted by agriculture and aquaculture. Front Microbiol 3:399. https://doi.org/10.3389/fmicb.2012.00399
82. Di Cesare A, Eckert E, Corno G (2016) Co-selection of antibiotic and heavy metal resistance in freshwater bacteria. J Limnol 75. https://doi.org/10.4081/jlimnol.2016.1198
83. Fechner LC, Gourlay-Francé C, Bourgeault A, Tusseau-Vuillemin M-H (2012) Diffuse urban pollution increases metal tolerance of natural heterotrophic biofilms. Environ Pollut 162:311–318. https://doi.org/10.1016/j.envpol.2011.11.033
84. Bonnard M, Barijhoux L, Dedrouge-Geffard O et al (2020) Experience gained from ecotoxicological studies in the Seine River and its drainage basin over the last decade: applicative examples and research perspectives. In: Flipo N, Labadie P, Lestel L (eds) The Seine River basin. Handbook of environmental chemistry. Springer, Cham. https://doi.org/10.1007/698_2019_384

River Basin Visions: Tools and Approaches from Yesterday to Tomorrow

Catherine Carré, Michel Meybeck, Josette Garnier, Natalie Chong, José-Frédéric Deroubaix, Nicolas Flipo, Aurélie Goutte, Céline Le Pichon, Laura Seguin, and Julien Tournebize

Contents

1 Introduction 382
2 Observing the Functioning of the Seine Hydrosystem Under Anthropogenic
  Pressures 383
  2.1 Strategies Devised by PIREN-Seine Scientists for the Spatiotemporal Observation
      of the River Basin 384
  2.2 Transforming Field and Laboratory Data into an Understanding of System
      Functioning 387
3 From Understanding Processes to Gaining Knowledge That Supports
  Decision-Making 389
  3.1 Numerical Models Simulate Hydrosystem Biogeochemical Functioning Across
      the Land–Ocean Continuum 391
  3.2 Models Are Used to Support Management Decisions 392
  3.3 Conceptual Models Interpret Relationships Between Society and the Hydrosystem
      Over the Longue Durée 395
  3.4 Models Enable Exploration of Socio-hydrosystem Scenarios and Their
      Consequences 398
  3.5 Advantages and Limitations of Models for Researchers and Practitioners 401
4 Participatory Devices Generated by PIREN-Seine Researchers and Their Effects
  on Public Action 403
  4.1 Participatory Experiments Conducted by PIREN-Seine Researchers 403

C. Carré (✉)
LADYSS, Université Paris 1 Panthéon-Sorbonne, Paris, France
e-mail: Catherine.carre@univ-paris1.fr

M. Meybeck, J. Garnier, and A. Goutte
SU CNRS EPHE UMR 7619 Metis, Paris, France

N. Chong and J.-F. Deroubaix
LEESU, ENPC, Université Paris-Est Créteil, AgroParisTech, Paris, France

N. Flipo
Centre de Géosciences, MINES ParisTech, PSL University, Fontainebleau, France

C. Le Pichon, L. Seguin, and J. Tournebize
Irstea, HYCAR Research Unit, Antony, France

Nicolas Flipo, Pierre Labadie, and Laurence Lestel (eds.), *The Seine River Basin*,
Hdb Env Chem (2021) 90: 381–414, https://doi.org/10.1007/698_2019_438,
© The Author(s) 2020, Published online: 3 June 2020

    4.2   Participatory Experiments Influence Collective Action 405
5   Conclusion 407
    5.1   Evolution of the Programme and Scientific Methods 407
    5.2   Scientific Knowledge and Action 410
References 412

Abstract The aim of this chapter is to provide a critical assessment of the approaches and production of tools within the PIREN-Seine programme over the past 30 years, as well as their use for river basin management and river quality improvement, and to analyse the challenges for the future. Three types of tools used in the PIREN-Seine programme are presented: metrology and fieldwork; model construction, simulation and their use in scenarios; and participatory science tools. These tools have been gradually built by the PIREN-Seine researchers and often developed together with the partners of the research programme, the main managers of the Seine River basin. Three issues raised by scientists and their partners are identified: (1) for metrology, how it has been improved to measure the state of waterbodies and to avoid their degradation; (2) for models, what they currently do and do not do and how they share common knowledge with practitioners; and (3) the place of researchers in the use of participatory devices in territories and their view of the effects of these tools to improve the quality of rivers and aquifers.

Keywords Long term, Metrology, Model, Participative tool, Quality trajectory, Scenario, Scientific production

1   Introduction

This analysis concerns the production of knowledge that is shared between researchers and the main managers of the Seine basin (drinking water producers and sewerage system operators, river basin agencies, watercourse unions, public institutions in charge of flood control, navigable waterway managers). Scientific knowledge is based on data produced from measurements, modelling and scenarios, while operators' knowledge is developed to jointly manage and improve the quality of the Seine River and of its aquatic environments. This way of doing science with the operational sector is similar to what Sheila Jasanoff described as a regulatory science: 'about how a fact is made and produced in this domain of scientific activity that serves public policy' [1]. As presented in the introduction of this book, since the third phase of the PIREN-Seine programme in 1998, researchers have developed a specific type of partnership research, not with laypersons but with stakeholders in charge of water management, elected officials, technicians, associations and professional groups. The tools discussed in this chapter, both numerical and conceptual, are anchored in the institutional reality of the Seine River basin, with groups of actors who exchange trust and legitimacy, and are therefore the product of this co-construction.

When the programme first started, the original focus was on hydrological processes (the link between river and alluvial aquifer), certain chemical pollutants (e.g. heavy metals) and certain biogeochemical methods of monitoring water quality (e.g. oxygen balance downstream of Paris, excessive algal growth). Today, the programme objective is to understand and quantify as far as possible the socio-ecosystem over a well-defined territory, a river basin on which multiple and varying social practices have been taking place for centuries, and to evaluate its possible future through prospective scenarios of basin evolution and its river response.

This chapter considers the way science is produced by using the paradigm developed by Gibbons et al. [2] on knowledge production and scientific discovery. 'Mode 1', which is very common, separates science and policy, and its sole purpose is the pursuit of scientific knowledge (science for the sake of science) and is defined by the domination of disciplinary science, strong internal hierarchy among disciplines and a strong sense of scientific autonomy. 'Mode 2', on the other hand, is described by Notwotny et al. [3] as socially distributed, application-oriented, transdisciplinary, diverse and heterogeneous in terms of sources and types of knowledge production, highly reflexive and subject to multiple accountabilities.

When applying Hatchuel's analysis of the position of scientists [4], it appears that PIREN-Seine knowledge production has evolved over 30 years from mono-disciplinary process-based studies in the field and/or laboratory, which are characteristic of the original environmental sciences (e.g. environmental chemistry), to deterministic modelling (e.g. Seine water quality within the Parisian sector with the ProSe model), enabling technical proposals to territorial problems, and to a pluri-disciplinary understanding in which the actors' rationales, decision-making and action procedures are a matter of knowledge in itself, eventually integrated into scenarios. These three stages, their potential and their limitations and perspectives in the PIREN-Seine programme are described here.

2 Observing the Functioning of the Seine Hydrosystem Under Anthropogenic Pressures

The scientific field studies of the PIREN-Seine do not aim only to describe the river state over the whole basin but also to fully identify and quantify the multiple hydrological, biogeochemical and ecological processes that control it. From the outset of the programme, the researchers considered these natural processes to be modified and sometimes regulated by present and past human activities and that their interactions with natural processes are always present in their research. Here, we describe the spatial and temporal scales used for this purpose.

2.1 Strategies Devised by PIREN-Seine Scientists for the Spatiotemporal Observation of the River Basin

2.1.1 Multiple and Nested Spatial Scales to Address River Basin Functioning

The first rationale of data acquisition is spatial and complements data from the Seine Water Agency, which is based on major rivers and urban areas in the 1980s. Scientists have combined a wide range of spatial scales from plots or specific point sources to the whole Seine basin up to the basin outlet (see Fig. 10 in [5]). The upstream–downstream Paris river profile, a stretch under maximum pressure, is still a permanent feature of the programme.

2.1.2 Multiple and Complementary Temporal Scales to Address River Basin Functioning, Regulatory Surveying and Reporting

This second approach overlaps with the previous one. Scientists addressing the functioning of the system select the most appropriate temporal scales to cover the dominant control factors and/or river processes (Fig. 1). The hydrological phenomena are reported on the daily scale, thanks to the regulatory hydrometeorological networks. Sub-daily observation is needed in headwater streams and for storm events in urbanized areas. At the beginning of the project, intensive fieldwork was carried out over short periods, as impact studies, quarterly longitudinal profiles and bimonthly samplings performed at key stations (e.g. near the laboratory).

Hydrologically driven variability also depends on the size of the river basin intercepted at the station: flood duration ranges from a few hours for headwater streams to a few weeks for large basins (100,000 km^2). Internal processes within the river, such as phytoplankton blooms and autochthonous formation of organic and/or inorganic particulates, last between a few days and weeks or more. Scientists are also interested in interannual variability (dry compared to wet years).

This temporal approach is largely finer than that of the regulatory chemical water quality surveys performed by river basin authorities, centralized on the monthly scale, whatever the size of the basin at the control station. Therefore, these surveys rarely capture sub-seasonal variability, and some of the river controlling factors are only taken into account statistically. Daily and sub-daily records are made by the water industries at the river intakes, on treated sewage and downstream of their outflows. These records, which are compiled for industrial monitoring or accident detection, are generally not included in river survey processing and annual or interannual state-of-the-environment reports (SOE; Fig. 1).

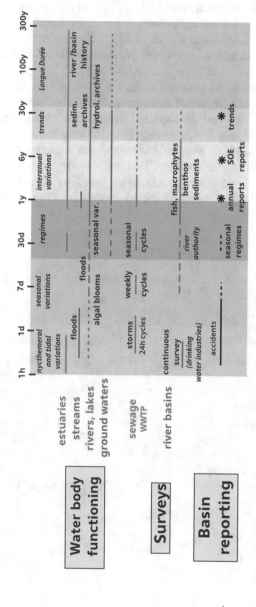

Fig. 1 Temporal scales of water quality observations considered in the PIREN-Seine programme (for waterbody functioning), of regulatory surveys and of basin reporting on various waterbodies

Over the past 10 years, the programme has developed a network of multiple chemical sensors that provide a continuous record of many analytes. They are mainly located in the Lower Seine, where they are used to focus on the impact caused by Paris (Box 1). Another continuous recording station has now been placed in the Orgeval experimental stream catchment (II on Fig. 1 in [5]), where it registers fine-scale variability in a drained agricultural basin. This monitoring generates an enormous amount of information on chemical and biogeochemical processes. The data then need to be assimilated in order to provide specific information on the real exposure of aquatic species to water quality variability.

At the opposite end of the temporal scale, the programme analyses trends over several decades (e.g. the impact of sewage treatment, the impact of phosphorus control on eutrophication and the expected hydrological changes due to climate change). Finally, the longue durée approach (50 years and more) is a specific feature of the PIREN-Seine, which aims to understand the interactions between river basins, their territories and their society. This is achieved by means of (1) historical research (e.g. [6]), (2) retrospective modelling based on data on past pressures established by scientists and on pressure–state models and (3) analysis of sedimentary archives [7].

Box 1

Simulated high-frequency measurements of river water chemistry and their use in the design of water quality surveys, using ammonia as an example.

The ProSe model generates high-frequency concentrations in the Parisian sector of the Seine River up to the estuary. This sector receives the discharges of more than 10 million inhabitants, including discharges from the Seine Aval wastewater treatment plant (SAV), Europe's largest. The upper distribution (90%) of NH_4 concentrations (left scale in Fig. 2) is simulated for the 2007–2011 period over 200 km, according to a longitudinal profile of the Seine from Paris to the Poses station, the Seine basin outlet [8, 9].

Different survey frequencies are simulated here, from 15 min to 3 months. The impact of the main sewage treatment plant, Seine Aval (SAV), is best simulated when the survey frequency is less than 3 days. The summer storm events within Paris centre generate combined sewer overflows (CSOs) that generate minor ammonia peaks, with limited impact at the interannual scale. The regulatory degradation of river water quality at the SAV facility is quantified and qualified by the colour scale derived from the European Water Framework Directive (WFD). Water quality is decreased by two colour codes at the SAV facility over 10 km and by one scale over 130 km, up to the Poses station.

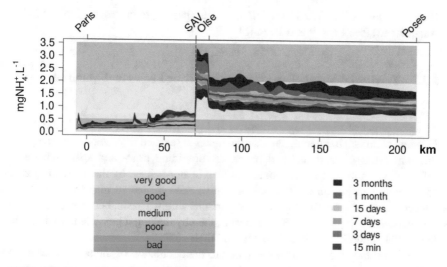

Fig. 2 Simulation of NH₄ distribution for the 2007–2011 period over 200 km from Paris to Poses, according to several measurement frequencies from 15 min; 3, 7 and 15 days; to 1 and 3 months (from Vilmin et al. [10])

2.2 Transforming Field and Laboratory Data into an Understanding of System Functioning

Targeted field studies make it possible to construct the hydrological, sedimentary and biogeochemical dynamics of the river along the stream order hierarchy and separate diffuse and point sources of material, natural and anthropogenic sources. These measurements contribute to the knowledge of the biogeochemical processes by examining the transfers between the different compartments of the Seine system and the transit times along the fluvial continuum.

2.2.1 Basic Processes and Their Controlling Factors

The river system is conceptualized first by a set of reservoirs (soil, plant, atmosphere, unsaturated zone, aquifers, river network) between which the material (gas, solute and particulate) is exchanged and transported (e.g. water, carbon, nitrogen). Once in the water runoff and/or in groundwaters, this material is routed from one stream order to the next and is exposed to another set of physical and biogeochemical processes (e.g. algal production, bacterial respiration, deposition/resuspension, solute–particulate exchange). Scientists measure amount in the environment, as well as rates, from which they can derive the main process parameters, which may vary spatially and temporally within the basin. These parameters integrated into conceptual frameworks are then used in the modelling at the basin scale

(Riverstrahler and CaWaQs model) or on a specific river reach (ProSe model, Anaqualand) (see Figs. 9 and 10 in [5]).

2.2.2   Territorial Metabolism: Comparing River Fluxes and Material Circulation in the Anthroposystem

A major target of many biogeochemical and pollutant transfer studies is the establishment of annual to interannual material fluxes at selected river stations intercepting different types of land use and pressures, providing time- and space-integrated information on this territory. Fluxes can then be normalized by the intercepted area to generate specific ones. Both specific fluxes and annual to interannual average concentrations can be used to make comparisons within a basin, between river-borne products, and comparisons between basins, thereby making it possible to understand the spatial variability of these transfers and their controls.

The river fluxes can also be compared to the circulation of the targeted element (e.g. nitrogen, cadmium, PCBs) within the anthroposystem (i.e. import–export, product manufacturing, transportation and recycling of material, etc.). In theory, a sustainable society should not release any waste into the environment, resulting in a zero level of contamination. In reality, there is a continuous leak of waste into soils, waterbodies or the atmosphere. The amount of material circulating over a given territory is difficult to assess, since statistical data were not originally designed for environmental objectives but rather for economic or fiscal purposes. The PIREN-Seine is one of the pioneers in France on the contamination of the river system by pharmaceuticals, particularly antibiotics [11]. This work has identified the sources and transfers of these products in the Seine basin, linked to human wastewater (hospitals, wastewater treatment plants) and to livestock farming. Moreover, the circulation of products should be established over basins or sub-basins, i.e. with hydrographic boundaries instead of administrative boundaries. Finally, the statistical data relate to manufactured goods, not their chemical composition, which is what would be needed for such a comparison, as in the case of individual metals, pesticides and nutrients [12, 13]. Only few flux–flow comparisons have been made for the Seine basin, providing precious information on the socio-hydrosystem metabolism and its evolution over time. The 'leakage rates' of the basin have been assessed for heavy metals, Cd, Cu, Hg, Pb and Zn, which ranged in the 2000s from 0.1 to 10%, depending on the metals [14] (see also [7] on sedimentary archives).

2.2.3   Aquatic Biota Contamination Across Trophic Levels: A Growing Concern

At the outset of the programme, the aquatic species (phytoplankton, zebra mussel, macrophytes) were studied as an important part of biogeochemical cycling or as indicators of the overall water quality and its restoration. Fish ecology and its disturbance by multiple human impacts (organic pollution, habitat degradation, river flow regulation and eutrophication) were considered as well. Harmful algal blooms observed in the outer estuary have also been addressed through dedicated models.

A new concern is the transfer of contaminants to aquatic species and from one trophic level to the next, as determined using nitrogen isotopic signatures. Primary producers (biofilm, leaf litter, macrophytes) have the lowest trophic levels, macroinvertebrates (*Gammarus* sp., *Lymnaea* sp., *Corbicula* sp., leech) have an intermediate trophic level, and fishes (roach, European perch, gudgeon, tench, bullhead, pumpkinseed and black bullhead) are at the top of the Seine River food web. For example, the transfer of polycyclic aromatic hydrocarbons (PAHs) and of phthalate metabolites between these trophic levels is presented in Box 2.

Box 2

Targeted monitoring of river biotic communities for PAHs, phthalates and their metabolites.

Polycyclic aromatic hydrocarbons (PAHs) and phthalates are the main organic micropollutants in the Seine hydrosystem and in three fish species [15] and prevent waterbodies from achieving good chemical status as defined by the Water Framework Directive (WFD). Biomonitoring of PAH and phthalates is complex, since they are rapidly metabolized in aquatic organisms, especially fish. Therefore, we monitored the levels of key PAH and phthalate metabolites in aquatic organisms caught in the river, from primary producers to macroinvertebrates and fish in order to assess the bioaccumulation and transfer of these pollutants and their metabolites across the trophic web [16].

The levels of PAH metabolites did not vary with trophic levels, likely due to the rapid elimination of these degradation products. We observed a trophic amplification of phthalate metabolites, probably since the excretion rate is slower (Fig. 3). This raises ecotoxicological concern, since phthalate metabolites also have deleterious effects.

3 From Understanding Processes to Gaining Knowledge That Supports Decision-Making

PIREN-Seine has developed a wide array of models, from numerical to conceptual.[1] Numerical models aim to synthesize knowledge on the functioning of the hydro-socio-system, on the whole basin, such as Riverstrahler, or on a river sector, such as ProSe. It enables to explore the possible futures of the basin over the next

---

[1]Details on PIREN-Seine models are given in the first chapter of the book [5] and in the corresponding chapters.

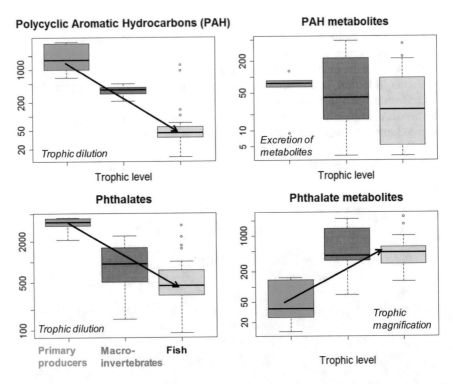

Fig. 3 Schematic behaviour of contaminants and their metabolites in the trophic food web of the Seine River: polycyclic aromatic hydrocarbons (PAHs), phthalates and their metabolites. Green, primary producers (biofilm, leaf litter, macrophytes); red, macroinvertebrates (*Gammarus* sp., *Lymnaea* sp., *Corbicula* sp., leech); and yellow, fishes (roach, European perch, gudgeon, tench, bullhead, pumpkinseed and black bullhead). Concentration in ng/g dry weight (log scales) [16]

50 years in the form of scenarios that can be used to support management decisions. Models and scenarios are conceived and developed in collaboration with PIREN-Seine partners. Their development has been presented regularly every year at the partner meetings since 1993. Conceptual models are generally developed to present a synthetic understanding of the society–basin–river evolution and interaction; their nonnumerical presentation makes them suitable for the communication of scientists' work to a lay audience. Conceptual and numerical models are also used in participatory scientific experiments, where they facilitate dialogue between water users and allow the transfer of scientific knowledge directly to the local level.

3.1 Numerical Models Simulate Hydrosystem Biogeochemical Functioning Across the Land–Ocean Continuum

The models developed within the PIREN-Seine are descriptions of the components of the hydro-socio-system of the Seine basin and their relationships, through the biogeochemical processes of the hydrosystem. For example, nitric contamination is presented as resulting from the opening of the terrestrial cycle of nitrogen linked to the development of agriculture and the modern agri-food chain. The fluxes of heavy metals transferred by the river network are seen as a consequence of the flows related to the production and consumption of these metals by industrial and domestic activity, as well as of regulations. Alongside the ecology of the aquatic environment, the industrial ecology of the Seine basin system has been also considered.

One of the main outputs of the PIREN-Seine models has been to provide a general model of the contribution of groundwater to the flow of the Seine. The CaWaQS/ MODCOU model performs the hydrological and hydrogeological coupling between seven aquifer units from the recent alluvial deposits to the lower cretaceous formations, with water residence times ranging from a few months to decades. The model has recently been focused on alluvial aquifer exchanges, and operates at the daily scale with a kilometric resolution over the whole river basin, which is well suited to certain needs of the basin authorities, and is completed by the modelling of nitrate reactivity and transfer from the unsaturated zone to the aquifer. This set of models, from the soil surface and unsaturated zone to aquifers, is well suited to addressing the evolution of water resources with climate change and the scenarios of variable water demand, particularly in the agricultural sector, as well for the study of diffuse pollutants (nitrates, pesticides, atmospheric pollutants) and their evolution with possible agro-economic measures.

Several families of numerical models have been generated. Their common characteristic is their continuous evolution: More processes are taken into account, their resolution is finer, and they are extended temporally (backcasting, scenarios of the basin futures), as well as spatially (to the estuary, to other basins). Sub-models are added and gradually connected to the original ones (as presented in Box 3).

Models work with data sets acquired at different time stages. They also provide an opportunity to check data consistency with the partners. 'The models appeared also very useful for confronting the various databases gathered by Water Authorities on human pressures and those on water quality. Many incoherent data could be detected during the process of informing the models, resulting in a significant improvement of the databases and hypotheses that have been hotly debated. As far as the Water Authorities were directly concerned with the results, a real dialog was established' [17].

Box 3

Coupling agrosystems and hydrosystems from headwater to estuary by a model cascade for nitrogen, phosphorus and carbon circulations.

GRAFS model calculates N, P and C budgets in agricultural systems [18]. From N surplus, it generates nitrate leaks into the unsaturated zone [19] and into the atmosphere (N_2O, a greenhouse gas; [20]). It operates from the farm scale (e.g. crop rotation, nitrogen uptake by crops, imports/exports of fertilizers, feed and food, recycling of organic residues). This diffuse nitrate is addressed in the Riverstrahler model, a semi-spatialized model organized by stream order that transfers the N products across the river continuum. Lateral groundwater river inputs of nitrate and local denitrification in the riparian zone are taken into account. In the river, the biogeochemical reactivity of C, N, P and Si species is handled by the RIVE model.

Ten-day river fluxes to the Seine estuary and to small coastal rivers are then used by the coastal zone biogeochemical model ECOMARS-3D, which was developed by marine scientists [21, 22]. This model simulates estuary functioning, from Poses to Honfleur and the Bay of Seine (Fig. 4). After crossing the estuarine filter, the impact at the coastal zone of the net nutrient fluxes can be evaluated in terms of eutrophication and harmful algal proliferation. It took almost three decades and millions of Euros for around a dozen scientists to build and operationalize such a chain of models, which has been conceptualized by biogeochemists since the 1990s.

The challenge in terms of modelling the biogeochemical functioning of aquatic environments is multiple: (1) to project at multiregional scales capable of including all contributory hydrosystems from headwaters to coastal zones by proposing a quantification of flows at sea and (2) to represent at various scales the microscopic processes operating in the aquatic continua that ultimately determine basic water quality (dissolved oxygen, ammonia, phosphorus, organic carbon, etc.). The scales considered by these models range from microscopic physiological processes (bacterial activity, algal uptake of nutrients, etc.) to large spatial scales from small catchments ($100 \ km^2$) to multiregional river basins (inputs to the North East Atlantic coast, $1,000,000 \ km^2$).

3.2 Models Are Used to Support Management Decisions

Primarily seen as research tools, models indeed allow for a better scientific understanding of the functioning of the Seine River basin while also supporting

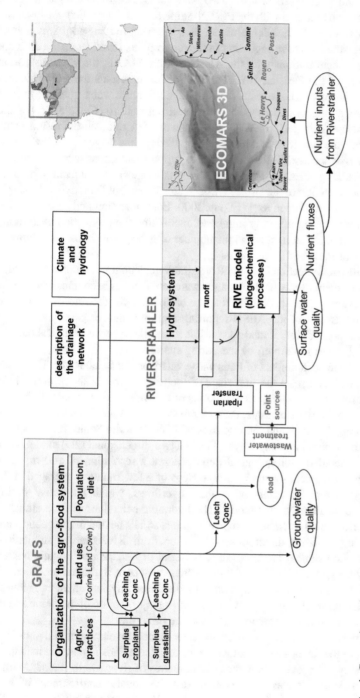

Fig. 4 Conception and spatial catenation of the biogeochemical models cascade across the whole river continuum from headwaters to the English Channel coastal zone. GRAFS: plot scale to stream. Riverstrahler: waterbodies and river network. RIVE: in-stream biogeochemistry. ECOMARS3D: coastal zone biogeochemical model of the English Channel

management and planning decisions. Two main tools, Seneque–Riverstrahler and ProSe, have been developed with the PIREN-Seine partners in explicit consultation.

The first use of a PIREN-Seine model as a decision-making tool was that of Seneque–Riverstrahler, at the beginning of the programme, to reflect on the different wastewater treatment options with the AESN and the SIAAP, the latter institution being in charge of the Parisian WWTPs. The model represents the major biogeochemical mechanisms of large watersheds involved in basic water quality (nutrient status, organic pollution and oxygen deficit, eutrophication). Calculations of the ecological processes are performed at a time scale of a few minutes, but results at the ecosystem scale are provided at a resolution of 10-day time stages.

Seneque–Riverstrahler (S–RS) was used to support water managers by simulating specific planning scenarios aimed at optimizing wastewater treatment in order to fulfil the implementation of the Urban Waste Water Treatment Directive in 1991 and the Water Framework Directive (WFD) in 2000. Eutrophication and oxygen depletion in the river were significantly improved, facilitating the production of drinking water and allowing for an increase in the number of fish species in the most human-impacted area of Paris and its outskirts.

To allow its use by technicians, the Riverstrahler calculation model is embedded in a GIS environment (Seneque; [23]), a software tool that has been transferred to the Seine Water Agency (AESN) and used at the basin scale for their needs. S–RS has also been used as part of the AESN reporting requirements for good-water-status data to the European Union. It made it possible to simulate physicochemical data for river waterbodies not monitored by measuring stations.

ProSe: The fine-scale impact of Paris wastewaters on the Seine River. The ProSe model was developed from the first stage of the programme, at the request of the SIAAP, on strongly human-impacted sections of the watercourse in order to be able to deal with transitional situations, such as rainstorms and the discharges from the Parisian combined sewers and the bypasses of WWTP in the Seine. It differs from the Riverstrahler model in many ways: (1) It is a two-dimensional (2D) model based on a specific, detailed morphological description at a resolution of 100 m; (2) it simulates highly transient events with time steps of a few minutes, averaged at the hourly scale. Dozens of combined sewage overflows, which are active during rainstorms, are localized in this model. The hydraulic behaviour of each stretch is dynamically calculated by Saint–Venant equations, and its upstream limit conditions are derived either from direct observations or from Riverstrahler simulations [17]. The biogeochemical processes (e.g. bacterial degradation of organic matter) are modelled through the RIVE model in the same way as in Riverstrahler. Today, it simulates basic water quality through more than a dozen physical chemical, biogeochemical and microbial indicators. It now covers a major section of the Seine and its tributaries within and downstream of Paris to the basin outlet.

ProSe is currently used to optimize the interannual development and sub-daily management of the sewage treatment works for the whole Paris region and its impact on water quality over several hundreds of kilometres of river length (Seine, Marne, Oise). The SIAAP operates the model to test the aquatic environment of new phosphorus and nitrogen treatments required by EU Directives for their WWTP and to manage their treatment facilities during repairs. It is also used by drinking water manufacturers using the river water supply. It can also be mobilized during crisis management in the case of accidental pollution, the last one being due to a fire that affected part of the Seine Aval WWTP in July 2019.

While Riverstrahler is well suited to diffuse pollution management over the whole basin and to the simulation of social responses over decades, ProSe has been designed for water quality restoration downstream of point sources of pollutants in urbanized and industrialized river sectors, from accidental pollution and rainstorm impacts to pluri-annual management. As such, both modelling approaches are fully complementary. Both models have been used by the river basin authority (*Comité de basin*) for the second general management scheme of the basin (SDAGE).

However, the tools themselves largely remain in the hands of researchers or practitioners with modelling expertise [24]. Only researchers have the ability to operate the model and change its code, while partners operate the model without being able to make changes in the code or benefit from the knowledge produced by modelling without operating the model or directly using its outputs. This strategy has the advantage of saving the substantial investment in time and resources required of researchers and practitioners in order to render them more accessible to nonexperts on the one hand and of developing the necessary internal expertise among practitioners on the other.

3.3 Conceptual Models Interpret Relationships Between Society and the Hydrosystem Over the Longue Durée

River basins and their societies coevolve over the longue durée, multi-decadal to multisecular. Since its second stage (see Figs. 9 and 10 in [5]), the PIREN-Seine has taken the view that the socio-hydrosystem was also dynamic at this temporality and that the current situation was the result of a long history of interactions between basin societies and the river. The history of the socio-hydrosystem can be presented through narratives and conceptual schematic interactions and the trajectories of the system and by extending models from the past. Compared with the factual content of an evolution, the notion of trajectory includes an interpretation of the elements selected to answer the question posed. Thus, in the example developed in Box 4, the long-standing relationship between the quality of watercourses and major European cities has been understood through the perception of pollution and its effects, as well as the measures taken by the water and sanitation services to address them [25] (Table 1).

Data to trace long-term trajectories on river state are not provided by watercourse managers, either because they do not exist or because they have not yet been exploited; they are reconstructed through (1) backward reconstruction of water

Table 1 Technical response in terms of water and sanitation for Paris compared with Berlin, Brussels and Milan

	Paris	Berlin	Brussels	Milan
1850s to 1890s	Drinking water supply plant Wastewater treated in sewage fields/excess into the Seine River	Drinking water supply plant. Installation of sewage fields/excess discharge rarely into the Spree River	Drinking water supply plant Wastewater discharge to river	Drinking water supply plant Wastewater reused for agriculture/ excess in canals
1900s to 1950s	Wastewater treatment plants initiated in 1940 (Achères–Seine Aval)	Construction of wastewater treatment plants and technical innovation	No change	No change
1960s to now	Gradual development and innovation in WWTPs. Unguaranteed performance of the system	West Berlin continuous innovation Lagging in East Berlin until 1989	After 2000, construction of the first WWTP. Erratic performance of the system	After 2000, construction of the first WWTP (high performance)

quality, using pressure–state models; (2) reconstruction of historical archives of water quality since the 1880s; (3) reconstruction of past contamination from sedimentary archives; (4) the construction of a unique and still growing set of data on the river and its internal and external control factors (climate and hydrology, land use, population, wastewater treatment, aquifers, urbanization and industrialization, material flow, water quality, river habitat, fish population, etc.) at fine resolutions and over 50–150 years; (5) the study of river institutions and water regulations over the past 150 years; and more (as in Box 4).

Box 4

The compared responses of four major European cities, Paris, Berlin, Milan and Brussels, to basic water quality issues faced during their urban expansion (1850–2015).

Each city has developed its own management strategy depending on water availability, policy choice, politics (when Berlin was split into two parts during the Cold War) and administrative division (the Zenne basin is managed today by three different Belgian regions). Berlin can be regarded as a precursor in terms of safeguarding its local water resources, lakes and surficial aquifers. At the end of the nineteenth century, the same water service was managing both the drinking water supply and sewage collection and treatment, both for the city and its suburban area, unlike other cities. A wide network of sewage collectors was built, and wastewater was treated on farmland around the city, which in turn was benefitting from the nutrient inputs. Three wastewater

(continued)

treatment plants were built in the 1920s when it became clear that the capacities of the sewage farms had been exceeded, whereas the Parisian sanitation organization waited until 1940 to start its first WWTP, Achères–Seine Aval. The level of treatment at Berlin's WWTPs has been constantly improved, with tertiary treatment starting in the 1980s, while in the Paris area, it was not until 1993 that all effluents were treated. Brussels water supply was far up in the Walloon region, and Milan was relying on medium-depth groundwaters, which were potentially contaminated by its own industries. Neither Milan nor Brussels had a WWTP until they were forced to do so by EU regulations at the beginning of the 2000s.

Trajectories in these four cities, which were exposed to maximum relative pressures (city population/river discharge), show a period of intense degradation in the late nineteenth century due to the discharge of untreated wastewater and the related alteration to physical environments. The rapid population growth was not followed by adequate technical, organizational or financial means sufficient to treat domestic and industrial discharges. The Zenne River was covered and transformed into a sewage collector across its urban sector. The urban impact of these cities on the quality of their rivers lasted over more than a century. The restoration of the Seine River and the Spree River was initiated several decades before that of the Zenne and Lambro rivers. European environmental policy played a major role in the converging restoration of all socio-hydrosystems in the early 2000s (Fig. 5) [25].

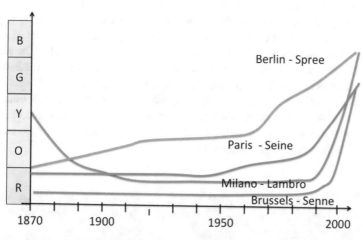

Fig. 5 Comparative trajectories (1870–2010) of the efficiency of the social response to reduce the organic pollution generated by four European cities in their rivers, using WFD standards and colour coding, from blue, very efficient policy (full river restoration), to red, inefficient measures, resulting in major degradation (summer oxygen levels occasionally below 2 mg O_2 L^{-1})

3.4   Models Enable Exploration of Socio-hydrosystem Scenarios and Their Consequences

Models have been spatially extended and gradually developed to incorporate more processes and interactions. On the basis of an analysis of trajectories of the major feature in the Seine basin, Riverstrahler made it possible to reconstitute the past evolution of water quality in the drainage network. A further useful feature of models is the exploration of the future of the hydro-socio-system through scenarios of sectorial changes in combination (or not) with climate change scenarios and simulation of the related river response [26]. These scenarios may simulate 'business as usual' (same trends in control factors), stepwise gradual changes or drastic changes ('what if' scenarios), which generate different sets of spatialized results on river state.

The model can be operated to feed scenarios, or the scenarios provide a new set of controlling parameters to the model and input data set. Two major approaches have been applied: (1) The preventive approach explores the possible changes in pressures on the basin, e.g. diffuse sources from agriculture; and (2) the curative approach tests the impact of river remediation measures.

3.4.1   Preventive Approach: Simulation of Agricultural Changes in the Seine Basin

PIREN-Seine scientists have explored the transition from a conventional intensive cropping system, which has dominated in the basin for several decades, to generalized organic production in response to new food demand for a 'demitarian' diet (reduction by half of animal product consumption). The impact of this new type of agriculture is based on an experimental agronomic field study with a set of farmers changing their agricultural practices (for more details, see Sect. 4.1). The nitrogen fluxes in soils and unsaturated areas were measured in a dozen farms over 3–6 years. These results were then incorporated in the models and upscaled to the whole basin. A realistic scenario of a deep structural agricultural change that is autonomous and organic, with a reconnection of crop, livestock and a demitarian diet, has been elaborated at the scale of the Seine basin and compared with a scenario based on good agricultural practices (see Box 5).

Box 5

> Models to explore river responses to agricultural scenarios and territorial management in the Seine basin.
>   'We have named "territorial biogeochemistry", the branch of science that describes and tries to understand the functioning of such complex systems,

(continued)

their internal and external exchanges of material, and the (physical, chemical, biological, or social-economic) mechanisms controlling these exchanges. Here we open the way to a comprehensive and, why not, citizen-oriented way of practicing science, helping to clarify the societal choices to which we are confronted to address the threat of global change' [27].

Two contrasting scenarios have recently been developed by Billen et al. [28] for 2040. The first one represents a continuation of the current trends of opening and specialization (O/S scenario) into cropping systems based on chemically synthesized inputs, as is the case in the Seine basin. The second explores a shift to autonomy via organic farming and a reconnection of crop and livestock farming while reducing animal protein in human diets by half, the so-called demitarian diet (A/R/D scenario). These scenarios both involve the strict application of regulations on WWTPs.

The results (Fig. 6) show that the basic water quality in terms of nitrate contamination could be reduced in the upstream rivers but worsened in the downstream ones with the O/S scenario. The A/R/D scenario would considerably reduce the level in nitrate everywhere in the river network. Such a reduction in nitrate levels and fluxes would make it possible to reduce eutrophication in the coastal zone of the Bay of Seine (see [26]). Interestingly, this A/R/D scenario would reduce greenhouse gases from agriculture by 36%, compared with 16% for the O/S scenario [20].

The aim of the discussion was not to define a possible future, but to encourage reflection on the possible, and sometimes conflicting, trends, such as the continuation of market openness and specialization, of agricultural territories and a sober and virtuous future with a shift to organic farming and agricultural autonomy accompanied by a change in our diet.

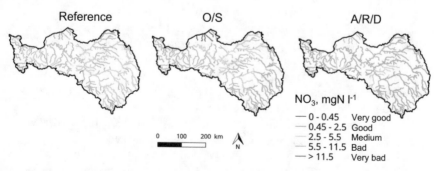

Fig. 6 Scenario outputs: spatial distribution of annual average nitrate pollution levels over the Seine Normandie river networks for the reference situation and for two contrasting scenarios for 2040: O/S, agriculture opening and specialization; A/R/D, autonomy, livestock reconnection and demitarian diet. Colour coding from blue, excellent, to red, very poor, according to the WFD (from [26])

3.4.2 Curative Approach: Simulated Fish Response to the Restoration of Degraded River Habitats

The freeware Anaqualand has been under development since 2002 at Irstea as a tool to support the development of a 'riverscape ecology' approach that explicitly considers connectivity in aquatic landscapes [29]. It was also designed to provide helpful results to bridge research and management preservation. Longitudinal and lateral connectivity for fish in rivers are modelled in 2D, through the calculation of hydrographic or functional distances (expressed in functional km). Functional connectivity is calculated according to the 'least cost' approach for which the environment facilitates or prevents the movement of fish from one life cycle habitat to another. This approach could be engaged to compare scenarios for the restoration of physical and chemical barriers that fragment river networks, thus making it possible to select priority actions and evaluate the relevance of restoration schemes for ecological continuity [30].

Anaqualand was used in the context of a SAGE (water development and management plan). Outputs from the ProSe hydraulic model were used as input parameters in Anaqualand, providing a map of fish migration probabilities based on participants' management decisions. This tool helps the local water commission (CLE) to define water resource management measures that take into account the different uses of the river. Similarly, several scenarios for restoring the ecological continuity of the Seine estuary to the Risle River for sea trout have been discussed with the Fishing Federation of the department of Eure to prioritize their actions (see Box 6).

Box 6

Modelling ecological continuity for sea trout under contrasting restoration scenarios (Anaqualand model).

The Risle River is the most downstream tributary of the Seine estuary, close to the Atlantic Ocean, and has a high potential for migratory species. It is subject to tidal influence, and no physical barriers impede the migration of fish up to Pont-Audemer, where there are a number of impassable obstacles. The Fishing Federation of the Eure department wanted to compare the actual ecological continuity of the Risle River (245 barriers) with 3 management scenarios for 66 barriers distributed along the main course and 2 tributaries (whereby 46 become passable and 20 are removed). In the short term, 21 initiatives are in preparation; in the medium term, 34 additional initiatives are under discussion, and in the long term, 11 more are planned.

Anaqualand models the accessibility of the Risle River, from mouth to headwaters (Fig. 7). The strong barrier effect for sea trout at Pont-Audemer is clearly shown in the present state, as well as in the short-term scenario, since no initiatives are envisaged in the near future for these most downstream

(continued)

barriers. In the medium-term scenario, the gain becomes significant, since these most downstream barriers are planned to be managed, with a total of 55 barriers becoming passable. The long-term scenario slightly improves accessibility, approaching the 'ideal' case without barriers, except for tributaries for which no management of barriers is planned as yet in the proposed scenarios.

3.5 Advantages and Limitations of Models for Researchers and Practitioners

Modelling tools offer numerous benefits to researchers and practitioners. They facilitate interaction and collaboration among different actors by providing a

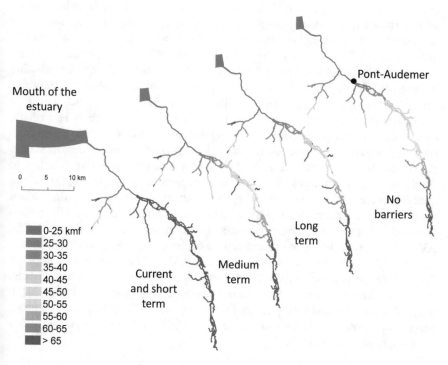

Fig. 7 Accessibility of the Risle River from the mouth of the estuary for sea trout under the different management scenarios measured by functional distances (classes of functional kmf). For the purposes of visualization, lines are enlarged and do not represent the actual width of the streams [31]

common reference. They not only provide a neutral frame for discussion, they also allow researchers and practitioners to achieve common goals without compromising individual objectives: The former conduct research to deepen their scientific under-standing, while the latter gain scientific evidence to support their decisions or justify their actions. With this in mind, the PIREN-Seine's choice of expert-based models can be understood as a way of maintaining the boundaries of science and practice, since emphasis on models as research tools subject to scientific validation and backed by the recognized expertise of PIREN-Seine researchers maintains a sense of credibility, legitimacy and trust.

But they pose just as many limitations and challenges. Scenarios are simplified representations of the world with uncertainties, despite steady progress over the past decades. The communication of modelling results must be clear about this, as the results generate passionate debates within society. While modelling results may be perceived as scientifically objective, the choice of what is modelled and how it is modelled (e.g. parameters considered, processes taken into account), as well as how the results are interpreted and used, is inherently subjective. Also, the generation of model knowledge and the simulation products potentially usable by practitioners need to be issued in a 'timely' manner in relation to the way the problems arise among the epistemic community. Those works that are not put on the policy agenda of the river basin authority might not be relevant and, even less, used. The co-construction of scenarios potentially explored by models is a good way to alleviate this difficulty, since it allows for collective reflection on possible futures.

The models have made it possible to envisage a paradigm shift (e.g. the transition to organic farming, source control and reduction of pollutant emissions, single vs. multiple sewage treatment plants in the Paris region), but they may encounter some socio-economic rationales. In some sectors, such as agriculture, models have not yet produced signals strong enough to generate drastic transforma-tions in practices or to overcome resistance to change. Also, model outputs should not contribute to the depoliticization of public decision by transferring decisions to purely technical committees. As a last point, dependence on expertise, the lack of available and/or reliable data, changing regulatory requirements demanding added functionality from the models and issues of propriety remain fundamental limitations to modelling. As a result, the Seneque model is no longer being used directly by the AESN, which now prefers to use an à la carte approach requesting modelling results from PIREN researchers when needed. The SIAAP, by contrast, has expressed a desire in recent years to move towards methods of real-time control and artificial intelligence involving the development of complex assimilation technologies [32]. It is up to researchers and their partners to see how they will develop their common patterns of practice, while the majority of operational partners still consider models to be important support tools.

4 Participatory Devices Generated by PIREN-Seine Researchers and Their Effects on Public Action

The participatory imperative is not specific to water management – participatory devices are foreseen for all policies – but it is one of the principles at the heart of the WFD (Article 14). The 1992 French water law already instituted basin committees for each major river basin district and local water commission by watershed. However, all water management procedures, such as catchment supply areas to produce drinking water, needed to integrate a participatory device henceforth.

The initiatives carried out over the past 10 years have taken several forms of partnership research, with researchers in multidisciplinary teams combining environmental and social sciences. Participation is understood as a means to facilitate the connection between actors and the production of common meanings. By working on controversial topics such as modifying agricultural practices to reduce diffuse pollution, or removing river weirs to restore ecological continuity, participation should not become a tool for accepting new rules or overcoming resistance. On the contrary, it must help to collectively identify a problem and then explore various levers of action, without one solution being put forward over another. Participants do this on a voluntary basis. The facilitation methods are fundamental in order that each participant can find their place and, above all, express themselves.

4.1 Participatory Experiments Conducted by PIREN-Seine Researchers

The first form of partnership research is to build scientific knowledge on a territory. Some experiments are designed to feed scenarios specific to the PIREN-Seine, where participants are co-constructing the data that feed models and scenarios. Others aim to build a dialogue between actors on a controversial issue by questioning the production of scientific knowledge and its appropriation.

4.1.1 Co-constructing Data for Models and Scenarios (ABAC Project)

Between 2012 and 2018, the ABAC project and network involved farmers from various parts of the Île-de-France region, as well as water agencies, farmers' federations and chambers of agriculture. The project is participatory given that farmers were, as far as possible, involved in sampling operations and interpreting results. A steering committee was set up and met every year to discuss the results and possibly reorient the project. The farmers' network made it possible to compare nitrogen leaching on their farms according to their practice [33]. The database was used for scientific publications but also by regional and local partners. Verbal reports of the results were regularly produced during meetings and formed the basis for exchanges between the actors.

4.1.2 Exploring River Restoration Scenarios in Model-Supported Participatory Experiments (Grand Morin Project)

In France, the ecological management of watercourses according to the WFD partly involves weir and dam removal to restore fish and sediment circulation. This approach generates conflicts between some river users and public services. Between 2010 and 2013, a number of PIREN-Seine researchers proposed their models to the local river commission of the Grand Morin River (CLE), a tributary of the Marne to the east of the Paris, which includes public sector services, the river union, the fishermen's association, the canoe/kayak clubs, elected officials and mill owners. The expectations of the commission members at the beginning of the experiment partly reflected their divergence: for the public sector services, the experiment should facilitate weir removal by silencing the controversies over the interest in and effectiveness of levelling, whereas most members of the CLE oppose the limited and partial nature of the scientific information on the subject (ecological interest in suppressing weirs).

The ProSe model was chosen to simulate flow variations on the watercourse and the Anaqualand model for the ability of fish to move along the river. Several workshops and field trips made it possible to collectively define the question to be addressed (What happens to the river and the ecosystem by manipulating the weirs and varying the water level?) and the actors concerned. They made it possible to explain the information provided by the two models, as well as their limitations, so that the actors could evaluate for themselves the effects of the weirs on the hydraulic functioning of the river, on its uses (kayaking, mills) and the circulation of three species of fish.

The interactive platform made it possible to simulate management of the works by various actors according to objectives chosen by all the participants based on the river discharge, which was also chosen by the participants. Each participant can play a different role. The model simulates the effects of the decisions made on the chosen indicators (water level in the diverted reaches, crossing of structures by canoes or different types of fish species). The results were displayed and discussed according to the objectives of each operator. Management decisions were evaluated collectively, thereby allowing for new simulations [34].

4.1.3 Building Territorial Dialogue to Act on Agricultural Diffuse Pollution (BRIE'eau Project)

The BRIE'eau project (2016–2020) aims to make the particularly controversial issue of agricultural water pollution (nitrates and pesticides) in the Brie region of Seine-et-Marne, a subject of debate. In this territory of large cereal crops, agricultural activity

Fig. 8 An example of the cards used (1). Workshop 'Imagine together scenarios for the evolution of the territory' (2). Workshop around the role-playing game (3) (photo courtesy of IRSTEA)

has led to a sharp deterioration in water quality of the limestone groundwater of Champigny, the main groundwater resource in Île-de-France. Since 2005, AQUI'Brie, an association of users of the Champigny aquifer, and Irstea researchers have been developing local initiatives to reduce pesticide flows from drained agricultural land. Artificial wetlands have been established on agricultural plots as landscape interfaces with a hydrological buffer function, which improve water quality parameters [35].

The project consists of a joint evaluation process of these land cover devices on water quality and the improvement of biodiversity. The challenge of the participatory approach is to create the conditions for dialogue between local actors, particularly those in the agricultural and drinking water sectors. To improve water quality, two levers for action are placed at the heart of the process: changes in agricultural practices and landscape design. The participatory approach consists of three complementary steps, each of which involves a specific dialogue tool (Fig. 8): (1) a set of cards to share the diversity of perceptions on buffer zones; (2) a scenario-building tool integrating buffer zones as a possible solution [36]; and (3) a role-playing game to simulate individual and collective action to be implemented.

The cards were used during individual interviews to generate highly varying or even opposing discourses among the respondents based on a question common to all actors. The different visions were then presented in a workshop. Two contrasting scenarios were constructed and discussed: one based on maximizing farmers' gross margin and the other on minimizing pressure in plant protection products. Participants expressed what they found interesting or problematic in these scenarios and also identified priority elements for improvement. The role-playing game brought together farmers, agricultural advisors, cooperatives and elected officials responsible for drinking water in order to identify the levers and obstacles to the implementation of individual or collective initiatives.

4.2 Participatory Experiments Influence Collective Action

It was the researchers themselves who first benefitted from the results, with the production of new research themes, the acquisition of data and the effectiveness of the flow of knowledge produced towards the other actors.

In the ABAC network, the collaboration with farmers who have totally or partially converted their farms to organic farming, or others who are reluctant to do so, was extremely rewarding. The community of researchers was the first to change, in terms of its science and messages to partners, with another perspective from the National Institute of Agricultural Research (INRA) on organic agriculture and the difficulties of the socioecological transition. Beyond the unifying and participatory activities, participants were looking to build up a common vision on the challenges and the means to reconcile environmentally friendly agriculture with farmers' incomes while maintaining or improving water quality.

In the Grand Morin experiment, the evaluation was carried out within the framework of a thesis [37] based on observations made during the workshops, interviews with the participants and, finally, decisions taken by the management of river works on the Grand Morin River. Concerning the mobilization of scientific knowledge, the members of the local river commission (CLE) found that, compared with the data provided by consulting firms, working with the scientists and constructing knowledge through modelling allow them to truly assimilate the results. The participants said they had a better perception of the interactions between the various elements of the system, the complexity of the management of the sluices, their positive effects on fish and kayak movement or the uselessness of certain weirs. For some participants, this way of sharing knowledge would even be a means to change the modes of public action. 'The infused science delivered in meetings does not make it possible to appropriate things. And it is just how we can join projects. In thematic commissions in the CLE there are exchanges but no co-construction. One could imagine the life of a SAGE[2] with tools allowing a real implication' (Chamber of Agriculture).

In the BRIE'eau project, attention is also paid to questions such as 'How do we learn?' or 'What could have facilitated this learning?'. The underlying assumption is that some tools would facilitate certain types of knowledge; for example, the scenario-building approach would lead to the construction of knowledge and technical skills, while the role-playing game is more likely to facilitate political, relational and communicational learning. These learning effects are less expected results and are often made invisible when trying to evaluate them. However, they are likely to influence public action – the way in which the actors of a territory discuss and decide together.

These experiments make it possible to mobilize other types of data in the debates and to advance the discussions. 'The ABAC project has enabled us to fill the obvious gap in baseline data on organic agriculture and to interpret its environmental effects in terms of water quality' (said by a PIREN-Seine researcher). By instrumentalizing approximately 15 typical crop rotations on different pedo-climatic conditions, the scientists aimed to obtain, within a comparatively short period of time, scientifically original results that would also be useful from a societal point of view for regional policies. The Île-de-France region is mobilized to promote rural and peri-urban areas

---

[2]SAGE: water management scheme.

but also seeks to quantify the services provided by organic farming. While consumer demand for organic products is steadily increasing, organic farming is not yet meeting demand in the Île-de-France region. The region is therefore determined to increase the proportion of organic farming to promote healthy and tasty food, at both the individual and the collective level, especially for school catering.

In the Grand Morin project, by playing with the interactive platform, elected officials and public services accepted the idea of only removing certain river weirs (while at the beginning of the experiment, elected officials did not want to remove any). 'Compromise should be to have some sections free and to deal with black spots of pollution with locks' (Mayor on the Grand Morin River). However, once the experiment was over, the actors resumed their strategic position, and some elected officials went back on weir removal discussed during the experiment. Finally, the weirs to be removed were chosen by the state services on the basis of criteria that were external to the local actors. Here we see the limits of participatory mechanisms to act on European regulatory standards imposed on local actors and the possibility of territorial innovation. Nevertheless, in 2014, the CLE of the Grand Morin River amended its watercourse management rules by recognizing both the heritage character of mill structures and its effects on river flow continuity and by asking property owners to manage them in winter in coordination with the public services.

It seems that this type of approach is able to create interest or a commitment from certain actors, which would then lead them towards a more operational reflection. However, in order to promote such effects, it is important to ensure continuity between this new participatory research space and the usual decision-making arenas in the region. For instance, in the BRIE'eau system, the main actors are the local community responsible for the catchment area, and the AQUI'Brie association is responsible for its animation. It is this latter actor – who has worked closely together on this applied research project – who will then be able to learn the lessons of this approach in the pursuit of his work of territorial animation.

5   Conclusion

5.1   Evolution of the Programme and Scientific Methods

The PIREN-Seine scientific practices have gradually evolved over 30 years. The programme was initially created at the request of river basin institutions to produce regulatory science to improve the quality of the Seine River. Scientists soon understood that basic scientific knowledge was lacking (e.g. on river processes, on water circulation) and that local issues could not be fully understood and resolved without a consideration of the whole basin territory. Hydrologists, geochemists and biogeochemists later realized that both the river network and the river basin that they were studying at time t were actually dependent on the past and present interactions of multiple actors and institutions and that history, geography, sociology, environmental policies and other disciplines needed to be part of the programme. Participatory social experiments were developed together with actors in the later phases, usually at the local level, using some of the tools (models, decision support systems) developed by the programme.

5.1.1 Seine System Design

The analysis of the Seine River system evolved from a consideration of the river hydrosystem (physical, chemical and ecological, including the connected waterbodies delimited by the basin boundaries) to a hydro-socio-ecosystem. It includes the river network, from headwater to the coast – the land–ocean continuum – the drainage basin, all human activities, institutions, regulations and the circulation of material, goods and products on the whole basin. Researchers were forced to recognize that this system is wide open to regional and global (transcontinental) effects (e.g. economics, regulations, consumer habits, production methods, import and export of products). For many products that characterize river quality and river fluxes, their circulation within the Seine basin is between one and two orders of magnitude higher than their export via the Seine River.

5.1.2 Research Methods and the Position of Scientists

The programme now includes three modes of research (question-oriented, fundamental and action research), with their various modes of valorization and exchange between three different communities (academics, river and basin stakeholders, local river users and managers) (Fig. 9). The PIREN-Seine has therefore developed a complex set of relationships with these communities on different levels: local with the general public, river associations and water commissions, basin-wide with institutions and global with peer scientists. Several types of products (scientific publications, reports to stakeholders, models and software, databases, leaflets and brochures, conferences, participatory meetings, etc.) are now generated by the programme, each adapted to a different audience. The current position of scientists with regard to the three communities is schematically presented in Fig. 9.

5.1.3 Programme Governance

From its inception, the project has been more on a bottom-up basis – conceived and elaborated by scientists with their counterparts at the institutions in question – than on a top-down basis, agreement between a dozen research institutions and as many stakeholders; therefore, its current governance is the result of 30 years of collaboration between a group of individual scientists and their individual counterparts among stakeholders: The research agenda is mostly generated by scientists, then discussed at the basic level and, finally, proposed and adopted by the funding partners. The current links between scientific financial partners, local actors and the scientific

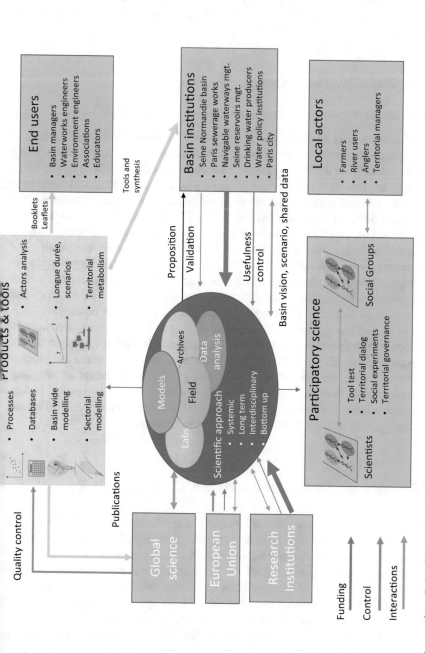

Fig. 9 Present-day relationships between the PIREN-Seine programme and its scientific institutions (left side) and partners (right side), as well as its knowledge transfer

community show the growing complexity of a programme of this kind (Fig. 9). As such, the PIREN-Seine has survived many institutional transformations in the field of environmental research and is now one of the oldest continuous programmes in this field.

5.1.4   Uses of Programme Outputs

The scientific outputs of the programme are now used in multiple ways:

- Models are used by the river authority to produce missing water quality data, to simulate future trends of many water quality indicators (e.g. basic chemistry, fish circulation) with different scenarios for each waterbody, as defined by the WFD. They are also used for river management planning at the local and basin-wide levels. Land-to-ocean models are also used as input to coastal zone models. The PIREN-Seine's model cascade of N, P, C and Si, from the plot scale to the ocean and from the past 70 years to the next 50 years, has no equivalent in other basins.
- Specific studies have been done on basic water quality on the whole river network, and chemical and microbiological impacts of Paris megalopolis, as requested by institutional partners 30 years ago. Their outputs can now be used to question preventive vs. curative environmental solutions, and production vs. consumption strategies (e.g. for agrochemicals, cosmetics).
- The project has numerous pioneering studies on both a national and an international level. In addition to the model cascade, the following investigations are particularly worthy of note: (1) the basin-wide coupling of surface and groundwaters within the context of climate change; (2) the detection of emerging contaminants such as antibiotics and microplastics; (3) the basin-wide history and circulation of persistent contaminants from sediment archives; and (4) longue durée greenhouse gas budget for a river basin territory.

5.2   Scientific Knowledge and Action

Scientific knowledge and action are linked but in a complex way. Action can be taken without the production of knowledge, and knowledge does not always trigger action. Actors must be regularly informed about scientific progress in a suitable way to generate action. Scenarios built up by scientists and their partners must be discussed by members of the river basin committee (the deliberative level). Many water quality issues have not been, or could not be adequately, addressed in the past by scientists due to a lack of fundamental research (e.g. in the field of ecotoxicology and long-term effects of low-level contamination on human health); technical means (lack or limitation of analytical means for most micropollutants before the 1990s; lack of GIS before the 1990s); or available databases (e.g. the historical population database was only completed in the mid-1990s; most databases on material circulation are only available at the national level). Finally, the necessary multidisciplinary character of environmental sciences was largely lacking in France when the programme began.

5.2.1 Knowledge Issues to Be Addressed in the Future

The programme has identified many remaining issues that will need to be addressed in future phases, such as:

- Modelling the transfer of particulate microcontaminants in rivers
- Analysing transfer of contaminants across trophic levels
- Coupling physicochemical quality and ecological quality, as measured by the WFD, and its modelling
- Coupling high-frequency chemistry records with models
- Linking river/water quality and human health, now and in the past based on reconstituted archives
- Taking into account perception, evaluation and representation of the quality of the river and water resources, now and in the past, within and outside the regulatory domain
- Developing scenarios of future expectations of the public and water actors regarding the waterbodies and the basin in the context of global change and changing expectations, such as the increasing interest in the recreational use of rivers by the public, for example, river bathing in urban rivers

The continuity of the PIREN-Seine programme in the near future will therefore depend on several conditions: (1) ensuring up-to-date mono- and multidisciplinary research recognized at an international level; (2) continuing to develop tools and devices usable by the funding partners; (3) ensuring the dissemination of research to the various end users; (4) developing activities and research across scales, from local river issues to questions relating to the entire basin and the land–ocean continuum; and (5) continuing to consider socio-ecosystem trajectories over a wide time window (200 years and more) – reconstruction of the past, present observations and near-future scenarios.

5.2.2 PIREN-Seine Governance Challenges

Since the PIREN-Seine operates on a voluntary basis and at the individual/small-team level and not at the laboratory/research-institution level, the programme will need to seek suitable scientists willing to fill these gaps in the future, as has been done so far at each of the seven stages of the programme (see [38]). Ecologists will join the next stage of the programme (2020–2023) to study the effects of changes in water circulation as a result of human activities on terrestrial ecosystems and biodiversity dynamics. PIREN-Seine researchers also want to work more closely with river managers and river users. During the flooding of the Seine in June 2016, they called on local residents to provide water samples taken from flooded banks in order to monitor contaminants [39]. The systematization of this participatory scientific approach would be particularly useful during extreme events that measuring stations are unable to monitor. However, a network of individuals may be difficult to set up due to the episodic and random nature of events.

Another development for the next stage of the programme will be the formalization of researcher–user interaction to valorize on a local level the results of the measurement stations and the modelling outputs. This work has already been done with river managers to monitor the physical effects of weir removal on the Orge River [40]. These initiatives of co-construction and valorization of research results will be continued at selected experimental sites (urban rivers around Paris, the Bassée floodplain ($150 \ km^2$)).

It will also be the task of stakeholders to ensure that these issues meet the expectations of the partners, always driven by regulatory constraints (e.g. ecosystemic services, effects of restoration of aquatic continuity on biodiversity, priority pollutants, good ecological state, etc.) with new issues arising (river water warming, 2024 Olympic Games) and the involvement of new partners (see [41]). The challenge is to take into account as accurately as possible the multiplicity of uses and points of view on the Seine basin socio-ecosystem and the resulting management complexity, as well as to achieve a shared vision of the hydrosystem as a common asset subject to public action, within a co-construction involving all actors.

Acknowledgements This work is a contribution to the PIREN-Seine research programme (www. piren-seine.fr), which belongs to the Zone Atelier Seine part of the international Long-Term Socio-Ecological Research (LTSER) network.

References

1. Jasanoff S (1998) The fifth branch: science advisors as policy makers. Harvard University Press, Cambridge
2. Gibbons M, Limoges C, Nowotny H et al (1994) The new production of knowledge: the dynamics of science and research in contemporary societies. SAGE, London
3. Nowotny H, Scott P, Gibbons M (2006) Re-thinking science: mode 2 in societal context. In: Carayannis EG, Campbell DFJ (eds) Knowledge creation, diffusion, and use in innovation networks and knowledge clusters. A comparative systems approach across the United States, Europe, and Asia. Praeger, Santa Barbara, pp 39–51
4. Hatchuel A (2000) Intervention research and the production of knowledge. In: LEARN Group (ed) Cow up a tree. Knowing and learning for change in agriculture. Case studies from industrialised countries. INRA, Paris, pp 55–68
5. Flipo N, Lestel L, Labadie P et al (2020) The Seine River basin. In: Flipo N, Labadie P, Lestel L (eds) The Seine River basin. Springer, Cham. https://doi.org/10.1007/698_2019_437
6. Dmitrieva T, Lestel L, Meybeck M et al (2018) Versailles facing the degradation of its water supply from the Seine River: governance, water quality expertise and decision making, 1852–1894. Water Hist 10(2–3):183–205

7. Ayrault S, Meybeck M, Mouchel JM et al (2020) Sedimentary archives reveal the concealed history of micropollutant contamination in the Seine River basin. In: Flipo N, Labadie P, Lestel L (eds) The Seine River basin. Springer, Cham. https://doi.org/10.1007/698_2019_386

8. Raimonet M, Vilmin L, Flipo N et al (2015) Modelling the fate of nitrite in an urbanized river using experimentally obtained nitrifier growth parameters. Water Res 73:373–387

9. Vilmin L, Flipo N, de Fouquet C et al (2015) Pluri-annual sediment budget in a navigated river system: the Seine River (France). Sci Total Environ 502:48–59

10. Vilmin L, Flipo N, Escoffier N et al (2018) Estimation of the water quality of a large urbanized river as defined by the European WFD: what is the optimal sampling frequency? Environ Sci Pollut R 25(24):23485–23501. https://doi.org/10.1007/s11356-016-7109-z

11. Tamtam F, Mercier F, Le Bot B et al (2008) Occurrence and fate of antibiotics in the Seine River in various hydrological conditions. Sci Total Environ 393(1):84–95

12. Lestel L, Meybeck M, Thévenot D (2007) Metal contamination budget at the river basin scale: an original Flux-Flow Analysis (F2A) for the Seine River. Hydrol Earth Syst Sci 11 (6):1771–1781

13. Lestel L (2012) Non-ferrous metals (Pb, Cu, Zn) needs and city development: the Paris example (1815–2009). Reg Environ Change 12(2):311–323

14. Meybeck M, Lestel L, Bonté P et al (2007) Historical perspective of heavy metals contamination (Cd, Cr, Cu, Hg, Pb, Zn) in the Seine River basin (France) following a DPSIR approach (1950–2005). Sci Total Environ 375(1):204–231. https://doi.org/10.1016/j.scitotenv.2006.12.017

15. Teil MJ, Tlili K, Blanchard M et al (2014) Polychlorinated biphenyls, polybrominated diphenyl ethers, and phthalates in roach from the Seine River basin (France): impact of densely urbanized areas. Arch Environ Contam Toxicol 66(1):41–57

16. Goutte A, Alliot F, Budzinski H et al (2020) Trophic transfer of PAH, phthalates, and their metabolites in an urban river food web. Environ Sci Technol (submitted)

17. Even S, Billen G, Bacq N et al (2007) New tools for modelling water quality of hydrosystems: an application in the Seine River basin in the frame of the water framework directive. Sci Total Environ 375:274–291

18. Le Noë J, Billen G, Garnier J (2017) Nitrogen, phosphorus and carbon fluxes through the French Agro-Food System: an application of the GRAFS approach at the territorial scale. Sci Total Environ 586:42–55. https://doi.org/10.1016/j.scitotenv.2017.02.040

19. Anglade J, Billen G, Garnier J (2015) Relationships for estimating N_2 fixation in legumes: incidence for N balance of legume-based cropping systems in Europe. Ecosphere 3:37

20. Garnier J, Le Noë J, Marescaux A et al (2019) Long-term changes in greenhouse gas emissions from French agriculture and livestock (1852–2014): from traditional agriculture to conventional intensive systems. Sci Total Environ 660:1486–1501. https://doi.org/10.1016/j.scitotenv.2019.01.048

21. Lazure P, Dumas F (2008) An external-internal mode coupling for a 3D hydrodynamical model for applications at regional scale (MARS). Adv Water Resour 31(2):233–250. https://doi.org/10.1016/j.advwatres.2007.06.010

22. Romero E, Garnier J, Billen G et al (2019) Modeling the biogeochemical functioning of the seine estuary and its coastal zone: export, retention, and transformations. Limnol Oceanogr 64 (3):895–912. https://doi.org/10.1002/lno.11082

23. Ruelland D, Billen G, Brunstein D et al (2007) SENEQUE: a multi-scaling GIS interface to the Riverstrahler model of the biogeochemical functioning of river systems. Sci Total Environ 375 (1–3):257–273

24. Chong N, Bach PM, Moilleron R et al (2017) Use and utility: exploring the diversity and design of water models at the science-policy interface. Water 9(12):983

25. Lestel L, Carré C (eds) (2017) Les rivières urbaines et leur pollution. Quae, Paris

26. Garnier J, Billen G (2016) Ecological processes and nutrient transfers from land to sea: a 25-year perspective. In: Glibert PM, Kana TM (eds) Aquatic microbial ecology and biogeochemistry: a dual perspective. Springer, Cham, pp 185–197

27. Billen G, Le Noë J, Garnier J (2018) Two contrasted future scenarios for the French agro-food system. Sci Total Environ 637–638:695–705. https://doi.org/10.1016/j.scitotenv.2018.05.043

28. Garnier J, Marescaux A, Guillon S et al (2020) Ecological functioning of the Seine River: from long term modelling approaches to high frequency data analysis. In: Flipo N, Labadie P, Lestel L (eds) The Seine River basin. Springer, Cham. https://doi.org/10.1007/698_2019_379

29. Le Pichon C, Gorge G, Faure T et al (2006) Anaqualand 2.0: freeware of distances calculations with frictions on a corridor. https://www6.rennes.inra.fr/sad/Outils-Produits/Outils-informatiques/Anaqualand. Accessed 15 July 2019

30. Roy ML, Le Pichon C (2017) Modelling functional fish habitat connectivity in rivers: a case study for prioritizing restoration actions targeting brown trout. Aquat Conserv 27(5):927–937

31. Le Pichon C, Alp M (2018) Projet ANACONDHA. Analyse spatiale de la connectivité des habitats fonctionnels pour les poissons à l'échelle de l'estuaire. Rapport de recherche du programme Seine-Aval 5. https://www.seine-aval.fr/wp-content/uploads/2017/02/RR-SA5-ANACONDHA.pdf. Accessed 15 July 2019

32. Wang S, Flipo N, Romary T (2019) Oxygen data assimilation for estimating micro-organism communities' parameters in river systems. Water Res 165:115021. https://doi.org/10.1016/j.watres.2019.115021

33. Benoit M, Garnier J, Beaudoin N et al (2016) A participative network of organic and conventional crop farms in the Seine Basin (France) for evaluating nitrate leaching and yield performance. Agr Syst 148:105–113

34. Carré C, Haghe JP, De Coninck A et al (2014) How to integrate scientific models to switch from flood river management to multifunctional river management. Int J River Basin Manag 12 (3):231–249. https://doi.org/10.1080/15715124.2014.885439

35. Tournebize J, Chaumont C, Mander U (2017) Implications for constructed wetlands to mitigate nitrate and pesticide pollution in agricultural drained watersheds. Ecol Eng 103:415–425

36. Gisclard M, Chantre É, Cerf M et al (2015) Co-click'eau: une démarche d'intermédiation pour la construction d'une action collective locale? Nat Sci Soc 23:3–13

37. De Coninck A (2015) Faire de l'action publique une action collective: expertise et concertation pour la mise en œuvre des continuités écologiques sur les rivières périurbaines. Dissertation, Université Paris Est

38. PIREN Seine Seine Teams. https://www.piren-seine.fr/fr/les-equipes. Accessed 15 Sept 2019

39. Flipo N, Mouchel JM, Fisson C (2017) Les effets de la crue de juin 2016 sur la qualité de l'eau du bassin de la Seine, vol 17. PIREN-Seine, Paris. Available via PIREN-Seine. https://www.piren-seine.fr/fr/fascicules/les-effets-de-la-crue-de-juin-2016-sur-la-qualité-de-leau-du-bassin-de-la-seine. Accessed 15 Aug 2019

40. Bellot C (2014) Evolution du fonctionnement sédimentologique et biogéochimique d'un bief de rivière suite à l'effacement d'ouvrages hydrauliques. Dissertation, Université Pierre et Marie Curie

41. PIREN-Seine Partners. https://www.piren-seine.fr/fr/partenaires. Accessed 15 Sept 2019

Correction to: The Seine River Basin

Nicolas Flipo, Pierre Labadie, and Laurence Lestel

Correction to:
Nicolas Flipo, Pierre Labadie, and Laurence Lestel (eds.),
The Seine River Basin, **The Handbook of Environmental**
Chemistry, https://doi.org/10.1007/698_2019_379,
https://doi.org/10.1007/698_2019_380, https://doi.org/
10.1007/698_2019_381, https://doi.org/10.1007/
698_2019_382, https://doi.org/10.1007/698_2019_383,
https://doi.org/10.1007/698_2019_384, https://doi.org/
10.1007/698_2019_385, https://doi.org/10.1007/
698_2019_386, https://doi.org/10.1007/698_2019_392,
https://doi.org/10.1007/698_2019_393, https://doi.org/
10.1007/698_2019_396, https://doi.org/10.1007/
698_2019_397, https://doi.org/10.1007/698_2019_407

The below and on the next page mentioned chapters of this volume were published
online first with the copyright year 2019 instead of 2020. This has now been corrected.

The updated online versions of the chapters can be found at
https://doi.org/10.1007/698_2019_379
https://doi.org/10.1007/698_2019_380
https://doi.org/10.1007/698_2019_381
https://doi.org/10.1007/698_2019_382
https://doi.org/10.1007/698_2019_383
https://doi.org/10.1007/698_2019_384
https://doi.org/10.1007/698_2019_385
https://doi.org/10.1007/698_2019_386
https://doi.org/10.1007/698_2019_392
https://doi.org/10.1007/698_2019_393
https://doi.org/10.1007/698_2019_396
https://doi.org/10.1007/698_2019_397
https://doi.org/10.1007/698_2019_407

Nicolas Flipo, Pierre Labadie, and Laurence Lestel (eds.), *The Seine River Basin*,
Hdb Env Chem (2021) 90: C1–C2, https://doi.org/10.1007/698_2020_667,
© The Author(s) 2020

1. Ecological Functioning of the Seine River: From Long-Term Modelling Approaches to High-Frequency Data Analysis. DOI: 10.1007/698_2019_379
2. Changes in Fish Communities of the Seine Basin over a Long-Term Perspective. DOI: 10.1007/698_2019_380
3. Contaminants of Emerging Concern in the Seine River Basin: Overview of Recent Research. DOI: 10.1007/698_2019_381
4. Mass Balance of PAHs at the Scale of the Seine River Basin. DOI: 10.1007/698_2019_382
5. Aquatic Organic Matter in the Seine Basin: Sources, Spatio-Temporal Variability, Impact of Urban Discharges and Influence on Micro-pollutant Speciation. DOI: 10.1007/698_2019_383
6. Experience Gained from Ecotoxicological Studies in the Seine River and Its Drainage Basin Over the Last Decade: Applicative Examples and Research Perspectives. DOI: 10.1007/698_2019_384
7. How Should Agricultural Practices Be Integrated to Understand and Simulate Long-Term Pesticide Contamination in the Seine River Basin? DOI: 10.1007/698_2019_385
8. Sedimentary Archives Reveal the Concealed History of Micropollutant Contamination in the Seine River Basin. DOI: 10.1007/698_2019_386
9. Pluri-annual Water Budget on the Seine Basin: Past, Current and Future Trends. DOI: 10.1007/698_2019_392
10. The Seine Watershed Water-Agro-Food System: Long-Term Trajectories of C, N and P Metabolism. DOI: 10.1007/698_2019_393
11. The Evolution of the Seine Basin Water Bodies Through Historical Maps. DOI: 10.1007/698_2019_396
12. Bathing Activities and Microbiological River Water Quality in the Paris Area: A Long-Term Perspective. DOI: 10.1007/698_2019_397
13. Past and Future Trajectories of Human Excreta Management Systems: Paris in the Nineteenth to Twenty-First Centuries. DOI: 10.1007/698_2019_407

Index

A

Achères-Seine aval WWTP, 7, 195, 221, 276, 396, 397
AESN, 10, 18, 21, 165, 277, 278, 394, 402
Agricultural scenarios, 209, 398
Agriculture, 3, 4, 9, 21, 34, 43–45, 53, 93, 96, 98, 100–102, 104, 110, 111, 119, 121, 124, 135, 138, 157, 167, 208, 391, 396, 398, 399, 402, 403, 406
Agro-food system, 2, 94, 100, 110–111, 118, 207, 210
Alkylphenols (APs), 247, 248, 277, 279, 281, 282
Anaqualand, 18, 88, 400, 404
Anthropocene, x, 2–4, 10, 13, 53, 61
Anthropogenic, 3, 11, 31–36, 42, 51, 63, 82, 100, 164, 170, 175, 202, 218, 262, 309, 357
Anthropogenic pressure, ix, 67, 69, 71, 244, 286, 313, 317, 356, 364, 367, 383–389
Anthroposphere, 11, 275, 277, 293
Antibiotics, 279, 284–285, 291, 295, 357, 370–375, 388, 410
Antiobiotic resistance (clinical integron), 357, 370–374
Aquifers, 3, 50, 61, 95, 144, 191, 319, 383
ArchiSeine, 38–39
ARSeine, 95, 98, 99, 103, 147, 152
Atlantic Multi-decadal Variability (AMV), 62, 63
Atmospheric deposition, 98, 102, 167–171, 175, 179–181, 364
Atrazine, 142–146, 148–154, 156
Austerlitz gauging station, 61–63, 222

B

Baseflow, 68, 194
Bassée, 32–34, 47–50, 53
Bassée floodplain, 5, 45, 47–50, 280, 412
Bathing, 323–349, 411
Bathing regulation, 327, 329, 336
Bièvre, 34, 46
Binding site, 230–233, 236
Bioaccumulation, 250, 258–260, 263, 365, 366, 368, 389
Bioavailability, 219, 228–230, 235–236, 238, 244, 259–263, 365
Biodegradable, 204, 222, 223, 335
Biodegradable dissolved organic carbon (BDOC), 222, 223
Biogeochemical cycles, 10, 11, 143, 238
Biological index (BIX), 223, 225
Biomarkers, 245, 250–261, 264, 375
Biomonitoring, 245, 250–259, 264, 368, 369, 389
Bougival, 5, 132, 204, 245–250, 256, 257, 261–263, 364, 366, 368, 369, 372, 373
BRIE'eau project, 404–406
Buache, P., 40

C

Cadmium (Cd), 230, 236, 271, 275, 277–279, 283, 285–286, 289–293, 373, 388
Caging, 251, 369
Caméré, 41
Carbon dioxide (CO_2), 34, 92, 100–102, 192, 201–206
Carbon sequestration, 100–102

CaWaQS, 18, 20, 22, 61, 65–68, 82, 388, 391
Chanoine, 52
Circularity, 119, 121–129, 131, 133–137
CLE, 400, 404, 406, 407
Climate change, 3, 17, 22, 60, 61, 71, 73, 74,
    82, 100, 102, 118, 209–210, 317–319,
    374, 391, 398, 410
Coastal zone, x, 13, 17, 191, 207–209, 211,
    290, 392, 393, 399, 410
Complexation, 229–231, 233, 235, 236
Contamination, 11, 45, 93, 132, 142, 164, 191,
    219, 245, 271, 314, 330, 358, 388
Core dating, 272, 282
Cores, 5, 10, 68, 271–289, 291, 293–295,
    312–315
Cropping systems, 93, 95, 96, 98, 100, 149,
    154, 156, 207, 398, 399
Crop rotations, 9, 95, 98, 99, 145, 149, 151,
    152, 156, 392, 406
Cryptosporidium, 330, 343, 345, 346, 367–369

D
Dam impact, 274
Daphnia magna, 235–236, 238, 366
Database, 14, 18, 20, 23, 31, 38–39, 66, 95, 98,
    99, 103, 142, 143, 145–147, 149, 152,
    154–158, 165, 167–173, 175–179, 181,
    183, 193, 194, 205, 303, 341, 391, 403,
    408, 410
Data integration, 260
Deethylatrazine (DEA), 142–145, 148–152,
    154–156
Digestive enzymes, 253–256
Discharge, 3, 31, 60, 107, 122, 150, 166, 190,
    219, 245, 274, 306, 325, 361, 386
Dissolved organic carbon (DOC), 220,
    222–224, 227–229
Dissolved organic matter (DOM), 218, 219,
    221–238
Dissolved oxygen, 203, 306, 311, 313, 392
Ditches, 32–34, 44, 46, 53
Domestic and industrial releases, 167

E
Ecosystem respiration, 196, 197, 203
EEM fluorescence spectroscopy, 223, 225
Effluent dissolved organic matter (EfDOM),
    227–338
Emissions, 14, 18, 69, 71, 100–103, 111, 122,
    133, 165, 166, 169, 170, 178–180, 182,
    192, 196, 197, 203, 205, 206, 220, 223,
    285, 287, 289, 292, 356, 374, 402
Endocrine disruptors (EDs), 245–250, 260, 349

Enrichment factors (EFs), 274, 283, 285–287,
    291
Environmental fluxes, 181
Environmental risk assessment (ERA), 259
Erosion, 11, 31, 34, 42–46, 51, 103–104, 107,
    108, 167, 168, 174, 177, 178, 181–183,
    193, 201, 227, 275, 278, 288, 340
Escherichia coli, 329–343, 346–349, 367
Estuary, x, 3–5, 7, 22, 30, 31, 34, 45, 50, 52,
    131, 194, 197, 199, 202, 205, 211, 264,
    271, 280–283, 285–287, 294, 295, 306,
    311–315, 318, 359, 386, 389, 391, 392,
    400, 401
Evapotranspiration, 6, 65, 74, 76, 77
Excess load, 276, 291

F
Faecal indicator bacteria, 367, 375
Fish, 3, 32, 132, 195, 245, 294, 302, 349, 365,
    388
Fish habitat, 42–45
Fish index, 312, 316
Floodplain, 5, 10, 13, 30–34, 44–50, 52, 53,
    167, 168, 177–178, 274–275, 277–281,
    288, 291, 412
Forest filter effect, 168, 174, 175, 178, 181
4‰ initiative, 100, 102
Functional species traits, 309

G
Garonne, 3, 21, 271, 287, 289, 292, 293
Generalized representation of the agro-food
    system (GRAFS), 93–95, 97, 99, 100,
    103, 104, 194, 392, 393
Genotoxicity, 250, 256, 258, 260
Geological structure, 6
Giardia, 330, 345–347, 367–369
Grand Morin River, 5, 22, 221, 224, 404, 406,
    407
Greater Paris, 13, 20, 339, 357, 358, 360–363
Gross primary Production (GPP), 203, 204
Groundwater, 3, 4, 6, 18, 22, 42–46, 48, 50, 53,
    60, 64–66, 68, 69, 71–74, 78–79, 82,
    104, 105, 107, 110, 111, 142, 144, 147,
    148, 151, 153, 154, 156–158, 166, 177,
    205–207, 210, 219, 224, 387, 391, 392,
    397, 405

H
Hazard quotients (HQs), 260, 261, 263
Health risks, 329, 330, 344, 367
High-frequency analysis, 190–211

Historical cartographic documents/maps, 30,
    31, 36, 38, 39, 52
Historical ecology, 302
Historical water quality, 396
History of water use, 18
Human excreta, 118–138
Humification index (HIX), 223, 225
Hydrological regime, 3, 5, 79–83
Hydrophilic (HPI), 145, 223–225, 228, 229,
    231, 232, 234, 235, 238
Hydrophobic (HPO), 223–225, 228, 229, 234,
    235, 248
Hydrosphere, 3, 102, 275
Hydrosystem, 2, 10, 60, 61, 64–71, 74, 76, 78,
    82, 97, 104–109, 151, 206, 358, 361,
    362, 383–389, 391–392, 395–401, 408,
    412

I
ICEP indicator, 208, 209
Immunotoxicity, 257
Infiltration, 65, 66, 68, 69, 76–78, 80, 105, 152,
    166, 171, 173, 347–349
Inorganic carbon, 193, 194, 204
Institut Géographique National (IGN), 37–40,
    49–51, 165, 175, 193
Invertebrates, 245, 250, 251, 253–256, 365
In vitro bioassays, 245–250

L
Lagrené, 41
Land cover, 9, 72–73, 94, 156, 194, 405
Lead (Pb), 15, 230, 235, 271, 272, 279, 283,
    286, 289–293, 388
Leakage, 123, 274, 275, 290–291, 333, 346,
    358, 359, 388
Lithology, 5, 6, 205
Livestock systems, 95, 96
Longitudinal profile/bathymetric maps, 42,
    44, 197, 198, 205, 222, 340, 341,
    384, 386
Long-term, x, 4, 6, 7, 13–16, 61, 62, 64, 74, 82,
    93–111, 119, 137, 138, 142–158, 183,
    190–211, 219, 244, 248, 272–276, 286,
    292, 295, 296, 302–319, 324–349, 375,
    395, 400, 401, 409, 410
Long-Term Socio-Economic and Ecosystem
    Research (LTSER), x, 4, 21, 238
Long-term study, 143
Longue durée, 2, 10, 14, 16, 23, 30, 35, 52, 386,
    395–397, 409, 410

Lower Seine sector, 5, 7, 10, 22, 34, 35, 40, 44,
    45, 197–199, 204, 205, 276, 285, 295,
    306, 307, 309, 311, 318, 386

M
Macroplastics, 357–362
Marly Machine, 46, 47
Marnay, 5, 245–250, 256–258, 261–263,
    364–366, 368, 369, 373
Marne river, 5, 7, 34, 35, 205, 220, 221, 224,
    225, 280, 325–329, 331–333, 339–347,
    349, 360
Mass balance analysis, 167, 181
Mass flux, 358, 359
Mercury (Hg), 229–234, 271, 277, 278, 283,
    286, 289–293, 314, 388
Metabolism, 11, 93–111, 138, 191, 196,
    201–206, 210, 251–253, 291, 388, 409
Metals, 10, 132, 165, 228, 271, 314, 356, 383
Methane emission, 220
Metrology, 383
Microplastics, x, 357–362, 410
Micropollutants, 252, 256, 271–296, 356, 374,
    389, 410
Migratory fish, 51, 53, 305–309, 317, 318, 400
Mills, 32, 33, 35, 37, 42, 44, 52, 305, 362, 404
MODCOU, 18, 20, 22, 65, 74, 93, 95, 105–107,
    151–153, 194, 391
Model, 3, 53, 60, 95, 118, 142, 169, 192, 225,
    245, 274, 359, 383
Modeling, x
Monitoring, 10, 11, 22, 148, 153, 154,
    156–158, 183, 195, 203, 221, 227, 245,
    246, 248–249, 252, 254, 259, 264, 271,
    276, 278, 292, 293, 295, 296, 302, 329,
    330, 337–339, 345, 346, 360, 363, 365,
    367, 368, 375, 383, 384, 386, 389

N
National Archives/Archives Nationales, 38, 40
Navigation channel, 32, 34, 41, 42, 44, 48,
    50–52, 304, 311
Net ecosystem production (NEP), 197, 203, 204
Nitrate leaching, 95, 102
Nitrogen, 3, 9, 11–15, 100, 102–103, 108–110,
    119, 128, 131, 133, 150, 191, 195,
    207–210, 219, 230, 232, 238, 275, 335,
    387–389, 391, 392, 394, 398, 403
Nitrogen retention, 97
Nitrous oxide emission, 133, 192
Non-native fish species, 43, 303–305, 317, 318

North Atlantic Oscillation (NAO), 6, 62, 63, 66, 74, 82
Nuclear magnetic resonance (NMR), 223, 224
Nutrients, x, 9, 14, 22, 93, 95, 96, 104, 105, 107–111, 123, 124, 126, 133, 137, 138, 191–194, 199–202, 205, 207–211, 219, 253, 256, 291, 388, 392, 394, 396

O

Oestrogenicity, 245–248, 250
Organic carbon (OC), 100, 101, 149, 193, 194, 204–206, 219, 220, 223, 230, 231, 272, 392
Organic farming, 98, 99, 207, 399, 402, 406, 407
Organic matter (OM), 22, 95, 97, 100, 102, 126, 191, 192, 196, 197, 201–203, 210, 218–238, 335, 394
Organic pollution, 190, 191, 194–199, 205, 219, 246, 311, 313, 314, 388, 394, 397
Orge river, 5, 22, 245, 255, 288
Orgeval, 5, 22, 143–147, 150, 153, 155–157, 386
Orgeval basin, 143, 145–147, 157
Oxidative stress, 252
Oxygen, 7, 12, 132, 192, 195–201, 203–204, 210, 219, 252, 306, 310–314, 316, 383, 392, 394, 397

P

Parallel factor analysis (PARAFAC), 223, 225, 226
Paris, 3, 30, 60, 93, 118, 143, 164, 194, 221, 244, 275, 302, 324, 357, 383
Participatory experiments, 403–407
Particulate organic matter (POM), 218, 219, 222–224
Pathogenic protozoa, 357, 367–369, 375
Perfluoroalkyl substances (PFASs), 362–366
Persistent organic pollutants (POPs), 248, 252, 271, 280–283, 286–287, 362
Pesticides, x, 22, 104, 110, 111, 142–158, 207, 209, 210, 248, 252, 314, 366, 388, 391, 404, 405
Phosphorus, 3, 9, 11, 13, 15, 103–104, 108–110, 119, 190–192, 194, 195, 199, 204, 207, 208, 210, 275, 286, 335, 386, 392, 394
Phosphorus accumulation, 104
Phthalates metabolites, 252, 389, 390
Phytoplankton, 2, 192, 193, 199–202, 210, 384, 388
Phytosanitary practices, 142, 143, 145–147, 149

PIREN-Seine, 3, 38, 60, 93, 118, 142, 164, 191, 220, 244, 272, 358, 382
Plastic litter, 358, 359, 362
Polarity, 223, 224, 229, 250
Pollutant fate, 228–232, 237
Pollutants, x, 2, 12, 14, 16, 42, 45, 151, 165, 169–171, 174, 176, 178, 180, 182, 207, 218–238, 245, 248, 250–252, 256, 259, 271, 272, 274–276, 280–283, 290, 296, 314, 319, 362, 367, 368, 383, 388, 389, 402, 412
Pollution, 3, 45, 119, 170, 190, 219, 244, 274, 306, 326, 357, 388
Polybrominated diphenyl ethers (PBDEs), 245, 247, 248, 252, 277, 279, 281, 282
Polychlorinated biphenyls (PCBs), 14, 245, 247, 248, 252, 271, 272, 277, 279, 281, 282, 293–295, 314, 388
Polycyclic aromatic hydrocarbons (PAHs), 12, 164–183, 229, 230, 236–238, 247, 248, 252, 257, 258, 263, 271, 277, 279, 281, 286–288, 356, 363, 389, 390
Ponds, 3, 31–34, 37, 41, 43, 44, 46, 52, 193, 194, 203, 303, 310, 317, 370
Ponts et Chaussées, 14, 33, 39, 40, 49, 50
Populations, 3, 33, 72, 91, 119, 166, 191, 221, 245, 276, 305, 325, 358, 396
Poses, 4, 5, 22, 33, 45, 50–52, 197, 200, 202, 205, 206, 279, 306, 386, 387, 392, 402
Precipitations, 64, 68, 74–76, 80, 81, 133
Pressures, x, ix, 3–10, 16, 23, 52, 60, 62, 67, 69, 71, 118, 146, 164, 165, 204, 221, 229, 236–238, 244, 264, 272, 276–277, 286, 288, 291, 293, 311, 313, 317, 356, 364, 367, 370, 372, 383–389, 391, 396–398, 405
Projections, 72–76, 83, 210
ProSe, 3, 18, 20, 22, 192, 196, 197, 201, 383, 386, 388, 389, 394, 395, 400, 404
ProSe model, 22, 192, 197, 201, 383, 386, 388, 404
Pumping, 69, 73, 78, 81, 152

Q
Quality trajectories, 16, 295

R
Rainfall, 5, 6, 60, 64, 65, 67–69, 73, 76–79, 83, 101, 166, 171, 196, 339, 364, 368
Reanalysis, 61, 74–77
Research programme, 3, 60, 82, 143, 172, 238, 245

Reservoirs, 3–5, 7, 11, 15, 16, 22, 31, 35, 36, 43–45, 47, 48, 53, 60, 63–66, 82, 144, 152, 167, 168, 177, 178, 183, 190, 193, 194, 222, 271, 272, 274, 278, 286–288, 330, 387, 409
Rhône, 3, 21, 30, 271, 288, 289, 292–295
Riparian wetlands, 105, 107–109
River-aquifer exchanges, 81, 82
Riverbank, 42
River dredging, 168, 174, 178
River Grand Morin, 5, 22, 221, 224, 404, 406, 407
River network, 5, 38, 42, 44, 45, 53, 66, 68, 69, 71, 82, 105, 107–109, 205, 206, 210, 291, 387, 391, 393, 399, 400, 407, 408
River Risle, 400, 401
River state, 16–17, 37, 44, 395
Riverstrahler, 3, 18, 20, 22, 93, 95, 105, 107, 192–194, 197, 198, 201, 202, 205–208, 211, 388, 389, 392–395, 398
Riverstrahler model, 22, 95, 107, 192–194, 197, 198, 392, 394
River systems, 2–4, 10, 11, 16, 30, 40, 45, 93, 95, 107, 108, 110, 111, 156, 167, 172, 180, 181, 191, 194, 196, 201, 205, 209, 210, 219, 220, 238, 275, 278, 387, 388, 408
Runoff, 11, 34, 64, 105, 166, 191, 227, 275, 325, 359, 387

S
SAGE, 400
Salmon, 44, 51, 305, 306, 318
Sandpits, 29, 32, 34, 35, 44, 47–50, 53, 190
Sanitation, 3, 14, 21, 129, 137, 138, 173, 195, 325, 329, 333, 339, 357, 358, 395–397
Scales, 4, 30, 60, 95, 134, 142, 165, 190, 258, 274, 302, 348, 356, 383
Scenarios, 17, 71, 111, 135, 157, 191, 318, 348, 374, 382
SDAGE, 395
Sediment, 10, 30, 165, 192, 245, 271, 330, 365, 404
Sedimentary archives, 13, 23, 271–296, 388
Sediment transport, 30, 35, 36, 275, 276
Seine, 3, 30, 60, 93, 118, 142, 164, 191, 219, 244, 271, 302, 325, 356, 382
   basin, 3–10, 13, 14, 17, 23, 30–53, 60–83, 94, 96, 97, 100–106, 108, 142, 143, 154–157, 181, 194, 197, 200, 202, 205, 207, 218–238, 244, 245, 271, 277, 302–319, 382, 384, 386, 388, 391, 398, 399, 408, 412

estuary, 5, 50, 211, 282, 294, 359, 392, 400
river, 4, 30, 61, 93, 119, 153, 164, 191, 219, 245, 272, 302, 325, 356, 382
river basin, x, 2–23, 32, 39, 40, 45–52, 82, 93, 104, 135, 142–158, 164–183, 220, 223–227, 244–264, 276–281, 285–287, 289–291, 293–294, 296, 303–311, 317–319, 356–375, 391, 392
river system, 4–10, 16, 30, 107, 110, 164, 181, 191, 196, 210
Seneque-Riverstrahler, 394
Sewage, 7, 18, 121, 122, 125–129, 131–134, 136, 137, 195, 196, 209, 219, 275, 277, 281, 289, 306, 330, 333, 335–337, 384, 386, 394, 396, 397, 402
Sewer deposit, 167, 169, 173
Sewer systems, 130, 137, 165, 167, 169, 171–173, 178, 180–182, 196, 313, 333, 336, 337, 346
Shad, 51, 305, 306, 308, 318
Silica, 3, 190, 192, 200–202, 207–209
Societal response, 16–17, 272, 294
Socioecological research, 17–21
Socioecological trajectories, 23, 119
Soil, 11, 45, 61, 93, 121, 142, 165, 218, 275, 326, 368, 387
Source separation, 134–138
Speciation, 217–238
Specific UV absorbance at 254 nm ($SUVA_{254}$), 223–225
Spectral slope ratio (S_R), 223, 224
STICS, 18, 93, 95, 99–101, 103, 105–107, 149–153, 158, 194
STICS-Pest, 149–151, 153
Storage, 11–13, 21, 32, 42, 44, 45, 50, 53, 61, 66, 93, 97, 100–105, 108–110, 123, 124, 168, 177–178, 195, 248, 336, 347, 349
Stream-aquifer interface, 65, 69, 71, 191
Stream order, 5, 22, 31, 33, 34, 45, 52, 53, 191, 193, 194, 209, 387, 392
Surface water, 2, 18, 37, 43, 44, 53, 65, 66, 74, 78–79, 93, 104, 108, 111, 121–123, 130, 132, 142, 148, 152, 157, 158, 165, 191, 210, 220, 223–225, 228, 229, 238, 246–248, 250, 324, 330, 333, 334, 336, 344–346, 363, 367, 370, 375
Suspended particulate matter (SPM), 274, 275, 277, 278, 280, 281, 287, 288, 292, 295
Syndicat Interdépartemental pour l'Assainissement de l'Agglomération Parisienne (SIAAP), 18, 128, 165, 171–173, 195, 196, 203, 204, 227, 331, 332, 334, 336, 339, 357, 394, 402

T

Timber rafting, 31–33, 41, 42, 44, 45, 47
Trace elements/trace metals, x, 228–236, 238,
     271, 277, 293, 356, 373
Trajectories, x, 2–23, 31, 34, 39, 45–53, 60, 61,
     71–82, 93–111, 118–138, 157, 196,
     272–276, 291, 293–296, 302–303,
     309–311, 359, 395, 397, 398, 411
Transphilic (TPI), 223, 224, 228, 229, 234
Treatments, 3, 49, 107, 121, 145, 165, 194, 219,
     247, 276, 306, 325, 368, 386
Triel, 5, 166, 245–250, 256–258, 261–263, 341,
     364–366, 368, 369, 373

U

Urban fluxes, 167, 169–174, 179, 183, 360
Urban sludge, 97, 167, 168, 174, 176, 178, 278
Urine, 122, 126, 134–136, 138, 252
UV/visible absorbance spectroscopy, 223

V

Vadose zone, 64, 108, 109, 152, 154
Variability, 5, 6, 22, 46, 61–64, 68, 74–76, 82,
     98, 146, 152, 157, 165, 172, 203,
     218–238, 250, 253, 258, 337, 374, 384,
     386, 388
VCN30-yearly minima of 30-day running
     average window, 63, 64, 80
Versailles, 33, 45–47, 52, 53
Vesle basin, 143–145, 152, 153, 156, 254
Vesles River, 5, 144
Viruses, 330, 343–346, 349
Vuillaume, 40, 41, 44, 51

W

Wastewater, 3, 104, 119, 165, 194, 219, 247,
     270, 306, 325, 359, 386
Wastewater treatment plants (WWTPs), 7, 13,
     16, 121, 122, 129–134, 136, 137,
     165–167, 169, 173–174, 178, 180–183,
     194–197, 199, 202–210, 219, 221,
227–229, 232, 233, 235, 238, 247, 276,
     313, 325, 333, 335, 336, 340–342,
     345–349, 359, 361, 371–373, 386, 388,
     394–397, 399
Water, 2, 31, 60, 93, 119, 142, 164, 190, 219,
     244, 271, 302, 324, 356, 382
balance, 35, 65, 66, 70, 72, 152, 153
budget, 9, 12, 60–83
column, 246–248, 250, 259, 260, 263, 364,
     365
quality, x, 2, 7, 14, 16–18, 23, 35, 61, 65,
     93, 104, 108, 110, 132, 137, 157, 190,
     191, 193–202, 205, 207–210, 219, 220,
     227, 238, 255–257, 264, 271, 277, 278,
     289, 293, 295, 306, 311, 313, 316–319,
     324–349, 362, 363, 367, 383, 385, 386,
     388, 391, 392, 394–396, 398, 399, 405,
     406, 410, 411
resources, 2, 13, 17, 34, 44–46, 52, 60, 67,
     69, 76–83, 157, 319, 391, 396, 400, 405,
     411
uptake, 73–74, 78, 79, 82
withdrawal, 60, 65, 67, 68, 71, 73, 74
Water Framework Directive (WFD), 34, 165,
     210, 244, 260, 276, 278, 295, 302, 357,
     386, 389, 394, 397, 399, 403, 404, 410,
     411
Weirs/locks, 22, 33–36, 40–44, 50, 51,
     133–135, 305, 311, 403, 404, 406, 407,
     412
Wetlands, 3, 9, 14, 30, 33, 34, 37, 42–44,
     47, 49, 50, 53, 105, 107–109, 147,
     157, 405

Y

Yield-fertilisation relationship, 98–100

Z

Zinc (Zn), 15, 229–230, 232, 234, 236, 271,
     279, 283, 289–293, 373, 388
Zone Atelier Seine (ZA Seine), x, 4, 21

Printed in the United States
by Baker & Taylor Publisher Services